Trends in Biological Processes in Industrial Wastewater Treatment

Online at: https://doi.org/10.1088/978-0-7503-5678-7

Trends in Biological Processes in Industrial Wastewater Treatment

Edited by
Maulin P Shah
Chief Scientist and Head, Industrial Wastewater Research Lab,
Division of Applied and Environmental Microbiology,
Enviro Technology Limited, Ankleshwar, Gujarat, India

IOP Publishing, Bristol, UK

ISBN   978-0-7503-5678-7 (ebook)
ISBN   978-0-7503-5676-3 (print)
ISBN   978-0-7503-5679-4 (myPrint)
ISBN   978-0-7503-5677-0 (mobi)

DOI   10.1088/978-0-7503-5678-7

Version: 20240601

IOP ebooks

British Library Cataloguing-in-Publication Data: A catalogue record for this book is available from the British Library.

Published by IOP Publishing, wholly owned by The Institute of Physics, London

IOP Publishing, No.2 The Distillery, Glassfields, Avon Street, Bristol, BS2 0GR, UK

US Office: IOP Publishing, Inc., 190 North Independence Mall West, Suite 601, Philadelphia, PA 19106, USA

Contents

7 Mycoremediation of wastewater: sustainable approaches   **7-1**

Maitri Nandasana and Sougata Ghosh

8 Advances in the remediation of xenobiotics using microbes   **8-1**

Anne Bhambri, Santosh Kumar Karn and Navneet Joshi

9 Fungi-based biosensing platforms for detection of heavy metals: focus on the eukaryotic system   **9-1**

Ankur Singh, Vipin Kumar and Sarika

Editor biography

Maulin P Shah

Maulin P Shah is a microbial biotechnologist with diverse research interests. A group of research scholars is working under his guidance on areas ranging from applied microbiology, environmental biotechnology, bioremediation, and industrial liquid waste management to solid state fermentation. His primary interest is the environment, the quality of our living resources, and the ways that bacteria can help to manage and degrade toxic wastes and restore environmental health. Consequently, he is very interested in genetic adaptation processes in bacteria, the mechanisms by which they deal with toxic substances, how they react to pollution in general, and how we can apply microbial processes in a useful way (e.g., bacterial bioreporters). He has more than 300 research publications in highly reputed national and international journals. He directs the research program at Enviro Technology Ltd, Ankleshwar. He has guided more than 300 postgraduate students in various disciplines of the life sciences. He is also a reviewer in various journals of national and international repute, and has edited more than 175 books in wastewater microbiology, environmental microbiology, bioremediation, and hazardous waste treatment.

List of contributors

Amera Adel Abaza
Department of Industrial Biotechnology, Genetic Engineering and Biotechnology Research Institute, University of Sadat City, Sadat City, Egypt

Dina Y Abdelghani
Department of Special Food and Nutrition, Food Technology Research Institute, Agricultural Research Center, Giza, Egypt

Manish Agrawal
Rungta Educational Campus, Kurud Road, Kohka, Bhilai, Chhattisgarh, India

J Anandkumar
National Institute of Technology Raipur, Chhattisgarh, India

G Aydogdu
Environmental Engineering Department, Faculty of Civil Engineering, Istanbul Technical University, Maslak, Istanbul, Turkey

Anne Bhambri
Department of Biochemistry and Biotechnology, SardarBhagwan Singh University Balawala, Dehradun, 248161, India
and
Department of Biotechnology, Shri Guru Ram Rai University, Patel Nagar, Dehradun, Uttarakhand, India

Grzegorz Boczkaj
Gdansk University of Technology, Faculty of Civil and Environmental Engineering, Department of Sanitary Engineering, Gdansk, Poland
and
EkoTech Center, Gdansk University of Technology, Gdansk, Poland

Vani Chandrapragasam
Division of Biotechnology, School of Agricultural Sciences, Karunya Institute of Technology and Sciences (Deemed to be University), Coimbatore, Tamil Nadu, India

E Cokgor
Environmental Engineering Department, Faculty of Civil Engineering, Istanbul Technical University, Maslak, Istanbul, Turkey

Joydeep Das
Department of Chemical Engineering, National Institute of Technology Agartala, Tripura, India

Ashraf El-Baz
Department of Industrial Biotechnology, Genetic Engineering and Biotechnology Research Institute, University of Sadat City, Sadat City, Egypt

Sougata Ghosh
Department of Physics, Faculty of Science, Kasetsart University, Bangkok, Thailand

Levin Anbu Gomez
Division of Biotechnology, School of Agricultural Sciences, Karunya Institute of Technology and Sciences (Deemed to be University), Coimbatore, Tamil Nadu, India

Jyoti Gulia
Department of Microbiology, Maharshi Dayanand University, Rohtak, Haryana, India

Shihu Hu
Australian Centre for Water and Environmental Biotechnology, The University of Queensland, St Lucia, Australia

Zhetai Hu
Australian Centre for Water and Environmental Biotechnology, The University of Queensland, St Lucia, Australia

G Insel
Environmental Engineering Department, Faculty of Civil Engineering, Istanbul Technical University, Maslak, Istanbul, Turkey

Farooque Ahmed Janjhi
Gdansk University of Technology, Faculty of Civil and Environmental Engineering, Department of Sanitary Engineering, Gdansk, Poland

Navneet Joshi
Department of Biosciences, Mody University of Science and Technology, Lakshmangarh, District-Sikar, Rajasthan, India

Santosh Kumar Karn
Department of Biochemistry and Biotechnology, SardarBhagwan Singh University Balawala, Dehradun, India

Sarita Khaturia
Mody University of Science and Technology, Lakshamangarh, Rajasthan, India

Divyajeet Kumar
Department of Chemical Engineering, National Institute of Technology Agartala, Tripura, India

Prasann Kumar
Department of Agronomy, School of Agriculture, School of Bioengineering and Biosciences, Lovely Professional University, Phagwara, Punjab, India

Vipin Kumar
Laboratory of Applied Microbiology, Department of Environmental Science and Engineering, Indian Institute of Technology (Indian School of Mines), Dhanbad, Jharkhand, India

Amit Lath
Centre for Biotechnology, Maharshi Dayanand University, Rohtak, Haryana, India

Soma Nag
Department of Chemical Engineering, National Institute of Technology Agartala, Tripura, India

Archana Nair
Department of Biotechnology, U. V Patel College of Engineering, Ganpat University, Mehsana, Gujarat, India

Maitri Nandasana
Department of Microbiology, School of Science, RK. University, Rajkot, Gujarat, India

G Ozyildiz
Environmental Engineering Department, Faculty of Civil Engineering, Istanbul Technical University, Maslak, Istanbul, Turkey

Rajakumar S Rai
Department of Mechanical Engineering, School of Agriculture and Biosciences, Karunya Institute of Technology and Sciences (Deemed to be University), Coimbatore, Tamil Nadu, India

Yashika Rani
Department of Microbiology, Maharshi Dayanand University, Rohtak, Haryana, India

Saloni Sahal
Mody University of Science and Technology, Lakshamangarh, Rajasthan, India

Biju Prava Sahariah
National Institute of Technology Raipur, Chhattisgarh, India

Anita Rani Santal
Department of Microbiology, Maharshi Dayanand University, Rohtak, Haryana, India

Sarika
Ashoka Trust for Research in Ecology and the Environment, Royal Enclave, Srirampura, Jakkur, Bengaluru, India

Ritu Shepherd
School of Liberal Science, Nehru Arts and Science College, Tirumalampalayam, Coimbatore, Tamil Nadu, India

Yousseria Shetaia
Department of Microbiology, Faculty of Science, Ain Shams University, Cairo, Egypt

Ankur Singh
Laboratory of Applied Microbiology, Department of Environmental Science and Engineering, Indian Institute of Technology (Indian School of Mines), Dhanbad, Jharkhand, India

Harlal Singh
Mody University of Science and Technology, Lakshamangarh, Rajasthan, India

Joginder Singh
Department of Microbiology, School of Bioengineering and Biosciences, Lovely Professional University, Phagwara, Punjab, India

Nater Pal Singh
Centre for Biotechnology, Maharshi Dayanand University, Rohtak, Haryana, India

D Soylu
Environmental Engineering Department, Faculty of Civil Engineering, Istanbul Technical University, Maslak, Istanbul, Turkey

J J Thathapudi
Division of Biotechnology, School of Agricultural Sciences, Karunya Institute of Technology and Sciences (Deemed to be University), Coimbatore, Tamil Nadu, India

Hameed Ul Haq
Gdansk University of Technology, Faculty of Civil and Environmental Engineering, Department of Sanitary Engineering, Gdansk, Poland

Vishruth Vijay
Division of Biotechnology, School of Agricultural Sciences, Karunya Institute of Technology and Sciences (Deemed to be University), Coimbatore, Tamil Nadu, India

Vijaylakshmi
Department of Biochemistry and Biotechnology, SardarBhagwan Singh University Balawala, Dehradun, India

G E Zengin
Environmental Engineering Department, Faculty of Civil Engineering, Istanbul Technical University, Maslak, Istanbul, Turkey

IOP Publishing

Trends in Biological Processes in Industrial Wastewater Treatment

Maulin P Shah

Chapter 1

Surfactants and bioremediation

J Anandkumar, Manisha Agrawal and Biju Prava Sahariah

Detergents, soaps, personal care products, liquid detergents, toiletries, and emulsifiers are used daily in modern societies and are common surfactant products. Surfactants are surface active agents that can reduce the interfacial tension of the interface. Due to their broad physical and chemical properties, this group of chemicals has a wide range of uses in industries such as personal care products, pharmaceutical, agriculture, food processing, textile, laundry fungicides, pulp and paper processing, paint industries, softeners, antistatic agents, detergents, and metal treatment industries. There are four broad categories of surfactants, namely anionic, cationic, zwitterionic, and nonionic, which are well studied and have growing utilization. The elements present in surfactants can significantly alter the quality of aquatic bodies and the soil. They can also induce a variety of problems in wastewater.

Microbial bioremediation is a successful option for various xenobiotics when applied with efficient microbes in their optimum environment. Furthermore, metabolic activity can transform complex surfactants to simpler forms. This chapter discusses the chemistry of dominant surfactants and their fate in the environment. Some possible bioremediation techniques will then be suggested for a sustainable and healthy environment.

1.1 Introduction

Surfactants are chemicals, more precisely organic substances (e.g., the regular used detergents or soaps, with the capacity to reduce the surface tension of a liquid, and hence facilitate high spreading and wetting properties), that are used for cleaning and softening of fabrics. The presence of one or more hydrophilic and hydrophobic groups enables surfactants to form micelles. The intrusion and dispersion properties of surfactants increase their use, not only in laundry but also in textile and perfume industries for dye penetration and perfume dispersion. Surfactants are also used as corrosion regulators, ore floaters, oil flow enhancers in porous materials, and aerosol production.

1.1.1 Chemistry of surfactants

Surface active agents that are used to decrease the surface tension of water are termed 'surfactants.' These are chemical compounds with a micelles creating property via self-assembled molecular clusters in a solution. The unique molecular structure of surfactants is responsible for their amphiphilic nature and are comprised of two parts: the first part is made by a water loving hydrophilic group that is soluble in water, such as hydroxyl, carboxyl, amino, and other polar groups; the second part is a lipophilic or hydrophobic group, such as alkyl, aryl, and other non-polar groups, which are water hating in nature and insoluble in water but soluble in lipids or oils. This property facilitates them to concentrate at the interfaces between bodies or droplets of water and hydrophobic substances such as oil or lipids, and perform as an emulsifying or foaming agent (figure 1.1).

On the basis of the nature of the hydrophilic group, surfactants are classified into four types:

(a) **Anionic surfactants:** Anionic surfactants contain negative charges on their hydrophilic head in solution and are the most commonly used surfactants, which are used in food, pharmaceutical, cosmetic, detergent, and other industries. Anionic functional groups are mainly sulfates (SO_4^{2-}), phosphates (PO_4^{2-}), and carboxylates ($RCOO^-$). Some examples of anionic surfactants are sulfates, sulfonates, and gluconates, such as Sodium dodecyl sulfate (SDS) $C_{12}H_{25}NaO_4S$ and Potassium oleate $C_{18}H_{33}KO_2$. They cover 50% of overall industrial production.

Uses

Textile industry: Oiling agents, dyeing auxiliaries, bleaching agents, and soft processing agent.

Pesticides: Spraying agents, emulsifying agents, and dispersants.

Highway: Asphalt emulsifier.

Civil engineering: Water reducing agent for cement.

Petroleum industry: Emulsifiers and fuel additives.

Polymer industry: Emulsifiers and antistatic agents, metal cleaning agents, mining flotation agents, and water treatment agents.

(b) **Cationic surfactants:** Cationic surfactants contain negative charges on their hydrophilic end. Alkyl ammonium chlorides and Quaternary Ammonium salt (N^+R_{4-}) are common examples of cationic surfactants. Cationic surfactants dissociate to positively charged ions in solution, and are used as fungicides, softeners, and antistatic agents. In general, the dosage of the cationic surfactant is less than that of the anionic surfactants. The hydrophobic group of cationic surfactants is composed of an alkyl group and a functional group containing nitrogen, phosphorus, sulfur, and so on. At present, most industrial cationic surfactants are composed of nitrogenous

a)

| Polar Hydrophilic Head, Water |

| Non- Polar Hydrophobic Tail, Water Hating |

b)

Br-

Hydrophilic Head

Cetyl trimethyl ammoniumbromide
(Cationic Surfactant)

Hydrophobic Tail

c)

O Na

O=S=O

Hydrophilic Head

Sodium dodecyl sulphate
(Anionic Surfactant)

Hydrophobic Tail

d)

Hydrophilic Head

OH
n

Triton X-100
(Non-ionic Surfactant)

Hydrophobic Tail

e)

Cl

OH

Hydrophilic Head

Coco-betain(CB)
(Zwitterionic Surfactant)

Hydrophobic Tail

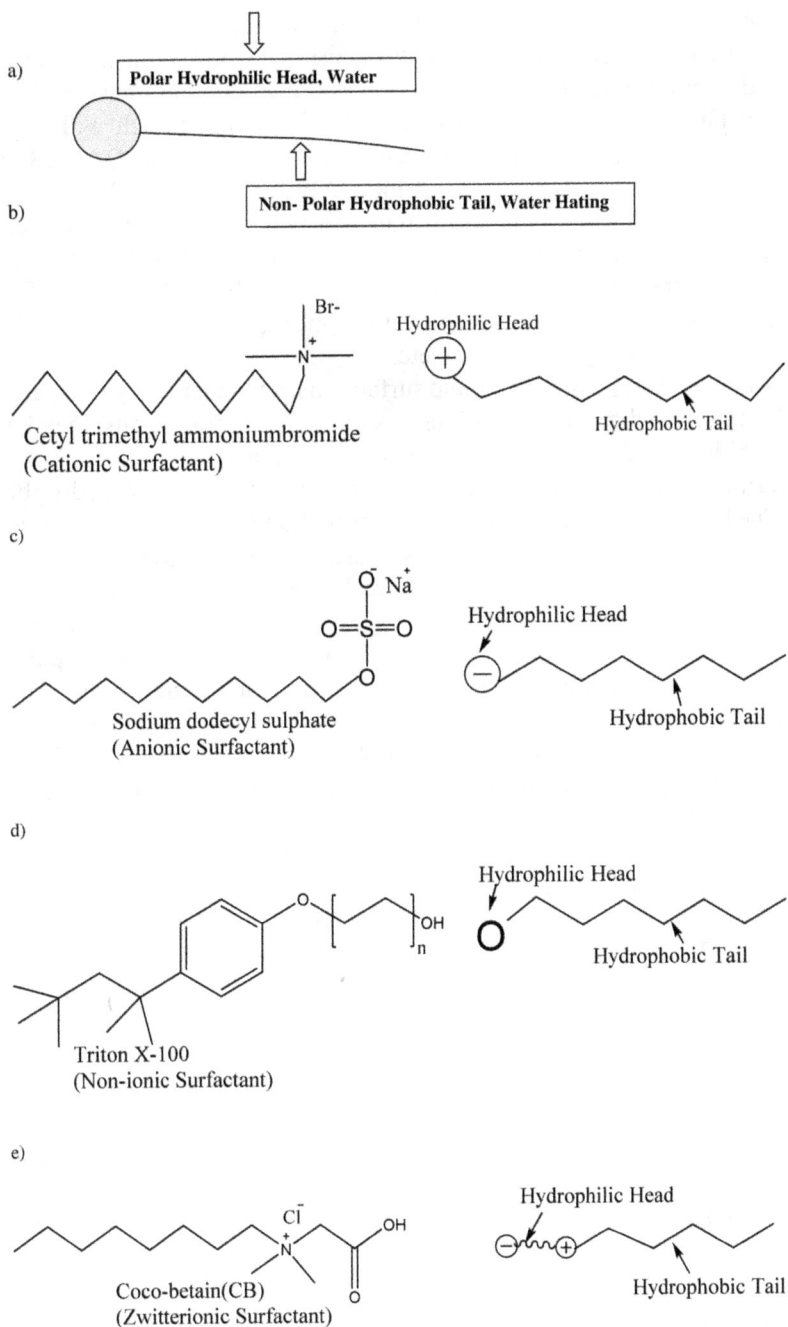

Figure 1.1. (a)–(e) Some examples of the structure of surfactants.

compounds, which are classified into amine, quaternary ammonium, heterocyclic, and rhodium salts. Quaternary ammonium salts (QACs) are widely used in cationic surfactants because of their solubility in both acids and bases. There are four functional groups that attach with a covalent bond to the positively charged central nitrogen atom in QACs. QACs are basically organic compounds. A long hydrophobic alkyl chain is present in functional groups, along with other short chains such as methyl or benzyl groups. The ingredients of ester cationic surfactants include (for example) diethyl ester dimethyl ammonium chloride (DEEDMAC). The QACs are mostly used in pharmaceuticals (antiseptic agents), softeners (fabric softeners), fungicides, preservatives, etc.

(c) **Nonionic surfactants:** Nonionic surfactants, as the term suggests, are neutral in nature and cannot dissociate in solution to produce ions, they have high stability, and they are the most commonly used type after anionic surfactants. They do not contain any charges on their hydrophilic end. The hydrophilic part possesses covalently bonded oxygen-containing groups such as hydroxyl and ether bonds, which are bonded to hydrophobic parent structures. Nonionic surfactants are less sensitive to water hardness than anionic surfactants. There is no or insignificant effect of acid and alkali on nonionic surfactants, which make them highly suitable for applications in washing, dispersion, foaming, solubilization, and in industries such as food, medicines, textiles, and paints. Examples of nonionic surfactants are ethoxylates, alkoxylates, alkylphenol polyoxyethylene ether (APEO), high carbon aliphatic alcohol polyoxyethylene ether (AEO), fatty acid polyoxyethylene ester (AE), and cocamide. They comprise 45% of overall industrial production, where they are used as wetting agents in coatings, food ingredients, cell staining, DNA extraction, and pharmaceutical experiments. The major ingredients of alkylphenol polyoxyethylene ether (APEO) are octylphenol polyoxyethylene ether and nonylphenol polyoxyethylene ether, composed of a polyoxyethylene chain and an alkyl chain attached to a benzene ring. They are widely used as insecticides, emulsifiers, and solvent enhancers. Other examples are Triton X-100 ($C_{14}H_{22}O(C_2H_4O)$ $n(n=9-10)$) and Tergitol ($C_{12-14}H_{25-29}O[CH_2CH_2O]xH$).

Uses:

Textile industry: Cationic surfactants are widely used as fabric softeners, as well as the antistatic agents for chemical fiber oils.

Metal industry: Cationic surfactants are used as metal corrosion inhibitors thanks to their high efficiency, low toxicity, easy production, low price, and the waste liquid can be used as a metal oil cleaning agent and rust remover after cleaning.

Paper industry: Cationic surfactants can improve the strength of paper, and the retention rate of the filler and fine fibers.

Coating industry: Cationic surfactants are mainly to evenly distribute and disperse pigment particles.

Oil industry: Cationic surfactants are used for oil viscosity reduction, oil well fixation, oil pipeline protection and sterilization.

Water purification industry: Cationic surfactants are used to treat domestic sewage and industrial wastewater.

(d) **Amphoteric surfactants:** These surfactants are also known as zwitterionic surfactants, whose amphoteric properties are independent of pH over a wide pH range. They have both positive and negative charges on their hydrophilic end with a net charge of zero. Betaines and amino oxides are common examples of this type of surfactant. Amphoteric surfactants include amphoteric surfactants (amphoterics) and amphoteric ionic surfactants (zwitterionics). The amphoteric surfactants are comparatively mild surfactants and are used in daily necessities and cosmetics. Nevertheless, the high cost and complex synthesis limits the production of amphoteric surfactants. There are three general types of amphoteric surfactants, namely amino acid amphoteric surfactants, betaine type amphoteric surfactants, and imidazoline surfactants. The cationic part of betaine type amphoteric surfactants is composed of quaternary ammonium salt, which can be divided into carboxylate betaine, sulfobetaine, and phosphate betaine according to different anions. The structure of the betaine type surfactants is based on the substitution of methyl or carboxyl groups of trimethylglycine with different substituents, such as betaines ($C_5H_{11}NO_2$) and amphoacetates ($RC(O)NH(CH_2)2N(CH_2CH_2OH)CH_2COONa$).

Uses:

These are expensive surfactants and are generally used in cosmetics, shampoos, shower gel, hair conditioners, and toothpaste because of their mildness, safety, and lack of irritating effects on skin and eyes.

The surfactant market is largely dominated by the nonionic surfactants, followed by the anionic surfactants. The cationic surfactants are generally expensive, and hence have a small share of the market. The demands for surfactants are increasing day by day, due to their wide range of domestic and industrial applications in many facets of daily life. Global market data about manufacturing of surfactants shows that surfactant market size was USD 42.1 billion in 2020 and is predicted to will reach to USD 52.4 billion by 2025 because of growing population and awareness about the importance of cleanliness. It is noticeable here that surfactants are an important ingredient in sanitizers and soaps. Frequent use of hand wash and soaps due to COVID-19 led to a nearly ten times increase in demand and production of surfactants and detergents. The brand site EVALED suggests that the formation of 800ppm detergent leads to 20ppm discharge of surfactants into wastewater. Generally, surfactants are present in detergent industry effluents and other industrial effluents. In addition, the residues of surfactants are discharged into the environment, which creates many problems, e.g., they hinder the transfer of oxygen from the atmosphere to the water in the process of aeration and they are accumulated in the body, which can lead to long term problems such as carcinogenicity and loss of fertility.

1.1.2 Toxicity of surfactants

The unique properties and diversified functions of surfactants have led to their widespread use in industries such as textiles, fibers, paints, polymers, plant protection, cosmetics, mining, oil recovery, pulp and paper, agriculture, pharmaceuticals, food, and daily chemicals. Their extensive use is directly related to the increased release of surfactants to the environment at levels that are sufficient to cause detrimental effects to the soil and aquatic ecosystems. A buildup of surfactant compounds in the soil can severely interfere with the structure and functioning of soil microbes, resulting in several harmful effects, e.g., on photosynthesis and damage to the root system of plants. In the case of aquatic organisms, the accumulation of surfactant compounds in the gills and viscera of fishes result in asphyxiation leading to death. Meanwhile, exposure to surfactants in the environment can lead to human diseases such as dermatitis, bronchopulmonary dysplasia, gingivitis and periodontitis, and oral ulcers. In addition, chemical species of surfactants can interact with bodily fluids through the food chain. The metabolites nonionic surfactants APEO are identified as environmental endocrine disruptors that can interfere with normal hormonal function and health status in animals, and are responsible for a decrease in sperm count in males, breast cancer, testicular cancer, and so on. Many surfactants have been shown in experiments to evade the interaction with the protein units involved in immune system, and are hence detrimental to health.

1.2 Treatment of surfactants

In view of the steady accumulation and negative impact of surfactants on the environment and population health, the removal of residual surfactants from the environment or conversion to non-toxic elemental compounds are important tasks. Eliminating surfactants during production or removing residual surfactants are required to maintain the environment and reduce potential health hazards.

The treatment process generally focuses on altering or breaking the structure of harmful compounds in the surfactant with the help of physical–chemical–biological forces into less harmful, elemental, and environmental-friendly compounds. Successful processes for the removal of surfactants from the environment include adsorption, oxidation, electrolysis, electrochemical oxidation, photolysis, sonication, and biodegradation (Berna *et al* 2007, Christopher *et al* 2021).

1.2.1 Physical–chemical process

The small pores in adsorbent materials such as activated carbon, alumina, and silica gels provide effective room for the adsorption and removal of surfactant molecules. The parent material of the activated carbon, and the electrostatic potential of the interface of the adsorbate and adsorbent influence the adsorption efficiency of pollutants (Adak *et al* 2005, Koner *et al* 2012, Siyal *et al* 2020). Strong oxides such as ozone, hydrogen peroxide, and ferric salt are capable of oxidizing surfactant molecules through oxidation reduction reactions dissociating them into non-toxic and harmless substances. In the Fenton oxidation process, hydrogen peroxide (H_2O_2) and ferrous sulfate ($FeSO_4$) are employed to form a strong oxidizing agent (hydroxyl radicals), which has high oxidation potential (2.8 V) compared to ozone (2.07 V). Photolysis of UV-excited hydrogen peroxide for oxidation of surfactant, the application of non-toxic and efficient semiconductor TiO_2, the ultrasonic degradation of surfactant in the presence of high sound pressure, and high temperature and pressure oxidation are used for surfactant removal. Electrolysis treatment is renowned for the efficient mineralization and removal of surfactants due to the application of strong oxidizing substances produced on the surface of the electrodes, such as PbO_2 and boron-doped diamond (BDD). For example, galvanostatic electrolysis using a Ti–Ru–Sn ternary oxide and a BDD anode for sodium dodecyl benzene sulfonate and car wash wastewater; and porous graphite as anode and cathode, with the effective CuO–Co_2O_3–PO_4^{3-} modified kaolin catalyst in a single undivided cell provided efficient anionic surfactant removal.

Though effective treatment results are achieved, oxidation, photocatalytic, and electrochemical treatment are relatively costly methods and either require secondary waste handling or consume more energy, and some of the processes with no secondary waste are associated with low treatment rate.

1.2.2 Bioremediation

Bioremediation involves mechanisms such as biosorption, biodegradation, bioaccumulation, and bioleaching for pollutant removal. Bioremediation includes phytoremediation (rhizodegradation, phytoextraction, phytovolatilization, phytodegradation, rhizofiltration, and rhizostabilization), mycoremediation, microbial remediation (bacteria, fungi, and microalgae), and phyco-remediation.

In biosorption, and bioaccumulation, pollutants are immobilized inside or outside a microbial cell equipped with extracellular polysaccharide substances (EPS) and through the secretion of various organic acids. Similarly, bioleaching involves numerous organic acids and enzyme secretions for pollutant extraction and processing according to the properties of the pollutants and the microbes. The biodegradation of surfactants biodegradation agents that derive their required life sustaining energy and nutrients from dissociating the chemical bonds in the surfactants while converting the latter to non-toxic elements. Biodegradation is highly prominent in the case of microbial, algae, and a few plant species (phytoremediation). Biotic and abiotic factors of the environment play a key role in the bioremediation technique. Biotic factors incorporate genetic and enzymatic molecules and properties that govern the physical and functional interactions of organisms and surfactants. Cell wall component orientation and secretion of specific enzymes, organic acids, EPS, and formation of biofilm are principal features that influence the performance efficiency of biotic factors. The abiotic factors mainly include temperature, pH, moisture, and bioavailability of surfactants together with nutrients ratios. Electron acceptors, mostly oxygen in aerobic biodegradation and nitrate/sulfate in anoxic conditions, significantly influence treatment efficiency.

In this chapter, the microbial process is emphasized for surfactant biodegradation. A number of microbes, which are tiny unicellular organisms associated with biodegradation, form a heterogeneous colony or mixed culture of different species in nature. In complete treatment, the mineralization of surfactants to carbon dioxide and water occurs via microbial metabolism, which is a factor of species involved in the biodegradation, environmental conditions, and property of the surfactants to be removed. A few frequently noted microbial species for surfactant biodegradation are *Acinetobacter*, *Aeromonas*, *Comamonas*, *Dechloromonas*, *Desulfovibrio*, *Geobacter*, *Holophaga*, *Parvibaculum*, *Pseudomonas*, *Sporomusa*, *Stenotrophomonas*, *Variovorax*, and *Zoogloea*.

Owing to the complex chemical structure of surfactants, degradation may follow a series of phases where the initial or parent surfactant compound is changed while losing its original structure, and hence surface activation properties. These intermediates are then processed by another group of microbes for rapid and complete mineralization, resulting in water, CO_2, mineral salt, and other inorganic substances, while avoiding secondary pollution.

The dissociation property of polar group defines the classification of surfactants as cationic, nonionic, and amphoteric, while the hydrophobic group determines the biodegradability to a great extent. It is also common to note that different surfactants have utterly different biodegradability.

1.3 Biodegradation mechanism of surfactants

The prominent ways to degrade surfactants with microorganisms are ω-oxidation, β-oxidation, α-oxidation, and benzene ring oxidation under the catalysis reactions of various enzymes. The microbes consider the pollutants (here surfactants) as a carbon source for their survival. In ω-oxidation, terminal alcohol is generated from alkyl chains via ω-oxidation in the presence of alkane monooxygenase. This is followed by sequential production of terminal aldehyde and the carboxylic acid in the presence of two kinds of dehydrogenases enzymes.

In the case of β-oxidation, the carboxyl groups generated by ω-oxidation of the alkyl chain influence β-oxidation in continuation of degradation of the alkyl chain. The significant enzymes are coenzyme A, dehydrogenase, and hydrolytic enzyme.

Benzene ring oxidation, especially the links, is considered as the speed control step of surfactants containing benzene rings. The most influential enzyme is oxygenase. The intermediates are catechol, hexadiene diacid, acetic acid, succinic acid, 2-hydroxyhexadiene semialdehyde acid, and finally formic acid, acetaldehyde, and pyruvic acid.

1.3.1 Bioremediation of anionic surfactants

The common anionic surfactants are linear alkylbenzene sulfonate (LAS), alkyl sulfonate (AS), and alcohol ether sulfate. LAS is a widely used surfactant, comprised of 20 or more homologues isomers and accounts for more than quarter share among the synthetic surfactants. The toxicity of LAS is generally mild and shows no accumulation in organisms. The biodegradation of LAS is highly efficient at all of the basic environment of anaerobic/aerobic or facultative conditions and relevant to its molecular structure for biodegradation. In anaerobic and aerobic conditions, LAS follows a degradation pathway with aromatic ring cleavage, desulfonation, beoxidation reactions, and ueoxidation reactions. The existing distance of the benzene ring with the central carbon atom determines the stability of the structure, i.e., when the distance is larger, there is more instability in the structure, which comprises the feasibility of degradation. The degradation of LAS by microorganisms is initiated by the breakdown of alkyl chain, which results in the generation of carboxylic acid by ω-oxidation of the terminal methyl group. This is followed by formation of acetyl-coenzyme A through β-oxidation, and hence enters the tricarboxylic acid cycle releasing CO_2 and mineral salts, which are the final products of the benzene rings. Owing to the complexity of the degradation of the benzene ring, it is regarded as the rate limiting element of LAS biodegradation. The literature recognizes more than 47 genera of microbes that are capable of LAS degradation, with almost 30 genera that are capable of aromatic ring cleavage, often denoted as aromatic compounds degraders. Desulfonation reactions are achieved by the species of seven genera, namely *Achromobacter*, *Acinetobacter*, *Aeromonas*, *Comamonas*, *Desulfovibrio*, *Hydrogenophaga* and *Pseudomonas*, and species of five prominent genera, namely *Geobacter*, *Magnetospirilium*, *Parvibaculum*, *Pseudomonas* and *Synergistes*, which are recognized for beoxidation reactions. *Parvibaculum* and *Pseudomonas* genera includes a few species that are capable of ueoxidation reactions. Several microbes from *Acinetobacter*, *Aeromonas*, *Comamonas*, *Dechloromonas*, *Desulfovibrio*, *Geobacter*, *Holophoga*, *Pseudomonas*, *Sporomusa*, *Stenotrophomonas*, *Variovorax* and *Zoogloea* genera are recognized for LAS degradation (Okada *et al* 2014). Aerobic treatment systems include a supply of molecular oxygen. The activated sludge process can provide more than 90% or even complete removal of LAS. Similarly, anaerobic processes (absence of oxygen) in the upflow anaerobic sludge blanket also exhibit very efficient treatment efficiency for LAS. A few significant strains that are capable of degrading LAS are *Pseudomonas nitroreducens*, *Pseudomonas aeruginosa*, *Pantoea agglomerans*, and *Serratia odorifera 2*. Each strain has specific preferences for degradation conditions, and hence exhibit different degradation times and rates. Therefore, achieving successful treatment depends on the selection of efficient species and the providision of their favorable conditions.

A few more of the commonly utilized anionic surfactants, namely AS and SDS, are also swiftly biodegradable. The biodegradation of AS is accomplished by devulcanization of sulfate esterase and β-oxidation of aliphatic alcohols. *Pseudomonas putida R1*, *Acinetobacter calcoaceticus*, and *Pantoea agglomerans* can completely degrade SDS into water and carbon dioxide by microorganisms.

1.3.2 Cationic surfactants

In the biodegradation of cationic surfactants, QACs are governed by the molecular structure of the same, the concentration of the surfactants, and the resistance of microorganisms to the surfactants. During QAC degradation, microbes generally follow three distinct pathways depending on the strain and surfactant, initiated with the hydroxylation of the alkyl chain, and then goes through ω-oxidation and β-oxidation, which results in acetyl coenzyme A for the tricarboxylic acid cycle. In the second pathway, there is partial hydroxylation of alkyl chain, which is trailed with partial cleavage of the C-N bond under the action of the enzyme, and finally the degradation is completed by β-oxidation. In the third pathway of cationic surfactant degradation, hydroxylation long chain alkyl is accompanied with demethylation of methyl carbon. A few QAC degradation efficient strains are *Pseudomonas* sp., *Xanthomonas* sp., *Thalassospira* sp., and *Aeromonas* sp. The strong germicidal effect and soothing absorbance of QACs in the sediment, minerals, or organic matter with positive charge in wastewater makes biodegradation more challenging.

1.3.3 Nonionic surfactants

The length and complexity of the alkyl chain, and the number of elements in the benzene ring and polyoxyethylene chain mostly influence the biodegradation of APEO, which is an ideal nonionic surfactant. The branched chain structure and higher units in polyoxyethylene chain add to the difficulties of biodegradation. Microorganisms break the long chains to short polyoxyethylene chains, octylphenol, and nonylphenol. The higher toxicity of the intermediates/metabolites, such as octylphenol and nonylphenol, than the parent compound APEO demand complete biodegradation. A few strains efficient in degradation of APEO are *Pseudomonas* sp., *Moraxella osloensis, Cupriavidus* sp., and *Brevibacterium* sp. *TX4*. Oxidation of terminal hydroxyl group, a single unit of polyoxyethylene chain, to carboxylic acid is followed by its removal, which is the initiation of APEO biodegradation.

Alkylphenols are highly resistant to microbial degradation. Two different mechanisms are recognized by different microorganisms to degrade it. The first mechanism includes hydroxylation under the action of microbial enzymes, resulting in cleavage of alkyl chains and phenols, followed by subsequent degradation by microorganisms. The second mechanism includes formation of alkyl catechol through phenol hydroxylase, followed by the cleavage of an aromatic ring by the action of dioxygenase, which is readily biodegradable.

The higher toxicity of the intermediates generated from APEo result in a gradual decrease or banning and increasing demand of aliphatic alcohol polyoxyethylene ether (AEO). AEO consists of numerous congeners, containing an alkyl chain of 12–18 carbon atoms connected to a polyoxyethylene chain of different unit count, which is widely used in decontamination, cleaning, care, and other industries. Strains such as *Pseudomonas aerugirosa, Pseudomonas stutzeri*, and *Flavobacterium* sp. are able to degrade AEO, where initially the ether bonds between the alkyl chain and the polyoxyethylene chain are broken to generate fatty acids and polyethylene glycols, and a few other metabolites for degradation.

1.3.4 Amphoteric surfactants

Betaine type surfactants are relatively easy to biodegrade and their primary biodegradation rate can reach 100%. Strains capable of degrading betaine surfactants include *Desulfobacterium, Eubacterium limosum,* and *Sporomusa ovata*.

1.4 Factors affecting the biodegradation of surfactants

The length and branching degree of alkyl chains, the chemical structures of surfactants, along with the number of units of polymers, benzene rings, and their location specifically govern the biodegradability of a surfactant. In addition, the type and population strength, as well as diversity highly influence biodegradation. High adaptability, mixed culture, or high diversity are favored for high degradation of surfactants in favorable environmental conditions.

1.5 Conclusions

Surfactants have many uses, such as domestic cleaners and soap, and in the pharmaceutical, agricultural, food, medicine, and other industries. Despite their use for hygiene and as detergents, the complex structure of surfactants can cause them to become toxic in the environment when they buildup. Fortunately, there are efficient treatment processes to handle waste surfactants. Bioremediation can overcome the drawbacks of these processes, such as generation of secondary waste generation, in physical-chemical processes and is facilitated by numerous microorganisms equipped with different types of cellular molecules and enzymes that are specifically required for breaking the chemical bonds in surfactants. Microbes follow different pathways for biodegradation for different surfactants, as well as type of microbes. Nature has many microbes whose potential is yet to be recognized, and many pollutants and chemicals are released everyday whose removal and degradation is yet to be understood. Therefore, analysis by biologists, bioinformaticians, and chemists will help to develop solutions for these environmental hazards.

References

Adak A, Bandyopadhyay M and Pal A 2005 Removal of anionic surfactant from wastewater by alumina: a case study *Colloids Surf.* A **254** 165–71

Berna J L, Cassant G, Hager C D, Rehman N, López I, Schowanek D, Steber J, Taeger K and Wind T 2007 Anaerobic biodegradation of surfactants—scientific review *Tenside Surfact. Deterg.* **44** 312–47

Christopher J M, Sridharan R, Somasundaram S and Ganesan S 2021 Bioremediation of aromatic hydrocarbons contaminated soil from industrial site using surface modified amino acid enhanced biosurfactant *Environ. Pollut.* **289** 117917

Koner S, Pal A and Adak A 2012 Cationic surfactant adsorption on silica gel and its application for wastewater treatment *New Pub: Balaban* **22** 1–8

Okada D Y, Delforno T P, Etchebehere C and Varesche M B A 2014 Evaluation of the microbial community of upflow anaerobic sludge blanket reactors used for the removal and degradation of linear alkylbenzene sulfonate by pyrosequencing *Int. Biodeterior. Biodegrad.* **96** 63–70

Siyal A A, Shamsuddin M R, Low A and Rabat N E 2020 A review on recent developments in the adsorption of surfactants from wastewater *J. Environ. Manage.* **254** 109797

Chapter 2

Advanced perspectives in industrial wastewater treatment: a novel approach for a sustainable environment

Jyoti Gulia, Yashika Rani, Amit Lath, Nater Pal Singh and Anita Rani Santal

Environmental challenges have been a major concern not only for society and public authorities over the past 30 years but more crucially the entire industrial world, particularly those related to the chemical and biological poisoning of water. Domestic, agricultural, and industrial processes all generate wastewater and effluents that may include harmful chemicals. Consequently, managing water supplies is an ongoing priority. The removal of both insoluble particles and soluble contaminants from effluents is the primary goal of conventional wastewater treatment, which typically employs a number of physical, chemical, and biological processes and activities. In this chapter, we will take a quick look at the various wastewaters that can be produced, describe a conventional and advanced wastewater treatment system, and study the limitations of the various technologies.

2.1 Introduction

The modern industrial sector has expanded significantly. The manufacturing sector generates a lot of trash, and among its byproducts is wastewater, which is of tremendous interest for reuse because of the shortage of drinkable water in most nations. However, there are serious ecological concerns related to wastewater disposal. Distinct products are being made with them, and unique and tenacious wastes are released from the production process (Obaideen *et al* 2022). Therefore, if only convective wastewater systems are capable of handling these contaminants, then the world's already acute potable water crisis may be aggravated. Therefore, it is essential to discover novel approaches to lessen the harm that wastewater produces to the globe. However, clean drinking water is increasingly hard to get due to pollution from industrialization (Saravanan *et al* 2021). So far, over a hundred methods, using the chemical, physical, and biological realms, have been documented for treating organic and inorganic wastewater. Industrial effluents pose a significant threat to public health and the environment, whereas wastewater treatment aims to eliminate these risks (Goh *et al* 2022). As with other forms of slow-rate land treatment, such as irrigation, the reuse of wastewater is an effective means of wastewater disposal. Raw urban wastewater normally requires some degree of treatment before it can be used for agricultural or landscape irrigation or aquaculture (Maryam and Büyükgüngör 2019). The quality of treated effluent used in agriculture has a direct effect on the efficiency and effectiveness of wastewater, soil, plants, and aquaculture systems. To meet the recommended microbiological and chemical quality requirements, effluent treatment methods should be both cost-effective and low maintenance so that they can be reused even in agriculture (Krzeminski *et al* 2019). For a variety of reasons, including cost and the difficulty of properly operating complex systems, the simplest treatment level is the best option in poor countries. In many cases, it is preferable to design the reuse system to accept a lower quality of wastewater rather than to rely on sophisticated treatment techniques to create reclaimed sewage that consistently meets a high-quality standard. The efficacy of various wastewater treatment systems must be evaluated because there are locations that call for stronger effluent. Wastewater treatment plants are often planned to reduce organic and suspended particle loads (Capodaglio 2017). Only pathogen eradication has been appropriate as a critical concern in choosing and implementing treatment systems for agricultural wastewater reuse (Al-Sa'ed 2007). Treating wastewater can remove components that could be harmful or injurious to crops, aquatic plants (macrophytes), and fishes, but is usually not an economically viable solution. Unfortunately, both the effluent quality standards crucial for agricultural use and the efficiency of wastewater treatment plants in developing nations are little understood. Short-term changes in wastewater flows observed by municipal treatment plants exhibit diurnal patterns. Base flow is mostly infiltration and insignificant volumes of sanitary wastewater early in the morning when water demand is at its lowest (Ofori *et al* 2021). The first peak of flow occurs in the late morning when wastewater from the highest water use period of the day reaches the treatment plant. The second peak of flow occurs in the early evening. Peaks occur at different times and in different amounts for different countries because of differences in population and the capacity of their sewer systems. Smaller communities with less extensive sewer systems have a much greater peak flow-to-average flow ratio (Penn *et al* 2012). Although the amplitude of peaks is reduced as wastewater runs through a treatment plant, the daily fluctuations in flow make it impracticable, in most cases, to irrigate with effluent straight from the plant. Efficient irrigation with reclaimed water calls for either flow equalization or short-term storage of treated effluent, with the latter offering additional benefits (Pettygrove *et al* 2018).

2.2 Wastewater and its characteristics

Water that has been tainted, typically by human activity, is known as wastewater. Domestic, commercial, agricultural, and stormwater runoff all contribute to the

99.9% water content of wastewater (Khan *et al* 2022). The composition and flow rate of the wastewater being treated is taken into account during the construction of each plant. According to Davis and Cornwell (2008), the concentrations of the most important physical, chemical, and biological constituents of wastewater are used to calculate its pollutant load. The characteristics of wastewater are determined by the quality of the water utilized by the community, the population's conservation habits and culture, the kind of industries in the area, and the treatment that those industries and their wastewater receive. The aforementioned factors often influence one another. Temperature influences many aspects of wastewater, including the quantity of dissolved gases and the activity of microorganisms. The particles in wastewater, which can either be floating or sedimentary, or suspended or soluble, are among its most distinguishing physical features (Banasiak *et al* 2005). Additional characteristics include heat, color, and clarity. In addition, both organic and inorganic substances are included in terms of their chemical properties. Carbon, hydrogen, oxygen, and sometimes nitrogen make up organic compounds. Many inorganic indicators of wastewater are crucial to the creation and regulation of wastewater quality standards, especially in relation to organic compounds (Crittenden *et al* 2012). Biochemical oxygen demand (BOD), chemical oxygen demand (COD), total solids, volatile solids, total nitrogen, total phosphorus (TP), pH, and alkalinity, among others, are often measured chemical properties of wastewater. Natural evaporation removes some water from wastewater but it leaves behind inorganic debris, which increases the concentration of those compounds (Young and Lipták 2018).

2.3 Various parameters of wastewater

2.3.1 pH

The pH of wastewater is determined by the raw water's chemical composition. Furthermore, the pH of a water supply can be affected by the many different chemicals and compounds found in industrial effluent and sewage. Bases such as sodium hydroxide (caustic soda) or acids such as hydrochloric acid can be introduced to the treatment process to adjust the pH level (Koul *et al* 2022). The effluent from a wastewater treatment plant must fulfill stringent discharge quality criteria, hence it is crucial that the pH of the wastewater be closely monitored and controlled at all times. Several operational stages of the treatment process are extraordinarily pH dependent. Certain suspended solids can be precipitated out at different pH values for removal, coagulants, and flocculants operate in specific pH ranges; bacteria used to degrade organic material will not survive if the pH is too high or low; and chlorine-based disinfectants are less effective under basic conditions. The pH value of different wastewaters varies depending on the chemicals present in them. For example, dairy effluent has an alkaline pH (6.6–12.2) whereas distillery wastewater has acidic pH (3.0–4.5) (Chandra *et al* 2017). The pH was found to range from 6.2 to 6.9 at the sewage treatment plants inlet during batch reactor investigation, and from 7.1 to 2.5 at the plant's exit. The technique of treating sewage with cyclic-activated sludge led to this elevation in pH (Kulkarni *et al* 2016). In another membrane bioreactor investigation (Ashok *et al* 2018), the pH was found to be constant between 6.5 and 8.5 at both the reactor's inlet and exit.

2.3.2 Temperature

Some industries discharge wastewater with temperatures in the 40 °C–50 °C range. For optimal efficacy during treatment, wastewater should be allowed to cool to room temperature. Ahsan *et al* (2005) investigated how temperature affected wastewater treatment using waste materials (refuse concrete and refuse garbage) and natural material. He found that increasing the temperature did not result in a significant improvement in removal efficiency for nitrate ions and ammonium ions. An activated sludge process is sensitive to temperature changes in organic carbon and nitrogen removal. Effluent concentrations are analyzed throughout a wide temperature range of 15 °C–35 °C using a steady-state simulation with changing kinetic parameters obtained from different temperature coefficients. The most often used temperature range in a wastewater treatment plant in India was 25 °C–30 °C. Temperatures below 20 °C and above 30 °C are also found to be in violation of the effluent limitations from the standard values (Tejaswini *et al* 2019).

2.3.3 Biological oxygen demand

BOD quantifies the quantity of oxygen needed by aerobic bacteria to decompose organic waste in water. Living bacterial organisms, which require oxygen to carry out their decomposition, stabilize or render the waste organic matter unobjectionable (Pasco *et al* 2000). BOD is a common water quality indicator that is used in wastewater treatment facilities. There is a significant quantity of organic material in residential sewage. The BOD was determined to be 198.67 mg l^{-1} at the entrance of a sewage treatment plant and 30–20 mg l^{-1} at the exit (Patil *et al* 2018). According to another study that evaluated a biological strategy for treating effluent, the BOD was 225 mg l^{-1} at the intake and 9 mg l^{-1} at the exit (Rajkumar 2016).

2.3.4 Chemical oxygen demand

The amount of oxygen needed to oxidize the organic molecules in water through chemical reactions involving oxidizing chemicals is known as the COD (Zhao *et al* 2004). The estimation of COD is of great importance for waters having unfavorable conditions for the growth of microorganisms, such as the presence of toxic chemicals. COD is always higher than the value of BOD. COD levels at the intake range from 315.1 to 365.6 mg l^{-1}, whereas those at the output range from 51.2 to 56.0 mg l^{-1}, according to research on the effectiveness of wastewater treatment plants (Dahamsheh and Wedyan 2017). Research on the effectiveness of an algae-based sewage treatment facility found that the input COD was 458.7 mg l^{-1} and the output COD was 208 mg l^{-1} (Mahapatra *et al* 2013).

2.3.5 Dissolved oxygen

The wastewater dissolved oxygen content is the amount of oxygen gas that has been dissolved in the water. Oxygen (O_2) can be taken in either from the air or from the waste products of the photosynthesis of nearby plants. In most cases, the detection of BOD in wastewater indicates that DO is not present. Microorganisms need dissolved oxygen, either naturally occurring or artificially produced, to breakdown the organic materials in wastewater. According to the research done on decentralized wastewater treatment, the DO at the intake was lower than the detection limit, but it was raised to 2 mg l^{-1} at the exit. Since the DO level in the aerobic reactors significantly influences the behavior and activity of the heterotrophic and autotrophic microorganisms living in the activated sludge, DO control is the most widely spread in practice (Mathew *et al* 2022). To facilitate the breakdown of organic matter and the conversion of ammonium to nitrate, the dissolved oxygen content in the aerobic portion of an activated sludge process should be high enough to deliver enough oxygen to the microorganisms in the sludge. However, if the DO is too high, then the airflow rate needs to be increased, which increases energy consumption and could reduce sludge quality. Denitrification efficiency is decreased due to high DO levels in the water that is used for internal recycling (Daigger and Littleton 2014).

2.3.6 Total dissolved solids

Total dissolved solids (TDS) is a measure of the molecular, ionized, or micro-granular (colloidal sol) suspended content of all inorganic and organic compounds present in wastewater. The concentration of TDS is often expressed as a ppm value. A digital meter can be used to monitor TDS levels in wastewater. It is a general indicator of the presence of numerous chemical pollutants and can be used to judge the visual quality of drinking water. According to the results of the experiments conducted to evaluate the operation of the sewage treatment plant, the TDS is 497.78 mg l^{-1} at the intake and 434.01 mg l^{-1} at the outlet (Negi and Sahu 2015). The intake TDS is 782 mg l^{-1} and the output TDS is 859 mg l^{-1} in the research of treatment efficiency of algae-based sewage treatment plants (Mahapatra *et al* 2013).

2.4 Types of various industrial wastewater

2.4.1 Distillery wastewater

Distillery effluent is characterized by its very dark color, high COD (110 000–190 0000 mg l^{-1}), high BOD (50 000–60 000 mg l^{-1}), large amounts of organic and inorganic salts leading to high electrical conductivity (30–45 dS m^{-1}), and TDS (90 000–150 000 mg l^{-1}), as well as an offensive odor (Mahimairaja and Bolan 2004) (table 2.1). Distillery effluent quality is mostly determined by the substrate utilized and the process parameters employed (Pant and Adholeya 2007). Due to the inclusion of complex recalcitrant organic contaminants and heavy metals, it is well known that the disposal of distillery effluent can pose major environmental problems (Santal *et al* 2011, 2016). High levels of BOD and COD are not the only

Table 2.1. Physico-chemical characteristics of distillery wastewater. Reprinted by permission from Springer Nature, Chandra and Kumar (2017), Copyright (2017).

Sl. No.	Components of wastewater	Values of distillery wastewater (mg l^{-1})	References
1	pH	3.0–4.5	Chandra and Kumar (2017)
2	Temperature	70 °C–80 °C	Mohana *et al* (2009)
3	Odor	Pungent	Santal *et al* (2016)
4	BOD	50 000–60 000	Chandra *et al* (2017)
5	COD	110 000–190 0000	Tripathi *et al* (2022)
6	TDS	90 000–150 000	Kumar *et al* (2020)
7	Color	Dark brown	Santal *et al* (2011)
8	Total solids	150 300	Tripathi *et al* (2022)
9	Total suspended solids	13 000–15 000	Chandra *et al* (2018)
10	Heavy metals (Cu, Zn, Cr)	1.1–1.5, 1.4–2.8, 0.8–2.3	Tripathi *et al* (2022)
11	Total nitrogen	5000–7000	Tripathi *et al* (2022)
12	Volatile suspended solids	200–7340	Chandra *et al* (2017)
13	Total nitrogen	100–200	Chandra *et al* (2017)
14	Total phosphorus	150–600	Chandra *et al* (2017)

negative effects on aquatic ecosystems that this waste can have (Chandra *et al* 2018, Shah 2020). Despite the difficulty, treating these streams is essential. Because of its high pollution load (COD 90 000 mg l^{-1}) and the existence of several nutrients in it (e.g., nitrogen, potassium, and phosphorus), the discharge of untreated effluent may promote the growth of several algae and other organisms, rendering the water bodies prone to eutrophication (Mathew *et al* 2022). The amount of sunlight that can penetrate rivers, lakes, and lagoons is diminished when highly colored components are present. Because of this, the photosynthetic activity of the aquatic microbiome decreases and the oxygen content of the water decreases along with it (Apollo and Kabuba 2022). The color of effluents is due to the presence of a dark brown color called melanoidin (Santal and Singh 2013). Distillery effluent, rich in organics and nutrients, is occasionally utilized for the irrigation of crops, particularly in dryland situations. However, inappropriate discharge of waste wash on lands may have detrimental effects on soil health, and should therefore not be considered a viable alternative in sustainable irrigation practice (Kalra and Gupta 2023). Distilleries often have difficulty in disposing of their effluent because of its hue.

2.4.2 Textile wastewater

Effluents from textile mills typically contain a wide range of contaminants, including toxic dyes and heavy metals. Dye additives can be categorized as either natural or synthetic. Synthetic dyes are more popular than natural dyes because they can be mass-produced and in a variety of colors, which are known for their fastness (Santal *et al* 2022, Shah 2021). Chemical composition and method of use are used to

Table 2.2. Physico-chemical characteristics of textile wastewater. Reproduced from Imtiazuddin *et al* (2012). CC BY 4.0.

Sl. No.	Components of wastewater	Values of textile wastewater (mg l^{-1})
1	pH	11.5
2	Temperature	49.2
3	Odor	Pungent
4	BOD	450
5	COD	821
6	TDS	2212
7	Color	Black
8	Heavy metals (Cu, Zn, Cr)	13.12,15.21
9	Total nitrogen	100–200
10	Dyes	10–200

categorize synthetic dyes into several categories. Color, pH, suspended particles, chemical oxygen demand, biochemical oxygen demand, metals, temperature, and salts are all elevated in dye effluents (Islam *et al* 2023). Various types of heavy metals are also found in discharged wastewater, which can be carcinogenic to human beings (Sharma *et al* 2023, Singh and Santal 2015). Therefore, prior to discharging the effluent to the receiving water body, it is essential to monitor and compare these parameters with the standard concentrations during the treatment procedures. Total organic carbon (TOC), ammonia-nitrogen (NH_4-N), nitrate-nitrogen (NO_3-N), and orthophosphate-phosphorus (PO_4-P) are just a few of the metrics that need to be monitored in relation to treatment effectiveness. Table 2.2 provides a summary of the typical characteristics of textile wastewater. Khandare *et al* (2013) point out that standard textile wastewater is hard to define because textile application methods vary, even within the same process.

2.4.3 Sewage wastewater

Water that has been contaminated by human waste is called sewage water. Wastewater is the water that is discharged from homes after being used for activities such as dishwashing, laundry, and toilet flushing. Pipes installed during the plumbing process carry the wastewater away from the homes. Sewage treatment (also known as domestic wastewater treatment or municipal wastewater treatment) is a form of wastewater treatment whose goal is to clean sewage so that it can be safely released into the environment or reused for another purpose (Shinde *et al* 2023). Sewage wastewater treatment comprises the physicochemical or biological breakdown of complex organic molecules in wastewater into stable, non-polluting chemicals. Discharging untreated wastewater into groundwater, surface water, or land has negative environmental effects. The decomposition of wastewater's organic materials can produce large amounts of malodorous gases. If untreated wastewater (sewage) is discharged into a river or stream, then it will consume the stream's dissolved oxygen to satisfy its BOD, causing fish kills and other undesirable effects (Anusha *et al* 2023).

2.5 Innovative treatment methods

Innovative procedures for treating industrial wastewater sometimes use toxicity reduction technologies to meet treatment criteria based on technological advancements. Industrial wastewater treatment options can be roughly categorized into physical, chemical, and biological methods. For example, the lagoon and the floating approaches are both physically demanding and relatively large in scale (Islam *et al* 2023). While chemical treatment is effective at removing pollutants, such as through ion exchange or chemical precipitation, or through advanced oxidation processes such as the Fenton process, it also has the potential to acidify the environment, increase reagent costs, and increase ferric sludge production. The use of microorganisms as a biological approach is better for the environment and has been employed to effectively treat industrial wastewater (Tripathi *et al* 2022). Although several new technologies are being implemented, they are still in the development stages.

2.5.1 Innovative physical treatment methods

Various methods are involved in physical treatment methods, such as membrane separation, ion exchange, ultrafiltration membrane, and nanofiltration membrane. Water emulsion separation remains a serious issue because of the high concentration of pollutants in wastewater. Membrane technology is highly sought after in the water separation sectors due to its simple operation and great effectiveness of separation (Ravichandran *et al* 2022). The hydrophobic properties of a membrane may aid fouling, shortening the membrane's life. In recent decades, scientists have investigated the feasibility of using carbon nanotubes, nanoporous graphene, and graphene oxide to create high-capacity membranes for wastewater desalination and treatment. There are two types of composite membranes that use graphene or graphene oxide: (i) those that are made entirely of graphene or graphene oxide, and (ii) those that are made of polymers or ceramics but have had their atomic structure altered to include graphene-based nanomaterials. It is possible to modify polymeric membranes by adding a nanomaterial to the membrane casting fluid before shaping the membrane. Water and permeate fluxes through nanomaterial membranes may be significantly higher than those using thin-film composite polyamide membranes (Mustafa *et al* 2022).

2.5.1.1 Ion exchange method

Ion exchange membranes are a crucial part of water desalination and electrolysis processes. As a result, their potential in cutting-edge energy conversion and storage systems, as well as effective desalination and wastewater treatment processes, has prompted extensive research and development. Dense polymeric membranes are characterized by fixed charges in the polymer matrix. It is possible that counter-ions (ion pairs with opposite charges) can pass through these membranes, whereas co-ions (ions with the same charge) are prevented from doing so (Nallakukkala *et al* 2022). The Donnan effect or exclusion (towards co-ions) is the name given to the counter-ion preferential permeability of ion exchange membranes, which was first observed by Donnan. Ion selectivity in membranes is essential for several commercial processes, including electrodialysis (ED), diffusion dialysis (DD), and electrolysis. In addition to the well-known uses in flow batteries and ED/DD, ion exchange membranes are finding new life in emerging applications such as membrane capacitive deionization (MCDI), reverse osmosis (RO), membrane fuel cells, and ion exchange membrane bioreactors. Low fixed charge density ion exchange membranes are also being investigated as chlorine-resistant membranes for wastewater treatment (Sedighi, *et al* 2023). Ion exchange can also eliminate all ions from wastewater. Therefore, the difference between deionization and total ionic pollution removal can be estimated. The choice between the two is determined largely by the nature of the cleaning solution and the level of contamination that must be removed (Bao *et al* 2022). Used galvanic baths and waste solutions from metal manufacturing typically contain high concentrations of dangerous ions. In many cases, only one type of ion is of sufficient toxicity or value to warrant removal from industrial wastewater (Thomas *et al* 2023). New forms of ion exchangers with specialized affinity to certain metal ions or metal groups allow for greater selectivity. It is important to note that ion exchange typically allows for the replacement of an undesired ion with one that has no discernible effect on the surrounding environment.

2.5.1.2 *Ultrafiltration membrane*
Ultra-filtration (UF) technology has become more common in wastewater treatment systems as a result of improvements in membrane performance and reduced costs. However, membrane fouling is the main problem with ultrafiltration technology. It is generally accepted that natural organic matter is safe for human ingestion (Huang *et al* 2022). However, when natural organic matter in effluent water combines with disinfectants, a number of hazardous and carcinogenic disinfection byproducts may be formed. Chlorinated water contains a wide variety of chlorinated dichloro benzo (p)phenyl compounds, including, but not limited to, trihalomethanes, haloacetic acids, haloacetonitriles, and various halo ketones. Due to the characteristics of organic matter, dichloro benzo (p)phenyl compounds can be generated in UF effluent. The role of pretreatment on dichloro benzo (p)phenyl compound formation following ultrafiltration has not been investigated. Ultrafiltration membrane performance is determined by membrane structure, which is governed by membrane fabrication methods, environmental factors, and the chemical constituents of the casting solution (Liu *et al* 2021). Several researchers have looked at how factors such as (i) membrane preparation technique, (ii) nature of the base polymer and its concentration in wastewater, (iii) nature of the solvent, (iv) nature of the non-solvent and its composition in the coagulation bath (v) type of additives, and (vi) relative concentration affect membrane morphology to achieve the desired membrane performance (Figoli *et al* 2014). The choice of membrane production technique is crucial in achieving the desired membrane properties and morphologies, and is one

of the influencing parameters. Several types of membranes are shown in figure 2.1. Phase inversion is a common method for synthesizing asymmetric polymeric ultrafiltration membranes, in which a liquid polymer is converted to a solid state in a controlled manner. The technique of phase inversion can be used to achieve either a porous or nonporous membrane shape, depending on how the initial condition of phase inversion is controlled. The phase inversion approach can be used to create a membrane using any polymer. The inability to properly select suitable solvents is the fundamental limitation of the phase inversion method (Hołda and Vankelecom 2015). Membrane morphology and performance can be optimized through careful choice of solvent to regulate polymer-solvent interactions. Several methods are used to precipitate polymer solutions using the phase inversion technique. These include (i) evaporation-induced phase inversion (EIPS), (ii) vapor-induced phase inversion (VIPS), (iii) thermally-induced phase inversion, and (iv) immersion-precipitation or non-solvent induced phase inversion (NIPS). NIPS is the most widely used phase inversion method for creating polymeric UFMs with the appropriate morphology, due to its solubility in the solvent and the simplicity of its synthesis method (Menut *et al* 2008, Rajabzadeh *et al* 2009). Guillen *et al* (2011) reported the mechanisms involved in the membrane's creation utilizing the NIPS approach. With this method, a uniform layer of CS composed of polymer and solvent is cast onto a glass plate at a specific depth. The polymer film hardens as it sinks into the non-solvent conduction band (CB), with solvent and non-solvent exchanging places until the system is thermodynamically stable.

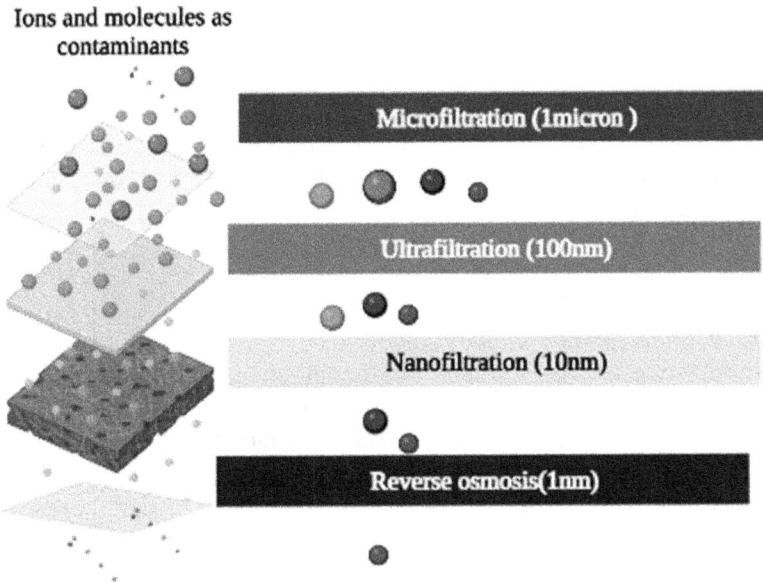

Figure 2.1. Membrane filtration method showing different processes on the basis of membrane thickness (Huang *et al* 2022). Reprinted from Saravanan *et al* (2021), Copyright (2021), with permission from Elsevier.

2.5.1.3 Nanofiltration method

Nanofiltration (NF) was discovered to be particularly appealing for a variety of uses, including water recycling, industrial wastewater treatment, and potable water production. Although RO and UF have found widespread use, their respective fields of application are constrained, making further expansion difficult (Warsinger *et al* 2018). In contrast, NF applications are rapidly growing and replacing existing membrane filtration methods. The manufacture of the NF membrane is flexible because it is made from a combination of components, including RO membrane polymers such as cellulose acetate and polyamide polymers, and other chemically resistant polymers (Nunes *et al* 2020). To endure high temperatures, NF membranes are also currently manufactured of ceramic materials. The versatility of NF preparation and the range of raw materials available will boost its use in a wide range of operations. Since NF can use a wide variety of raw materials and can easily be adapted to serve a variety of purposes, the research community would do well to devote greater resources to its advancement. Isotropic micro-porous, nonporous, dense, electrically-charged, asymmetric, ceramic, and liquid membranes are some of the many membrane types that have been identified based on membrane structure and pore shape (Abdel-Fatah 2018). NF is a technique whereby some of the feed is allowed to pass through a semipermeable membrane (figure 2.1). The incoming stream is separated into a filtered part called permeate and an unfiltered concentration called retentate or concentrate. NF has been shown to efficiently remove organic matter. Although microbial growth has been reported in NF distribution systems, this can be eliminated with chlorine treatment. NF membranes that are low in organic material retention and high in organic material removal can decrease microbial growth and create high-quality water (Choudhury *et al* 2018). The chemical industry, food, textiles, metal finishing, pulp and paper, pharmaceutical and biotechnology applications, and power generating are just some of the many industrial processes that frequently use membrane processes. The most commonly reported uses are desalination of the food industries; partial desalination of whey; desalination of textile dyes and optical brighteners; the removal of metal, nickel, and chrome plating from the metal finishing industries and purifications spent clean-in-place (CIP) chemicals.

2.5.2 Innovative chemical treatment methods

2.5.2.1 Advanced oxidation method

The advanced oxidation method includes the photocatalytic oxidation process and Fenton's oxidation process. Photocatalytic oxidation technology is an efficient and energy-saving green technology to treat organic wastewater. High efficiency and enhanced recycling are two of the main advantages of photocatalysis over more conventional wastewater treatment methods. This method can be used for the treatment of industrial wastewater, including pharmaceutical wastewater, dyeing wastewater, and other wastewater containing a variety of refractory compounds. Photocatalytic oxidation is frequently employed in combination with other advanced oxidation processes. The theory of photocatalysis is on the basis of the redox capability of photocatalysts in the presence of light (Liu *et al* 2020). In photocatalytic processes, semiconductors are typically utilized as catalysts to convert organic contaminants into carbon dioxide and water. Valence band and CB topologies can both be found in semiconductor photocatalysts (Wang *et al* 2014). The area between the C and V bands is known as the prohibited band. Semiconductor catalysts absorb photons of a particular energy and release electrons and holes with reduction and oxidation capabilities, respectively (Al-Nuaim *et al* 2023). Soon after, some of the created e/h will migrate to the surface of the catalyst, where it will take part directly in the degradation process of the pollutant and some will form reactive oxygen species to take part in the action.

2.5.2.2 Fenton's oxidation process

In the process of Fenton oxidation, Fe^{2+} and hydrogen peroxide (H_2O_2) undergo a chain reaction to produce catalytic OH. In general, OH is a kind of highly oxidizing radical that is effective at oxidizing a wide range of potentially harmful and inert chemical molecules. Fenton oxidation is recommended for organic wastewater that is difficult to biodegrade or treat with traditional chemical methods (such as landfill leachate) (Zhang *et al* 2019a). In the standard Fenton reaction, H_2O_2 is degraded to OH under the catalysis of Fe^{2+}. So, OH can oxidize organic materials and break them down into small molecules via electron transfer. The Fenton-like reaction, in contrast to the standard Fenton reaction, is not restricted to the interaction between Fe^{2+} and H_2O_2. Multiple free radicals can be involved in the breakdown of organic contaminants through Fenton-like reactions (Kurniawan *et al* 2006). Many different types of catalysts besides Fe^{2+} have been shown to be effective in Fenton-like reactions, including Fe(II), Fe(III), Cu(II), nano zero-valent iron, metal foam-based catalysts, oxygenated g-C3N4, and bimetallic organic frameworks. Heterogeneous Fenton reactions involve catalysts immobilized on porous supports, such as mesoporous silica, graphene oxide, metal–organic frameworks, carbon nanotubes, and so on.

2.5.3 Innovative biological methods

Trhe teatment of industrial effluents through the biological removal of dyes and heavy metals from wastewater shows promise over more conventional approaches (Zhu *et al* 2019). These can be partially degraded, mineralized, or transmuted by using microorganisms to immobilize their molecular structure. Significant progress has been achieved in recent years toward the use of biological techniques for the long-term and low-cost removal of dyes and heavy metals from wastewater. Despite these advances, there is still a need for a unified platform that incorporates biological approaches for remediating recalcitrant compounds present in wastewater. Biological treatment methods coupled with these other processes are more practical for treating industrial wastewater (Ahmed *et al* 2021). Because of its elevated mitigation and long-term viability, biological treatment processes are particularly well-suited for hybridization with cutting-edge approaches. Microorganisms may degrade many different substances, and consequently biological processes can handle a wide variety of wastes. Most industrial waste can be eliminated using microorganisms in the treatment process, which utilize complex organic molecules and some inorganic chemicals as a substrate and drive energy out of it. Different pollutants and types of microorganisms have different effects on biological treatments. The aerobic treatment uses oxygen-consuming bacteria to degrade organic molecules in wastewater. Oxidative degradation is the process by which organic molecules are broken down into simpler components, such as water, carbon dioxide, and biomass. Meanwhile, anaerobic degradation happens in the absence of oxygen, and it is the anaerobes that are responsible for breaking down complex organic compounds into simpler ones. Experimental research by Ledakowicz *et al* (2017) integrated the processes of ozonation and biological treatment before degradation, creating an enhanced method for removing industrial effluent from the textile industry. The purpose of this method is to enhance current methods used to treat wastewater. Bioreactors have been used to improve upon the conventional biological treatment method of activated sludge. Some of the processes that make up conventional biological treatment are nitrification and denitrification, composting, bioremediation, and biofiltration. Some novel methods have been developed to improve the effectiveness of biological degradation as part of non-traditional treatment processes.

Bacteria are the most widespread microorganisms, making up a sizable portion of the planet's total biota, and have been studied for their ability to take up contaminants. Bacterial membrane cell walls function as powerful heavy metal chelating mediators. Polysaccharide slime layers are present on bacterial membranes, and they include functional groups that aid in the binding of heavy metals present in wastewater. The binding of heavy metals to the outer surface of a bacterial cell wall typically involves two distinct steps. In the first step, heavy metal ions interacted with reactive groups on the cell surface. In the second step, successive heavy metal species accumulated at elevated concentrations. The glutamic acid carboxyl groups in peptidoglycan are the most common sites of heavy metal deposition. In addition, a major pathway for heavy metal uptake in a number of bacteria is metabolically independent biological adsorption. In other wastewater treatment bacterial isolates, *Alcaligenes faecalis* and *Paracoccus pantotrophus* showed 72.6% and 81.2% decolorization in distillery wastewater, respectively (Santal *et al* 2011, 2016). Various potential bacterial isolates have been found for degradation of textile dyes from wastewater. These microbes include *Lysobacter* sp. (Ranga *et al* 2015); *Nesterenkonia lacusekhoensis* (Prabhakar *et al* 2022); and *Bacillus* sp. (Santal *et al* 2022), a consortium including *Bacillus* sp., *Stenotrophomonas* sp., *Pseudomonas* sp., and *Alcaligenes* sp. (Shah and Bera 2021). Decolorization of 82% of mixed azo dyes and degradation of 56% of Cr (VI) metal were observed on the fifth day of the study at 37 °C and pH 8.5 in a moving bed biofilm reactor treating wastewater from synthetic textiles with 1200 ppm concentrations of mixed azo dyes—reactive red, reactive brown, and reactive black, and 300 ppm Cr (VI) metal (Biju *et al* 2022). Conventional and advanced biological treatment methods for industrial wastewater treatment process are shown in figure 2.2.

2.6 Microbial fuel cells for wastewater treatment

MFCs can be put to use in the production of both bioenergy and purified wastewater. These bio-products can be made by simply converting the organic and chemical energy present in wastewater into electrical energy. An MFC is a device that converts organic materials into energy using microorganisms as the biocatalyst. The three basic components of a standard MFC are electrodes, a separator, and electrogens. Each MFC has two electrodes, which are mechanically divided into either one or two chambers. These sections have completely mixed reactors (Gude 2016). Cation exchange membranes and proton exchange membranes have electrodes on both sides of the membrane. In this arrangement, the anode faces the chamber containing the liquid phase, while the cathode faces the chamber containing just air. The aforementioned literature cites carbon, graphite, and metal-based compounds as candidates for electrode components. Carbon, platinum, platinum black, carbon cloth, paper, felt, graphite granules, carbon mesh, and activated carbon can all be used as either single or multiple electrodes (Chauhan *et al* 2023). These electrodes must be stable and compatible with living tissue. It is also preferred if the material has high electrical conductivity and a large surface area. Electron acceptors can be added to the cathode in the form of permanganate, chromium hexacyanoferrate, azo dye, or oxygen. A separator, such as a cation exchange membrane or a salt bridge, keeps the chamber isolated (Verma *et al* 2021). By generating a potential difference between two chambers in a circuit, bioelectricity can be generated from the microbial breakdown of wastewater. MFCs have been proposed for use in wastewater treatment (Singh *et al* 2019). Many

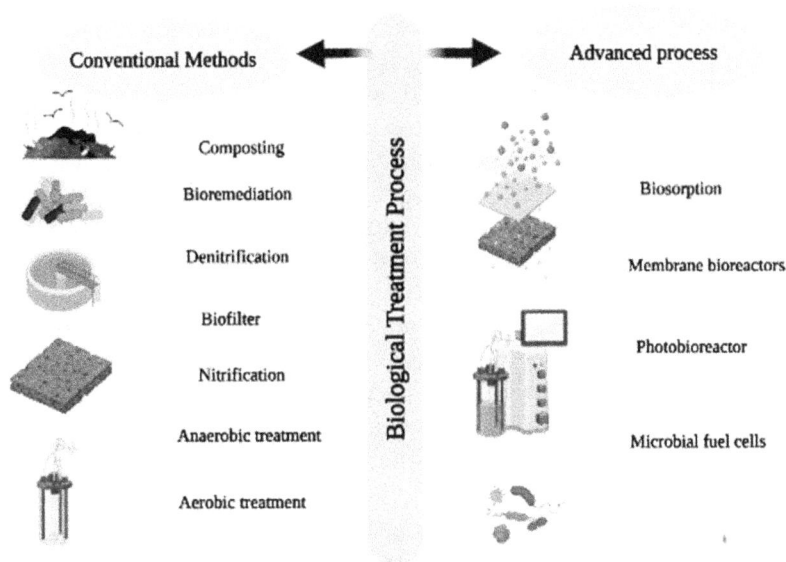

Figure 2.2. Conventional and advanced biological treatment methods for industrial wastewater treatment process. Reprinted from Ahmed *et al* (2021). Copyright (2021), with permission from Elsevier.

organic compounds found in municipal wastewater can be used to power MFCs. The amount of power created by MFCs in this process has the ability to slash in half the amount of electricity needed for aerating the activated sludge in a conventional wastewater treatment approach. MFCs reduce solid waste by 50%–90% when compared to conventional activated sludge treatment methods. Combining anaerobic digesters with aerobic sequencing batch reactors is a viable solution for sludge disposal issues. Complete oxidation of acetate, propionate, and butyrate leads to carbon dioxide and water. Because of the wide variety of organic compounds that can be biodegraded, a hybrid MFC including both electrophiles and anodophiles is optimal for wastewater purification. It is vital for the treatment of wastewater that some microbial fuel cells (MFCs) have been shown to have greater sulfide removal efficiency. By fostering the development of bio electrochemically active bacteria, MFCs can enhance the wastewater treatment process and bring it closer to a state of operational stability (Zhang *et al* 2019b). Continuous flow, single-compartment MFCs, and membrane-less MFCs are chosen for wastewater treatment due to the difficulty in scaling up alternative technologies. Sewage, food processing effluent, swine wastewater, and corn stover are all good sources of biomass for MFCs because they have a high concentration of organic matter (Rahimnejad *et al* 2015).

MFC technologies present a novel approach to wastewater treatment because instead of being a net consumer of electrical energy, the treatment process can now collect energy in the form of electricity or hydrogen gas. Kim *et al* (2016) first used bacteria in a biofuel cell as an indicator of water lactate concentration, which allowed energy generation. Although the technology's power output was low, its effectiveness in weakening wastewater was not known. The domestic wastewater could be treated to practical levels while producing energy, and a direct link between electricity generation using MFCs and wastewater treatment was established. The power output in this study ($26\ mW\ m^{-2}$) was lower than that found with other types of wastewater, but still significantly higher (Li *et al* 2014). Organic and inorganic matter in marine sediments could be used in a novel type MFC design, demonstrating the versatility of substrates, materials, and system architectures for harnessing electricity from bacterial biomass. Even with all of these features, power consumption was still generally modest. The recent finding that glucose can increase the power density produced by an MFC has stimulated broad interest in MFCs. This was an entirely biological process that did not necessitate any external chemical mediators or catalysts. The major objective is to develop a flexible MFC that may be used in a variety of wastewater treatment settings. The small sample of studies on MFC technology shows that the energy output per person is not particularly impressive compared to conventional wastewater treatment plants (WWTPs).

MFC offers significant energy reductions in aeration and solids handling (Choudhury *et al* 2017). Additionally, these resistant substrates can be used to generate electricity and remove high-concentration pollutants. The primary components of wastewater treatment operating costs are aeration, sludge treatment, and pumping. Some estimates place the cost of aeration at 50% of a WWTP's overall operating expenditures. Combining MFCs with other treatment technologies can mitigate this expense and WWTPs can become energy positive.

2.6.1 Limitations of MFC wastewater treatment

When comparing MFCs to currently used wastewater treatment technologies, there are various operational and functional benefits and drawbacks associated with MFCs. These include the ability to remove large amounts of organic pollutants (in the form of CODs) and to valorize bioenergy (in the form of electricity) (Hernández-Fernández *et al* 2015). Most people understand that the MFCs' bioenergy component is around their use of wastewater treatment to generate green or blue power. When electricity is generated from biomass and organic matter, chemical energy is directly converted to electrical energy. It has also been reported that the direct conversion of wastewater substrates to bioenergy accounts for the input during the thermal combustion of biogas (Ahmed *et al* 2021). The collection of electrical energy in an MFC significantly reduces the sludge yield in comparison to aerobic process sludge production. While hydrogen and methane are the primary byproducts of an anaerobic process, nitrous gases are also commonly present in the off-gas (Maktabifard *et al* 2018). Because the substrate's energy was traditionally channeled into the MFC's anodic chamber during processing, MFC off-gases have decreased economic viability. Since the anodic chamber gas in an MFC does not include any particularly strong chemicals or aerosols with dangerous or undesired bacterial components, it can be released straight into the atmosphere. While there have been significant advances in power production via MFCs, and even achievement of the primary power aim in small-scale systems, scaling up remains a critical difficulty and constraint on the general usage of these technologies. The expensive cost of cation exchange membranes, the potential of biofouling, and the accompanying high internal resistance limit their use in power generation and commercial applications. Domestic wastewater contains organic debris that contains around 10 times the energy needed for treatment (Nawaz *et al* 2022). Although there are several promising new approaches, no existing technology is yet able to fully recover all of the energy present in wastewater. When considering the inoculum, electrode materials, reactor design, and operating parameters (e.g., temperature, external loading rates, etc), the substrate type has the greatest impact on the time it takes for the MFC system to begin producing useful energy. When it comes to treating wastewater, MFCs' limited electrolyte buffer capacity is a major limitation. Some external mediators or chemical material may be required to keep and stabilize the hydrogen potential of the anodic and cathodic chambers. In the MFC context, this will improve wastewater treatment and encourage the valorization of bioenergy (Kataki *et al* 2021).

2.7 Conclusions

Wastewater has been shown in numerous studies to contain cytotoxic, genotoxic, and carcinogenetic chemicals that stay in the environment for long periods of time and have deleterious effects on the biota. This chapter has examined several physical, chemical, electrochemical, and biological treatment methods. Each method has its own set of benefits and drawbacks. Various filtration methods were also discussed. Due to the improved treatment efficiency of membrane bioreactors, low land area usage, and easy operation, this technique has become a popular technology. Although membrane bioreactors face significant obstacles, they are a promising treatment technique that may eliminate contaminants and reduce effluent concentration in large-scale applications. However, microbial fuel cells are also being deployed for wastewater treatment, which can maximize the benefits of energy generation using wastewater. The overall efficiency of the treatment process can be increased to roughly 90%. It has been found that the Fenton procedure is effective at removing COD and color from acidic wastewater.

References

Abdel-Fatah M A 2018 Nanofiltration systems and applications in wastewater treatment *Ain Shams Eng. J.* **9** 3077–92

Ahmed S F, Mofijur M, Nuzhat S, Chowdhury A T, Rafa N, Uddin M A and Show P L 2021 Recent developments in physical, biological, chemical, and hybrid treatment techniques for removing emerging contaminants from wastewater *J. Hazard. Mater.* **416** 125912

Ahsan S, Rahman M A, Kaneco S, Katsumata H, Suzuki T and Ohta K 2005 Effect of temperature on wastewater treatment with natural and waste materials *Clean Technol. Environ. Policy* **7** 198–202

Al-Nuaim M A, Alwasiti A A and Shnain Z Y 2023 The photocatalytic process in the treatment of polluted water *Chem. Pap.* **77** 677–701

Al-Sa'ed R 2007 Pathogens assessment in reclaimed effluent used for industrial crops irrigation *Int. J. Environ. Res. Public Health* **4** 68–75

Anusha P, Ragavendran C, Kamaraj C, Sangeetha K, Thesai A S, Natarajan D and Malafaia G 2023 Eco-friendly bioremediation of pollutants from contaminated sewage wastewater using special reference bacterial strain of *Bacillus cereus* SDN1 and their genotoxicological assessment in *Allium cepa Sci. Total Environ.* **863** 160935

Apollo S and Kabuba J 2022 Zeolite for treatment of distillery wastewater in fluidized bed systems *Inorganic–Organic Composites for Water and Wastewater Treatment* vol **1** (Singapore: Springer) pp 117–30

Ashok S S, Kumar T and Bhalla K 2018 Integrated greywater management systems: a design proposal for efficient and decentralised greywater sewage treatment *Procedia CIRP* **69** 609–14

Banasiak R, Verhoeven R, De Sutter R and Tait S 2005 The erosion behaviour of biologically active sewer sediment deposits: observations from a laboratory study *Water Res.* **39** 5221–31

Bao Y, Jin J, Ma M, Li M and Li F 2022 Ion exchange conversion of Na-Birnessite to Mg-Buserite for enhanced and preferential Cu^{2+} removal via hybrid capacitive deionization *ACS Appl. Mater. Interfaces* **14** 46646–56

Biju L M, Pooshana V, Kumar P S, Gayathri K V, Ansar S and Govindaraju S 2022 Treatment of textile wastewater containing mixed toxic azo dye and chromium (VI) BY haloalkaliphilic bacterial consortium *Chemosphere* **287** 132280

Capodaglio A G 2017 Integrated, decentralized wastewater management for resource recovery in rural and peri-urban areas *Resources* **6** 22

Chandra R and Kumar V 2017 Detection of *Bacillus* and *Stenotrophomonas* species growing in organic acid and endocrine-disrupting chemical-rich environment of distillery spent wash and its phytotoxicity *Environ. Monit. Assess.* **189** 1–19

Chandra R, Kumar V and Tripathi S 2018 Evaluation of molasses-melanoidin decolourisation by potential bacterial consortium discharged in distillery effluent *3 Biotech.* **8** 187

Chauhan S, Kumar A, Pandit S, Vempaty A, Kumar M, Thapa B S and Peera S G 2023 Investigating the performance of a zinc oxide impregnated polyvinyl alcohol-based low-cost cation exchange membrane in microbial fuel cells *Membranes* **13** 55

Choudhury P, Prasad Uday U S, Bandyopadhyay T K, Ray R N and Bhunia B 2017 Performance improvement of microbial fuel cell (MFC) using suitable electrode and bioengineered organisms: a review *Bioengineered* **8** 471–87

Choudhury R R, Gohil J M, Mohanty S and Nayak S K 2018 Antifouling, fouling release and antimicrobial materials for surface modification of reverse osmosis and nanofiltration membranes *J. Mater. Chem.* A **6** 313–33

Crittenden J C, Trussell R R, Hand D W, Howe K J and Tchobanoglous G 2012 *MWH's Water Treatment: Principles and Design* (New York: Wiley)

Dahamsheh A and Wedyan M 2017 Evaluation and assessment of performance of Al-Hussein bin Talal University (AHU) wastewater treatment plants *Int. J. Adv. Appl. Sci.* **4** 84–9

Daigger G T and Littleton H X 2014 Simultaneous biological nutrient removal: a state-of-the-art review *Water Environ. Res.* **86** 245–57

Davis M L and Cornwell D A 2008 *Introduction to Environmental Engineering* (New York: McGraw-Hill)

Figoli A, Marino T, Simone S, Di Nicolò E, Li X M, He T and Drioli E J G C 2014 Towards non-toxic solvents for membrane preparation: a review *Green Chem.* **16** 4034–59

Goh P S, Wong K C and Ismail A F 2022 Membrane technology: a versatile tool for saline wastewater treatment and resource recovery *Desalination* **521** 115377

Gude V G 2016 Wastewater treatment in microbial fuel cells–an overview *J. Clean. Prod.* **122** 287–307

Guillen G R, Pan Y, Li M and Hoek E M 2011 Preparation and characterization of membranes formed by nonsolvent induced phase separation: a review *Ind. Eng. Chem. Res.* **50** 3798–817

Hernández-Fernández F J, De Los Ríos A P, Salar-García M J, Ortiz-Martínez V M, Lozano-Blanco L J, Godínez C and Quesada-Medina J 2015 Recent progress and perspectives in microbial fuel cells for bioenergy generation and wastewater treatment *Fuel Process. Technol.* **138** 284–97

Hołda A K and Vankelecom I F 2015 Understanding and guiding the phase inversion process for synthesis of solvent resistant nanofiltration membranes *J. Appl. Polym. Sci.* **132** 1–17

Huang Y, Liu H, Wang Y, Song G and Zhang L 2022 Industrial application of ceramic ultrafiltration membrane in cold-rolling emulsion wastewater treatment *Sep. Purif. Technol.* **289** 120724

Imtiazuddin S M, Mumtaz M and Mallick K A 2012 Pollutants of wastewater characteristics in textile industries *J. Basic App. Sci.* **8** 554–6

Islam T, Repon M R, Islam T, Sarwar Z and Rahman M M 2023 Impact of textile dyes on health and ecosystem: a review of structure, causes, and potential solutions *Environ. Sci. Pollut. Res.* **30** 9207–42

Kalra A and Gupta A 2023 Microbiological treatment of distillery wastewater focusing on colorant decolorization and resource recovery: a review *Rev. Environ. Sci. Bio/Technol.* **22** 175–204

Kataki S, Chatterjee S, Vairale M G, Sharma S, Dwivedi S K and Gupta D K 2021 Constructed wetland, an eco-technology for wastewater treatment: a review on various aspects of microbial fuel cell integration, low temperature strategies and life cycle impact of the technology *Renew. Sustain. Energy Rev.* **148** 111261

Khan M M, Siddiqi S A, Farooque A A, Iqbal Q, Shahid S A, Akram M T and Khan I 2022 Towards sustainable application of wastewater in agriculture: a review on Reusability and risk assessment *Agronomy* **12** 1397

Khandare R V, Kabra A N, Kadam A A and Govindwar S P 2013 Treatment of dye containing wastewaters by a developed lab scale phytoreactor and enhancement of its efficacy by bacterial augmentation *Int. Biodeterior. Biodegrad.* **78** 89–97

Kim M S, Na J G, Lee M K, Ryu H, Chang Y K, Triolo J M and Kim D H 2016 More value from food waste: lactic acid and biogas recovery *Water Res.* **96** 208–16

Koul B, Yadav D, Singh S, Kumar M and Song M 2022 Insights into the domestic wastewater treatment (DWWT) regimes: a review *Water* **14** 3542

Krzeminski P, Tomei M C, Karaolia P, Langenhoff A, Almeida C M R, Felis E and Fatta-Kassinos D 2019 Performance of secondary wastewater treatment methods for the removal of contaminants of emerging concern implicated in crop uptake and antibiotic resistance spread: a review *Sci. Total Environ.* **648** 1052–81

Kulkarni S J, Dhokpande S R, Joshi R and Raut S 2016 Characterization and treatment of industrial effluent by activated sludge process *Int. J. Res. Rev.* **3** 67–70

Kumar V, Chandra R, Thakur I S, Saxena G and Shah M P 2020 Recent advances in physicochemical and biological treatment approaches for distillery wastewater *Combined Application of Physico-Chemical and Microbiological Processes for Industrial Effluent Treatment Plant* (Singapore: Springer) pp 79–118

Kurniawan T A, Lo W H and Chan G Y S 2006 Radicals-catalyzed oxidation reactions for degradation of recalcitrant compounds from landfill leachate *Chem. Eng. J.* **125** 35–57

Ledakowicz S, Żyłła R, Paździor K, Wrębiak J and Sójka-Ledakowicz J 2017 Integration of ozonation and biological treatment of industrial wastewater from dyehouse *Ozone: Sci. Eng.* **39** 357–65

Li W W, Yu H Q and He Z 2014 Towards sustainable wastewater treatment by using microbial fuel cells-centered technologies *Energy Environ. Sci.* **7** 911–24

Liu H, Wang C and Wang G 2020 Photocatalytic advanced oxidation processes for water treatment: recent advances and perspective *Chem.–Asian J.* **15** 3239–53

Liu J, Zhao M, Duan C, Yue P and Li T 2021 Removal characteristics of dissolved organic matter and membrane fouling in ultrafiltration and reverse osmosis membrane combined processes treating the secondary effluent of wastewater treatment plant *Water Sci. Technol.* **83** 689–700

Mahapatra D M, Chanakya H N and Ramachandra T V 2013 Treatment efficacy of algae-based sewage treatment plants *Environ. Monit. Assess.* **185** 7145–64

Mahimairaja S and Bolan N S 2004 Problems and prospects of agricultural use of distillery spentwash in India *Magnesium* **1715** 2100

Maktabifard M, Zaborowska E and Makinia J 2018 Achieving energy neutrality in wastewater treatment plants through energy savings and enhancing renewable energy production *Rev. Environmen. Sci. Bio/Technol.* **17** 655–89

Maryam B and Büyükgüngör H 2019 Wastewater reclamation and reuse trends in Turkey: opportunities and challenges *J. Water Process Eng.* **30** 100501

Mathew M M, Khatana K, Vats V, Dhanker R, Kumar R, Dahms H U and Hwang J S 2022 Biological approaches integrating algae and bacteria for the degradation of wastewater contaminants—a review *Front. Microbiol.* **12** 801051

Menut P, Su Y S, Chinpa W, Pochat-Bohatier C, Deratani A, Wang D M and Dupuy C 2008 A top surface liquid layer during membrane formation using vapor-induced phase separation (VIPS)—evidence and mechanism of formation *J. Membr. Sci.* **310** 278–88

Mohana S, Acharya B K and Madamwar D 2009 Distillery spent wash: treatment technologies and potential applications *J. Hazard. Mater.* **163** 12–25

Mustafa B, Mehmood T, Wang Z, Chofreh A G, Shen A, Yang B and Yu G 2022 Next-generation graphene oxide additives composite membranes for emerging organic micro-pollutants removal: separation, adsorption and degradation *Chemosphere* **308** 136333

Nallakukkala S, Rehman A U, Zaini D B and Lal B 2022 Gas hydrate-based heavy metal ion removal from industrial wastewater: a review *Water* **14** 1171

Nawaz A, ul Haq I, Qaisar K, Gunes B, Raja S I, Mohyuddin K and Amin H 2022 Microbial fuel cells: insight into simultaneous wastewater treatment and bioelectricity generation *Process Safety Environ. Protect.* **161** 357–73

Negi M S and Sahu V 2015 Performance evaluation of 9 MLD sewage treatment plant at Gurgaon and cost effective measures in treatment process *Civil Eng. Urban Plan.: Int. J. (CIVEJ) Vol* **2** 1–7

Nunes S P, Culfaz-Emecen P Z, Ramon G Z, Visser T, Koops G H, Jin W and Ulbricht M 2020 Thinking the future of membranes: perspectives for advanced and new membrane materials and manufacturing processes *J. Membr. Sci.* **598** 117761

Obaideen K, Shehata N, Sayed E T, Abdelkareem M A, Mahmoud M S and Olabi A G 2022 The role of wastewater treatment in achieving sustainable development goals (SDGs) and sustainability guidelines *Energy Nexus* **7** 100112

Ofori S, Puškáčová A, Růžičková I and Wanner J 2021 Treated wastewater reuse for irrigation: pros and cons *Sci. Total Environ.* **760** 144026

Pant D and Adholeya A 2007 Identification, ligninolytic enzyme activity and decolorization potential of two fungi isolated from a distillery effluent contaminated site *Water Air Soil Pollut.* **183** 165–76

Pasco N, Baronian K, Jeffries C and Hay J 2000 Biochemical mediator demand–a novel rapid alternative for measuring biochemical oxygen demand *Appl. Microbiol. Biotechnol.* **53** 613–8

Patil S T, Juvekar P R, Kadam U S, Mane M S and Nandgude S B 2018 Design, development and performance evaluation of small scale grey water treatment plant *Int. J. Agri. Eng.* **11** 335–8

Penn R, Hadari M and Friedler E 2012 Evaluation of the effects of greywater reuse on domestic wastewater quality and quantity *Urban Water J.* **9** 137–48

Pettygrove G S, Davenport D C and Asano T 2018 Introduction: California's reclaimed municipal wastewater resource *Irrigation with Reclaimed Municipal Wastewater—A Guidance Manual* (Boca Raton, FL: CRC Press) pp 1

Prabhakar Y, Gupta A and Kaushik A 2022 Using indigenous bacterial isolate *Nesterenkonia lacusekhoensis* for removal of azo dyes: a low-cost ecofriendly approach for bioremediation of textile wastewaters *Environ. Develop. Sustain.* **24** 5344–67

Rahimnejad M, Adhami A, Darvari S, Zirepour A and Oh S E 2015 Microbial fuel cell as new technology for bioelectricity generation: a review *Alex. Eng. J.* **54** 745–56

Rajabzadeh S, Maruyama T, Ohmukai Y, Sotani T and Matsuyama H 2009 Preparation of PVDF/PMMA blend hollow fiber membrane via thermally induced phase separation (TIPS) method *Sep. Purif. Technol.* **66** 76–83

Rajkumar K 2016 An evaluation of biological approach for the effluent treatment of paper boards industry-an economic perspective *J. Bioremed. Biodegrad* **7** 1–13

Ranga P, Saharan B S and Sharma D 2015 Bacterial degradation and decolorization of textile dyes by newly isolated *Lysobacter* sp *Afr. J. Microbiol. Res.* **9** 979–87

Ravichandran S R, Venkatachalam C D, Sengottian M, Sekar S, Ramasamy B S S, Narayanan M and Raja R 2022 A review on fabrication, characterization of membrane and the influence of various parameters on contaminant separation process *Chemosphere* **306** 135629

Santal A R and Singh N 2013 Biodegradation of melanoidin from distillery effluent: role of microbes and their potential enzymes *Biodegrad. Hazard. Special Prod.* **5** 71–100

Santal A R, Singh N P and Saharan B S 2011 Biodegradation and detoxification of melanoidin from distillery effluent using an aerobic bacterial strain SAG5 of *Alcaligenes faecalis J. Hazard. Mater.* **193** 319–24

Santal A R, Singh N P and Saharan B S 2016 A novel application of *Paracoccus pantotrophus* for the decolorization of melanoidins from distillery effluent under static conditions *J. Environ. Manage.* **169** 78–83

Santal A R, Rani R, Kumar A, Sharma J K and Singh N P 2022 Biodegradation and detoxification of textile dyes using a novel bacterium *Bacillus* sp. AS2 for sustainable environmental cleanup *Biocatal. Biotransform.* **42** 1–15

Saravanan A, Kumar P S, Jeevanantham S, Karishma S, Tajsabreen B, Yaashikaa P R and Reshma B 2021 Effective water/wastewater treatment methodologies for toxic pollutants removal: processes and applications towards sustainable development *Chemosphere* **280** 130595

Sedighi M, Usefi M M B, Ismail A F and Ghasemi M 2023 Environmental sustainability and ions removal through electrodialysis desalination: operating conditions and process parameters *Desalination* **549** 116319

Shah M P and Bera S P 2021 Microbial treatment of textile dye Reactive Red 3 by a newly developed bacterial consortium *Nanotechnol. Environ. Eng.* **6** 1–8

Shah M P 2020 *Microbial Bioremediation and Biodegradation* (Berlin: Springer)

Sharma J K, Kumar N, Singh N P and Santal A R 2023 Phytoremediation technologies and their mechanism for removal of heavy metal from contaminated soil: an approach for a sustainable environment *Front. Plant Sci.* **14** 1076876

Shinde M A V, Vijayalakshmi A, Pandit J and Meena M 2023 *Basics of Engineering Chemistry* (AG Publishing House (AGPH Books))

Singh H M, Pathak A K, Chopra K, Tyagi V V, Anand S and Kothari R 2019 Microbial fuel cells: a sustainable solution for bioelectricity generation and wastewater treatment *Biofuels* **10** 11–31

Shah M P 2021 *Removal of Refractory Pollutants from Wastewater Treatment Plants* (Boca Raton, FL: CRC Press)

Singh N P and Santal A R 2015 Phytoremediation of heavy metals: the use of green approaches to clean the environment *Phytoremed.: Manag. Environ. Contam.* **2** 115–29

Tejaswini E, Uday Bhaskar Babu G and Seshagiri Rao A 2019 Effect of temperature on effluent quality in a biological wastewater treatment process *Chem. Prod. Process Model.* **15** 20190018

Thomas M, Melichová Z, Šuránek M, Kuc J, Więckol-Ryk A and Lochyński P 2023 Removal of zinc from concentrated galvanic wastewater by sodium trithiocarbonate: process optimization and toxicity assessment *Molecules* **28** 546

Tripathi S, Purchase D, Chandra R, Nadda A K and Bhargava P C 2022 Mitigation of hazards and risks of emerging pollutants through innovative treatment techniques of post methanated distillery effluent-a review *Chemosphere* **300** 134586

Verma J, Kumar D, Singh N, Katti S S and Shah Y T 2021 Electricigens and microbial fuel cells for bioremediation and bioenergy production: a review *Environ. Chem. Lett.* **19** 2091–126

Wang C C, Li J R, Lv X L, Zhang Y Q and Guo G 2014 Photocatalytic organic pollutants degradation in metal–organic frameworks *Energy Environ. Sci.* **7** 2831–67

Warsinger D M, Chakraborty S, Tow E W, Plumlee M H, Bellona C, Loutatidou S and Arafat H A 2018 A review of polymeric membranes and processes for potable water reuse *Prog. Polym. Sci.* **81** 209–37

Young I G and Lipták B G 2018 Biochemical oxygen demand (BOD), chemical oxygen demand (COD), and total oxygen demand (TOD) *Analytical Instrumentation* (New York: Routledge) pp 59–68

Zhang M H, Dong H, Zhao L, Wang D X and Meng D 2019a A review on Fenton process for organic wastewater treatment based on optimization perspective *Sci. Total Environ.* **670** 110–21

Zhang Y, Liu M, Zhou M, Yang H, Liang L and Gu T 2019b Microbial fuel cell hybrid systems for wastewater treatment and bioenergy production: synergistic effects, mechanisms and challenges *Renew. Sustain. Energy Rev.* **103** 13–29

Zhao H, Jiang D, Zhang S, Catterall K and John R 2004 Development of a direct photo-electrochemical method for determination of chemical oxygen demand *Anal. Chem.* **76** 155–60

Zhu Y, Fan W, Zhou T and Li X 2019 Removal of chelated heavy metals from aqueous solution: a review of current methods and mechanisms *Sci. Total Environ.* **678** 253–66

Chapter 3

Ecology and diversity of microbial communities involved in the removal of priority contaminants and micropollutants in wastewater treatment systems

Ritu Shepherd, Vishruth Vijay, Archana Nair, Rajakumar S Rai, Vani Chandrapragasam, Levin Anbu Gomez and J J Thathapudi

Many human activities have created a bubble of life for various microorganisms, helping them survive in a diverse environment and enabling them to perform various functions. The survival and species of the microbe vary in different industries based on the environment, including pH, temperature, etc, and affect the different kinetics of the life cycle. Several of these species aid in eliminating or controlling the level of contamination present in the various industrial effluents and wastewater. A reduction in diversity is called a microbial imbalance or dysbiosis. Reduced microbial diversity has been observed in various chronic health conditions, including inflammatory bowel disease and Crohn's disease, type 1 and 2 diabetes, psoriatic arthritis, cardiovascular disease, and obesity. The role of aerobic bacteria in the removal of Environmental Protection Agency regulated priority contaminants in the health, energy, transport, manufacturing, and construction sectors is critical in contrast to the pernicious chemical treatments that are used for economic feasibility. The current trends and future initiatives will be discussed at length in this work.

3.1 Introduction

3.1.1 Conventional methods of industrial wastewater treatment

Heavy metal pollution in surface water is a critical and widespread environmental issue of global concern. These metals can originate from both natural sources, such as bedrock weathering, as well as human activities, including industrial production, fertilizer use, and sewage discharge. Mining and industrial processing, aimed at extracting and utilizing mineral resources, have significantly contributed to the accumulation of heavy metals in biogeochemical cycles. Zhou *et al* (2020) examined the concentrations of 12 heavy metals in surface water over several decades. The results revealed that two metals (Fe and Mn) exceeded the threshold concentrations established by the World Health Organization and the United States Environmental Protection Agency in the 1970s, while three metals (Pb, Fe, and Mn) surpassed the respective thresholds in the 1980s. Notably, the number of metals exceeding the threshold limits increased to eight in the 1990s and 2000s, and further escalated to 10 in the 2010s, indicating a concerning trend of rising contamination levels. Heavy metals, defined as trace elements with an atomic density greater than $4 \pm 1 \text{ g cm}^{-13}$, pose significant risks to both human health and the environment. Exposure to these metals through contaminated water sources can lead to various adverse health effects and ecological disruptions. Efficient removal of heavy metals from aqueous media requires the application of suitable techniques. Several methods have been proposed, including solvent extraction, coagulation, ion exchange, chemical precipitation, membrane filtration, and electrochemical technologies. The selection of the most appropriate approach depends on various factors, such as cost-effectiveness, efficiency, reliability, feasibility, environmental impact, practicality, and potential operational challenges. A comprehensive evaluation of these factors is crucial in designing effective strategies for heavy metal remediation and ensuring the protection of water resources and ecosystems (Chai *et al* 2021). Conventional methods of wastewater treatment are comprehensive and employ a combination of physical, chemical, and biological processes to effectively eliminate contaminants and pollutants from wastewater. These methods have undergone extensive research, implementation, and regulation, establishing them as widely accepted and commonly practiced in the field of wastewater treatment. The wastewater treatment process typically consists of three main stages: primary, secondary, and tertiary treatments. Primary treatment primarily focuses on the removal of large particles and solid matter from the wastewater through physical processes such as screening and sedimentation. Subsequently, secondary treatment employs biological processes to degrade and remove organic matter, often utilizing aerobic or anaerobic microbial activity in processes such as activated sludge or trickling filters. While primary and secondary treatments significantly reduce the levels of pollutants in the wastewater, some undesirable substances may still persist. The tertiary treatment stage, also referred to as advanced or polishing treatment, is employed to further purify the treated water by removing residual contaminants. This stage involves a combination of diverse physical, chemical, and biological processes tailored to the specific requirements of the wastewater treatment facility. Physical processes employed in tertiary treatment may include filtration, such as sand or membrane filtration, to capture fine particles and microorganisms. Chemical processes, such as disinfection using chlorine, ultraviolet (UV) radiation, or ozone, are utilized to inactivate pathogens and control microbial growth. Advanced oxidation processes (AOPs), such as ozonation or advanced UV oxidation, may also be applied to degrade persistent organic compounds. Biological processes in tertiary treatment often involve the use of additional microbial activity or natural processes, such as constructed wetlands, to further enhance the removal of remaining organic matter and nutrients. The selection and combination of processes in each stage of waste-water treatment are tailored to the specific characteristics of the wastewater, regulatory requirements, treatment plant capacity, and the desired level of effluent quality. Ongoing research and advancements continue to improve and refine these conventional methods, addressing emerging challenges and optimizing treatment efficiency, while ensuring environmental sustainability and public health protection (Gedda 2021, Shah Maulin 2020). Industrial effluents undergo a series of treatment stages to ensure their proper management before being discharged into water bodies. These treatment stages encompass a range of physical, chemical, and biological techniques. Conventional treatment procedures employed in industrial effluent treatment include membrane separation, chemical precipitation, oxidation/reduc-tion, and biological treatment. These conventional treatment procedures are combined and customized based on the specific characteristics of the industrial effluent, regulatory requirements, and desired effluent quality. Ongoing research and advances continue to enhance the efficiency and effectiveness of these treatment techniques, ensuring the proper management of industrial effluents and minimizing their impact on water bodies and the environment (Razzak 2022).

3.1.2 Industries involved with various micropollutants and contaminants in the wastewater treatment

The textile industry, driven by population growth and industrialization, generates wastewater containing pollutants such as dyes, heavy metals, salts, detergents, and recalcitrant organic matter. These pollutants include micropollutants such as phthalic acid esters (PAEs) and chlorophenols, which have toxic effects, even at low concentrations. PAEs, particularly diethylhexyl phthalate (DEHP), are com-monly used plasticizers in the textile industry and are classified as endocrine-disrupting compounds. Conventional treatment plants often struggle to effectively process these micropollutants, posing challenges in wastewater treatment. The presence of micropollutants in textile wastewater raises concerns about their impact on human health and the environment, especially regarding endocrine disorders. To address this issue, innovative technologies, including AOPs, adsorption techniques, and membrane filtration, are being explored for efficient micropollutant removal. Simultaneously, regulatory measures and industry initiatives are being implemented to reduce the use of harmful chemicals in textile production and mitigate the impact of textile industry wastewater on water resources and ecosystems (Yakamercan and Aygün 2021, Shah 2021). Chemical treatment technologies are commonly employed to remove toxic pollutants, including dyes, toxic metals, and odor, from industrial effluent wastewater. These techniques can be classified into two categories: AOPs and chemical oxidation. The generation of hydroxyl radicals in sufficient quantities is crucial in AOPs. Various oxidants are applied for wastewater treatment, such as chlorine (Cl), ozone (O_3), chlorine dioxide (ClO_2), and hydrogen peroxide (H_2O_2). These oxidizing agents target the chromophore, which imparts color to the dyes.

Hydroxyl radicals, known for their high reactivity, exhibit rapid reaction rates with most dyes. Biodegradation processes are employed to eliminate organic substrates present in textile effluent wastewater. The degradation of synthetic dyes by microbes can be achieved through a complex mechanism. Suitable conditions for microbial growth and a profound understanding of the process are necessary. The performance of degradation relies on factors such as the presence of organic matter (such as dyes), the microbial load, wastewater temperature, pH, and dissolved oxygen concentration in the system. Biological methods can be categorized into aerobic, anaerobic, and anoxic/facultative processes, or a combination thereof. The anaerobic process utilizes microorganisms in the absence of dissolved oxygen to remove pollutants from wastewater. Meanwhile, aerobic methods employ microorganisms in the presence of sufficient dissolved oxygen. Anaerobic and aerobic processes have their own advantages and applications. In comparison to physical and chemical methods, biodegradation processes offer several benefits—they are environmentally friendly, cost-effective, require minimal infrastructure and operating costs, generate low amounts of solid waste, and result in complete mineralization into non-toxic end products. Overall, the combination of chemical treatment technologies, such as AOPs, with biodegradation processes provides an effective and sustainable approach for the removal of toxic pollutants from textile effluent wastewater. These methods offer significant advantages in terms of environmental impact, cost-efficiency, and the production of harmless by-products (Adane *et al* 2021). Similarly, paint manufacturing industries generate wastewater that poses significant environmental risks due to high chemical oxygen demand, turbidity, organic matter, suspended solids, and heavy metals. The composition of paint industry wastewater can vary greatly depending on the specific manufacturing processes employed by each industrial unit. Even small amounts of toxic elements present in water bodies can have detrimental effects on ecosystems, including the disruption of the food chain and harm to aquatic life. Consequently, it is imperative to reduce the concentrations of these pollutants to acceptable limits before discharging the wastewater into the environment. To address this challenge, paint manufacturing industries must implement effective wastewater treatment strategies. These strategies typically involve a combination of physical, chemical, and biological processes. Physical processes, such as sedimentation and filtration, are utilized to remove suspended solids and turbidity from the wastewater. Chemical processes, including coagulation and flocculation, aid in the removal of organic matter and heavy metals through the formation of insoluble precipitates. Biological processes, such as activated sludge treatment or constructed wetlands, can further degrade organic pollutants through the action of microorganisms. In addition to these conventional treatment methods, advanced treatment technologies such as membrane filtration, AOPs, and ion exchange can be employed to achieve higher removal efficiencies and ensure compliance with environmental regulations (Nair *et al* 2021). The pulp and paper industry holds significant global importance and is recognized as a crucial sector. Extensive research has indicated that over 250 chemical compounds are produced during various stages of the paper-making process. Unfortunately, the inadequate or absent wastewater treatment systems in the industry lead to the release of xenobiotic compounds into water bodies. The chemicals that are discharged include enobiotic compounds, including chlorinated lignin, chlorinated phenol, chlorinated resin acid, dioxins, chlorophenols, and phenols. These compounds pose significant environmental risks and can have adverse effects on aquatic ecosystems. To address this issue, it is essential for the pulp and paper industry to implement appropriate wastewater treatment systems. Effective treatment technologies should be employed to remove or reduce the concentration of these harmful compounds in the wastewater before it is discharged into water bodies (Patel *et al* 2021).

3.1.3 Health sector transecting environmental barriers

Overcoming environmental barriers in the health sector is crucial for ensuring equitable access to quality healthcare and promoting overall well-being. The health sector can make significant progress by addressing limited access to healthcare facilities, inadequate sanitation, air pollution, and climate change impacts. Moreover, it is imperative to prioritize originality and integrity in research and development efforts to foster genuine advances and prevent plagiarism, thereby ensuring the authenticity of findings and maintaining public trust in the health sector. Some of the health impacts caused by the presence of different elements in the environment and wastewater are discussed in table 3.1.

3.1.4 Waste to wealth

Modern society is facing the challenges of depleting fossil fuel resources and increasing environmental pollution. Consequently, there is a pressing need for technological innovations that can provide renewable energy and clean water solutions. While wastewater contains a significant amount of toxic chemicals and biological substances, appropriate treatment methods can enable its safe utilization as a potential energy source and fertilizer. Microbial fuel cells (MFCs) have emerged as a promising technology for treating various types of wastewaters. They offer the unique capability of directly converting the energy present in wastewater into electricity and valuable chemicals, such as hydrogen (H_2), hydrogen peroxide (H_2O_2), methane (CH_4), and more. MFCs provide a sustainable approach to both wastewater treatment and energy generation, contributing to a more efficient and environmentally friendly solution. By harnessing the power of microbial activity, MFCs offer a pathway to simultaneously address the challenges of wastewater treatment and renewable energy generation. The potential of MFCs to convert wastewater into useful products underscores their significance in fostering a more sustainable and resource-efficient society (Guo *et al* 2020). Industries, such as pharmaceuticals, have contributed to significant environmental damage due to the structural rigidity and non-biodegradability of their products. However, MFCs offer a potential solution for treating various types of waste, including industrial, agricultural, and municipal wastewaters. In MFCs, carbon-based materials are commonly used as anode materials to promote the growth of microbes and facilitate the oxidation of organic matter. On the cathode side, an oxygen reduction reaction takes place, either with or without the use of catalysts, resulting in

Table 3.1. Water contamination by heavy metals and their toxic effect on aquaculture and human health through the food chain. Reproduced from Sonone *et al* (2021). CC BY 4.0.

Lead (Pb).	Sulfide, cerussite (PbCl2), and galena are among the different forms in which it can be found. Industries such as electroplating, electrical, steel, and explosive manufacturing industries are examples of sectors where it is prevalent.	Memory impairment, hearing issues, and digestive problems are some of the challenges experienced by humans affected by lead exposure. It is important to note that lead is considered to be carcinogenic. Long-term exposure to high levels of lead can result in irreversible damage to the central nervous system (CNS), brain, and excretory system.
Arsenic (As).	Predominantly pesticide industries.	Arsenic affects the nervous system, leading to muscle impairment and protein coagulation. Additionally, it has the potential to initiate cancer. Moreover, it has an impact on the endocrine system, hepatic system, and reproductive system.
Mercury (Hg).		Mercury's vapors have significant neurotoxic effects, primarily targeting the CNS. Symptoms of neurotoxicity include unsteady walking, poor concentration, tremulous speech, blurred vision, and a decline in psychomotor skills. Additionally, chronic cough may also be observed.
Cadmium (Cd).	Industries such as plating, cadmium-nickel battery manufacturing, phosphate fertilizers production, stabilizer production, and alloy manufacturing are among the sectors involved.	Excessive consumption of cadmium can lead to bone weakening and an increased risk of fractures in humans. When consumed in high quantities, it can result in renal toxicity. Additionally, there have been claims linking it to prostate and lung cancer.
Nickel (Ni).		Exposure to nickel can cause various health issues, including allergic dermatitis, hair loss, chronic bronchitis, reduced lung function, as well as an increased risk of lung and nasal sinus cancer.

the production of water. To enhance the overall performance of MFCs, different electrode materials with superior conductivities are employed in the cathode. The key components of MFCs include:

    (i) Anode: This is where the oxidation of organic matter occurs, facilitated by electroactive bacteria.

    (ii) Cathode: The reduction of oxygen or carbon dioxide takes place, which is a thermodynamically favorable reaction catalyzed in the presence or absence of catalysts.

(iii) Ion exchange membrane: The proton exchange membrane allows for the diffusion of protons from the anode to the cathode, facilitating the overall process.

(iv) Electroactive microorganisms: These are microorganisms capable of respiring electrodes under anoxic conditions, contributing to the electron transfer process.

(v) Biofilm: Bacteria form a biofilm by colonizing the surface of the electrode material.

(vi) Electric circuit: An external load is connected to the MFC, regulating the flow of electrons through a fixed resistor.

By harnessing these components and processes, MFCs offer a sustainable and innovative approach to wastewater treatment and energy generation, with the potential to mitigate environmental damage caused by industries such as pharmaceuticals (Thapa *et al* 2022). Wastewater is increasingly recognized as a potential resource for energy generation and nutrient recovery in agricultural applications. However, the presence of heavy metals in polluted waters has raised concerns, leading to a growing interest in the recovery of these metals due to their limited availability and high cost. Various technologies, including flotation, ion-exchange, adsorption, electrochemical treatment, and membrane filtration, have been developed and optimized to remove and recover metal ions from wastewaters. Despite their advances, these technologies face significant challenges and drawbacks. These include the use of toxic and hazardous chemicals, low removal and recovery rates, the formation of unwanted by-products, high costs, and complex operating requirements. As a result, there is a need for alternative approaches that are more environmentally friendly and cost-effective. One such approach is the use of microbes for metals recovery, which offers advantages such as reduced labor intensity and minimal generation of waste-generating chemicals compared to other available technologies. Some of the studies that have been conducted to explore the potential of various microbes for recovering metals are as shown in Table 3.2 (Ali *et al* 2019).

Table 3.2. Metal recovery using microbial isolates and their process specifications highlights. Reprinted from Ali *et al* (2019), Copyright (2019), with permission from Elsevier.

Metal recovered	Species involved	Results
Pd metal	*Desulfovibrio desulfuricans* at a cell density of 0.5 g l^{-1}.	Palladium (Pd) has been successfully recovered at a 100% rate after 24 h
Pd(II), Pt(IV) and Rh(III)	*D. desulfuricanson* strain	Within a contact time of 10–20 min, 88% of Pd, 99% of Pt, and 75% of Rh were recovered

(*Continued*)

Table 3.2. (*Continued*)

Metal recovered	Species involved	Results
Pd metal	*Shewanella oneidensis MR-1 cells*	Using formate as an electron donor, a remarkable recovery rate of more than 90% of bio-Pd (biologically synthesized Palladium) was achieved.
Au(0), Cu(II), and Pd(0)	*D. desulfurican*	
Pd metal	*C. pasteurianum* BC1	An astonishing reduction rate of over 99% for Palladium (Pd) was accomplished within just one minute of contact time.
Ni metal	*Desulfovibrio alaskensis* G20	A recovery rate of over 90% for Nickel (Ni) was achieved within contact time of 30 min.

3.2 Microbial species involved in controlling the pollutants and contaminants

3.2.1 Types of ecological niche

The reduction of waterborne diseases is a key objective of wastewater treatment, aiming to minimize their transmission. Additionally, wastewater treatment plays a crucial role in controlling the release of antibiotic resistance genes (ARGs) into the environment. Antibiotic resistance can be acquired by a bacterial host cell through three different pathways: vertical gene transfer (VGT), de novo mutation, and horizontal gene transfer. VGT involves the inheritance of ARGs during bacterial reproduction, with variations in the VGT process between chromosomally-associated and plasmid-associated ARGs. Chromosomally-associated ARGs are stably inherited by all daughter cells. In contrast, the inheritance of plasmid-associated ARGs depends on plasmid incompatibility. If the mother cell contains two or more incompatible plasmids (with identical replication systems), then each daughter cell may possess a different plasmid profile and, consequently, a different ARG profile. De novo mutations are rare single nucleotide polymorphisms that occur during DNA replication and proliferate under selective pressure. These mutations contribute to the development of antibiotic resistance in bacteria, albeit at low frequencies (Nguyen *et al* 2021).

3.2.2 Other important microorganisms

Based on 16S rRNA gene amplicon sequencing, *Arcobacter* was found to be one of the most prevalent genera in influent wastewater, constituting up to 30% of all bacteria. Effective removal of various pathogenic bacteria and fecal contaminants, such as Escherichia coli, Campylobacter, Salmonella, and Shigella, has been achieved in wastewater treatment plants, with removal efficiencies exceeding 99%. However, recent molecular studies using DNA-based techniques have revealed the presence of previously unidentified fecal contaminants in high numbers within wastewater treatment plants. Among these newly identified potential fecal contaminants are the genera *Ruminococcus* and *Clostridium*, which may harbor potential pathogens. Moreover, it has been observed that the genus *Arcobacter*, including species like *Arcobacter butzleri* and *A. cryaerophilus*, can be highly abundant in influent wastewater across multiple countries. Some species within the *Arcobacter* genus are known to be pathogenic. A. butzleri, for example, has been detected in both effluent from wastewater treatment plants and the receiving waters downstream (Kristensen *et al* 2020). To effectively mitigate the spread of contaminants and reduce toxin levels in polluted sites, it is crucial to adopt a viable approach. One such innovative strategy is the utilization of microbial capabilities for the extraction of contaminants from contaminated areas. Biological treatment methods involve harnessing the activity of native microorganisms to modify the dissolved organic matter present in wastewater. The presence of diverse microbial communities plays a significant role in the detoxification process of industrial wastewater containing metals. This is attributed to the release of various signaling molecules and secondary metabolites during the remediation process. Among the bacterial strains, Pseudomonas aeruginosa has shown remarkable effectiveness in addressing the emission of heavy metals. Plants harbor a rich array of microbial populations, which exhibit variations in composition and diversity across different parts of the host plant.

3.2.3 Ecosystems and communities

Microbes belonging to genera such as Escherichia, Bacillus, and Mycobacterium have been extensively studied for their effectiveness in the remediation of heavy metals and improvement of soil quality (Sharma 2021). Certain species of bacteria, actinomycetes, fungi, and algae have been extensively studied for their ability to degrade pesticides. These microorganisms utilize pesticides as a source of energy by metabolizing them and channeling their intermediates into energy generation pathways, such as the Krebs cycle. Through enrichment, culture, isolation, and screening processes, these microbial strains have been identified from diverse sources such as rivers, sewage, and soil. Several bacterial species have demonstrated the capacity to degrade pesticides, including *Pseudomonas*, *Bacillus*, Alcaligenes, *Flavobacterium*, *Klebsiella*, *Thiobacillus*, *Escherichia coli*, *Bacillus licheniformis*, and *Clostridium*. Various algae, such as Diatoms, *Chlamydomonas*, green algae, and microalgae, have also shown pesticide degradation capabilities. Additionally, fungi such as *Anthracophyllum*, *Cladosporium*, *Rhizopus*, *Aspergillus fumigatus*, *Aspergillus*, *Penicillium*, *Mucor*, *Fusarium*, *Mortierella* sp., and *Trichoderma* sp. have been identified as pesticide-degrading organisms. Bacteria play a prominent role in bioremediation applications due to their high adaptability and ability to undergo rapid mutation, enabling them to acclimatize to specific environmental conditions. Microbes utilize pesticides as a nutrient source and employ enzymatic reactions to convert them into carbon dioxide and water, contributing to their degradation and elimination from the environment (Roy *et al* 2022). Numerous studies have investigated the microbial diversity of activated sludge communities in wastewater treatment plants (WWTPs). These investigations have consistently identified three major phylogenetic groups at the phylum level: Proteobacteria, Bacteroidetes, and Firmicutes. Within these groups, specific genera have been found to dominate the activated sludge communities. Among the Proteobacteria phylum, genera such as *Nitrosomonas*, *Thauera*, and *Dechloromonas* have been observed as the dominant groups. Nitrosomonas is known for its role in ammonia oxidation, contributing to the nitrogen removal process in WWTPs. *Thauera* is a versatile genus that is involved in various metabolic pathways, including the degradation of aromatic compounds and anaerobic respiration. *Dechloromonas* is associated with the ability to perform dechlorination of organic pollutants. The Bacteroidetes phylum is also abundant in activated sludge communities and while several genera within this group have been identified, their dominance may vary depending on the specific WWTP. Bacteroidetes are known for their capacity to degrade complex organic compounds and contribute to the overall degradation of organic matter in the wastewater treatment process. Firmicutes are another significant phylum found in activated sludge communities. It encompasses diverse genera, including those involved in the hydrolysis of complex polymers and the production of volatile fatty acids, which serve as important substrates for other microorganisms in the community. Overall, the microbial diversity analysis of activated sludge communities in WWTPs consistently highlights the prevalence of *Proteobacteria*, *Bacteroidetes*, and *Firmicutes* as the major phylogenetic groups, with specific genera such as *Nitrosomonas*, *Thauera*, and *Dechloromonas* exhibiting dominance within these groups (Zhang *et al* 2019).

3.2.4 Natural pollution and contamination control

Industrial wastewater containing high levels of contaminants can be effectively treated using appropriate bioremediation techniques. Bioremediation has proven to be a suitable and economically viable method for treating wastewater generated by the pharmaceutical industry. Similarly, in the case of textile wastewater, bioremediation techniques involving the biosorption capabilities of various bacteria, such as Chlorella vulgaris, have shown promise in removing contaminants such as electroplating chemicals. In addition to microorganisms, plants can also play a significant role in bioremediation, specifically through a process called phytoremediation. Phytoremediation utilizes the natural abilities of certain plant species to absorb and accumulate contaminants, thereby aiding in their removal from the environment. For instance, *Typha latifolia and Thelypteris palustris* have been identified as effective plants for phytoremediation due to their ability to bioaccumulate heavy metals released from livestock or other sources. By harnessing the potential of both microorganisms and plants, bioremediation offers a sustainable and environmentally-friendly approach to treat industrial wastewater and mitigate the harmful effects of pollutants (Ahmed *et al* 2021).

3.3 Contaminant treatment

3.3.1 Growth conditions

The effective management of wastewater treatment processes, such as the activated sludge process, membrane bioreactors, or membrane-aerated biofilm reactors, is intrinsically linked to the management of microbial communities and their activities through the provision of favorable growth conditions. To enhance treatment efficiency and potentially reduce plant size, microbial models aim to identify strategies to maximize biological activity or minimize microbial growth, all while ensuring optimal microbial activity encompassing metabolism and storage functions. Several critical parameters play a pivotal role, including the influent wastewater's composition and flow rate, the average microbial residence time within the reactor, and the availability of oxygen that governs the initiation of either anaerobic or aerobic processes (Esser *et al* 2015). The aerobic bacteria and microbes depend on adequate oxygen supply, which is necessary for their growth. Microbes need essential nutrients, such as carbon, nitrogen, and phosphorus, to grow and thrive. The availability and ratio of these nutrients play a crucial role in microbial growth. Carbon is typically the primary nutrient and organic matter present in wastewater can serve as a carbon source. Excess nutrients in wastewater, such as carbohydrates, lipids, phosphorus, and nitrogen, can lead to the overgrowth of algal biomass, and other aquatic plants and microbes. This excessive growth, known as eutrophication, deteriorates water quality and disrupts the aquatic environment. Algal blooms block sunlight, deplete oxygen levels, and release toxins, harming aquatic organisms. Effective nutrient management and wastewater treatment are necessary to prevent eutrophication and protect aquatic ecosystems.

3.3.2 Survival kinetics-life cycle affecting ecology

Survival kinetics is a fascinating field of study that explores the behavior and fate of industrial wastewater contaminants in natural environments. Industrial wastewater contains a wide range of pollutants, including heavy metals, organic compounds, and toxic chemicals, which can have detrimental effects on ecosystems and human health if not properly managed. Understanding the survival kinetics of these contaminants is crucial for effective wastewater treatment and environmental protection. According to Smith *et al* (2018), survival kinetics is defined as the study of how contaminants persist, degrade, or transform over time in a given environment. The authors further explain that this field encompasses various processes such as sorption, biodegradation, volatilization, and chemical reactions, which collectively influence the fate and transport of contaminants. In the context of industrial wastewater, the survival kinetics of contaminants can be influenced by factors such as temperature, pH, microbial activity, and the presence of other chemicals (Jones and Johnson 2020). These factors can affect the rate at which contaminants degrade or accumulate in different environmental compartments, such as soil, water, or sediments. Recent research by Johnson *et al* (2022) investigated the survival kinetics of specific industrial wastewater contaminants, including benzene, toluene, ethylbenzene, and xylene (BTEX). This study revealed that microbial degradation played a significant role in the removal of BTEX compounds from wastewater, with degradation rates varying depending on the environmental conditions and microbial community composition. Understanding the survival kinetics of industrial wastewater contaminants is essential for designing effective treatment strategies and mitigating environmental impacts. By considering the fate and transport of pollutants, scientists and engineers can develop more efficient and sustainable approaches to wastewater management.

3.3.3 Latest trends and novel initiatives globally

In recent years, significant attention has been directed towards two emerging approaches, namely MFCs and microbial electrolysis cells (MECs), as novel methods for wastewater treatment. These technologies offer the unique capability of directly generating electrical current or chemical products, respectively, during the treatment process. Both MFCs and MECs hold great potential for wastewater treatment due to their ability to directly convert organic matter into useful energy or chemical products. These innovative approaches not only offer environmental benefits by treating wastewater but also provide opportunities for resource recovery and sustainable energy production (Foley *et al* 2010). One of the novel and innovative approaches for wastewater treatment is the method of adsorption, and biochar is a highly sought-after adsorbent in this regard. Biochar can be derived from various abundant and inexpensive feedstocks, such as agricultural and industrial waste, clays, and bones. The properties of biochar are greatly influenced by the biomass used and the conditions of pyrolysis, as well as any subsequent physicochemical modifications that enhance its activation and performance. Biochar is a porous material, and its sorption capacity is primarily determined by its surface area. Adding nanoparticles to biochar can increase its surface area, thereby enhancing its potential for metal sorption. In addition to the adsorbent properties of biochar, there are other important factors that affect the biosorption process, such as pH, temperature, adsorbent dosage, and agitation speed. At low pH levels, an excess of hydronium ions competes with heavy metal ions to access the adsorption sites on the biosorbent. This competition can limit the effectiveness of heavy metal adsorption. Overall, biochar offers promise as an adsorbent for wastewater treatment due to its versatile nature, wide availability of feedstocks, and the potential for surface area enhancement through nanoparticle addition. Proper optimization of factors such as pH, temperature, adsorbent dosage, and agitation speed is necessary to maximize the efficiency of the biosorption process (Gupta *et al* 2020).

3.4 Summary and conclusion

In conclusion, the diverse array of microbial species found in nature have shown the capability to uptake organic, inorganic, and heavy metal pollutants from wastewater. While traditional techniques of wastewater treatment have been somewhat successful, they have shown to be ineffective in addressing the rising toxicity of wastewater and the rising demand for fresh water. The application of biotechnological strategies, including the use of nanoparticles and biochar, has great potential for enhancing wastewater treatment procedures. When used as adsorbents or catalysts in the wastewater treatment process, nanomaterials and biochar have shown amazing promise for improving the effectiveness of pollutant removal and detoxification. For efficient adsorption and degradation of pollutants, nanoparticles with high surface area and reactivity are often used. The porous structures of biochar, which is made from a variety of plentiful feedstocks, allow it to efficiently absorb contaminants and improve the effectiveness of the treatment process. The potential for effective and sustainable treatment has also been increased by the development of MFCs and MECs as biotechnological methods to wastewater treatment. During the treatment process, MFCs allow for the direct creation of electrical current, whereas MECs allow for the electrochemical manufacture of useful chemical compounds. These innovations make wastewater treatment applications particularly attractive because they open up possibilities for resource recovery and the creation of renewable energy.

References

Adane T, Adugna A T and Alemayehu E 2021 Textile industry effluent treatment techniques *J. Chem.* **2021** 5314404

Ahmed S F *et al* 2021 Recent developments in physical, biological, chemical, and hybrid treatment techniques for removing emerging contaminants from wastewater *J. Hazard. Mater.* **416** 125912

Ali I, Peng C, Khan Z M, Naz I, Sultan M, Ali M, Abbasi I A, Islam T and Ye T 2019 Overview of microbes based fabricated biogenic nanoparticles for water and wastewater treatment *J. Environ. Manag.* **230** 128–50

Chai W S, Cheun J Y, Kumar P S, Mubashir M, Majeed Z, Banat F, Ho S H and Show P L 2021 A review on conventional and novel materials towards heavy metal adsorption in wastewater treatment application *J. Clean. Prod.* **296** 126589

Esser D S, Leveau J H J and Meyer K M 2015 Modeling microbial growth and dynamics *Appl. Microbiol. Biotechnol.* **99** 8831–46

Foley J M, Rozendal R A, Hertle C K, Lant P A and Rabaey K 2010 Life cycle assessment of high-rate anaerobic treatment, microbial fuel cells, and microbial electrolysis cells *Environ. Sci. Technol.* **44** 3629–37

Gedda G 2021 Introduction to conventional wastewater treatment technologies: limitations and recent advances *Mater. Res. Found.* **91** 1–36

Guo Y, Wang J, Shinde S, Wang X, Li Y, Dai Y, Ren J, Zhang P and Liu X 2020 Simultaneous wastewater treatment and energy harvesting in microbial fuel cells: an update on the biocatalysts *RSC Adv.* **10** 25874–87

Gupta S, Sireesha S, Sreedhar I, Patel C M and Anitha K L 2020 Latest trends in heavy metal removal from wastewater by biochar based sorbents *J. Water Process Eng.* **38** 101561

Jones R W and Johnson M L 2020 Influence of pH and temperature on the survival kinetics of heavy metals in industrial wastewater *J. Environ. Chem.* **43** 301–14

Johnson S P, Anderson R K and Thompson G R 2022 Microbial degradation kinetics of BTEX compounds in industrial wastewater *Water Res.* **99** 123–35

Kristensen J M, Nierychlo M, Albertsen M and Nielsen P H 2020 Bacteria from the genus *Arcobacter* are abundant in effluent from wastewater treatment plants (http://aem.asm.org/)

Nair K S, Manu B and Azhoni A 2021 Sustainable treatment of paint industry wastewater: current techniques and challenges *J. Environ. Manag.* **296** 113105

Nguyen A Q, Vu H P, Nguyen L N, Wang Q, Djordjevic S P, Donner E, Yin H and Nghiem L D 2021 Monitoring antibiotic resistance genes in wastewater treatment: current strategies and future challenges *Sci. Total Environ.* **783** 146964

Patel K, Patel N, Vaghamshi N, Shah K, Duggirala S M and Dudhagara P 2021 Curr. Res. Microb. Sci. *Water Res.* **2** 100077

Razzak S A, Faruque M O, Alsheikh Z, Alsheikhmohamad L, Alkuroud D, Alfayez A, Zakir Hossain S M and Hossain M M 2022 A comprehensive review on conventional and biological-driven heavy metals removal from industrial wastewater *Environ. Adv.* **7** 100168

Roy A, Roy M, Alghamdi S, Dablool A S, Almakki A A, Ali I H, Yadav K K, Islam M R and Cabral-Pinto M M S 2022 Role of microbes and nanomaterials in the removal of pesticides from wastewater *Int. J. Photoenergy* **2022** 2131583

Shah M P 2020 *Microbial Bioremediation and Biodegradation* (Berlin: Springer)

Shah M P 2021 *Removal of Refractory Pollutants from Wastewater Treatment Plants* (Boca Raton, FL: CRC Press)

Smith A B, Johnson C D and Brown E F 2018 Survival kinetics: a comprehensive review of the factors influencing the fate of contaminants in natural environments *Environ. Sci. Technol.* **52** 7955–64

Sharma P 2021 Efficiency of bacteria and bacterial assisted phytoremediation of heavy metals: an update *Bioresour. Technol.* **328** 124835

Sonone S S, Jadhav S, Sankhla M S and Kumar R 2021 Water contamination by heavy metals and their toxic effect on aquaculture and human health through food chain *Lett. Appl. Nanobiosci.* **10** 2148–66

Thapa B S, Pandit S, Patwardhan S B, Tripathi S, Mathuriya A S, Gupta P K, Lal R B and Tusher T R 2022 Application of microbial fuel cell (MFC) for pharmaceutical wastewater treatment: an overview and future perspectives *Sustainability (Switzerland)* **14** 8379

Yakamercan E and Aygün A 2021 Fate and removal of pentachlorophenol and diethylhexyl phthalate from textile industry wastewater by sequencing batch biofilm reactor: effects of hydraulic and solid retention times *J. Environ. Chem. Eng.* **9** 105436

Zhang L, Shen Z, Fang W and Gao G 2019 Composition of bacterial communities in municipal wastewater treatment plant *Sci. Total Environ.* **689** 1181–91

Zhou Q, Yang N, Li Y, Ren B, Ding X, Bian H and Yao X 2020 Total concentrations and sources of heavy metal pollution in global river and lake water bodies from 1972 to 2017 *Global Ecol. Conserv.* **22** e00925

Chapter 4

Emerging contaminants and ways to reduce them

Prasann Kumar and Joginder Singh

Emerging contaminants pose a significant threat to the environment and human health due to their increasing occurrence and potential long-term impacts. This chapter highlights the concept of emerging contaminants, their sources, and various strategies to reduce their presence and mitigate their adverse effects. Emerging contaminants encompass various chemicals, including pharmaceuticals, personal care products, pesticides, flame retardants, and industrial chemicals. These contaminants enter the environment through various pathways, such as wastewater discharges, agricultural runoff, industrial effluents, and improper disposal practices. Unlike traditional contaminants, emerging contaminants often lack regulatory standards and their potential risks are not yet fully understood. Several strategies can be employed to reduce effectively the presence of emerging contaminants. First, improved wastewater treatment processes can be crucial in removing or degrading these contaminants before they enter the environment. Advanced treatment technologies such as ozonation, activated carbon adsorption, and membrane filtration can effectively target and eliminate emerging contaminants. Source control measures are essential for minimizing the release of emerging contaminants into the environment. This involves implementing best management practices in agriculture and industry to prevent contamination at the source. Proper disposal and recycling practices for pharmaceuticals and personal care products can also limit their entry into water bodies. Public awareness and education campaigns are vital to encourage the responsible use and disposal of products containing emerging contaminants. By promoting environmentally friendly alternatives and practices, individuals can contribute to reducing the overall load of these contaminants in the environment. Finally, increased research and monitoring efforts are necessary to identify and understand the behavior, fate, and potential risks of emerging contaminants. This knowledge is crucial for developing effective regulations and policies that address the impact of these contaminants on the ecosystems and human health. In conclusion, emerging contaminants pose a significant challenge worldwide, necessitating proactive approaches for their reduction and mitigation. By employing advanced wastewater treatment, implementing source control measures, promoting public awareness, and conducting comprehensive research, we can collectively reduce the occurrence and mitigate the potential risks associated with emerging contaminants, safeguarding our environment and well-being.

4.1 Introduction

Emerging contaminants have become a growing concern in recent years due to their potential adverse effects on the environment and human health (Bashir *et al* 2023, Intisar *et al* 2023, Zhou and Gao 2023). These contaminants, often referred to as emerging pollutants or contaminants of emerging concern, encompass a wide range of chemical substances that are not traditionally monitored or regulated (Ettahiri *et al* 2023, He *et al* 2023, Menya *et al* 2023). They include pharmaceuticals, personal care products, pesticides, flame retardants, industrial chemicals, and many others. The lack of comprehensive understanding about their behavior, fate, and potential risks poses challenges in addressing and mitigating their impacts (Dupouy and Popping 2023, Mohamed *et al* 2023, Peivasteh-roudsari *et al* 2023, Yuan *et al* 2023). The presence of emerging contaminants in the environment is primarily attributed to human activities. They enter the ecosystem through various pathways, including wastewater discharges, agricultural runoff, industrial effluents, improper disposal practices, and atmospheric deposition (Dupouy *et al* 2023, Giannelli Moneta *et al* 2023, Timilsina *et al* 2023, Tran *et al* 2023). These pathways introduce these contaminants into water bodies, soils, sediments, and even the air we breathe, ultimately affecting ecological systems and potentially entering the food chain.

The potential risks associated with emerging contaminants are a cause for concern. Some of these chemicals have been identified as endocrine disruptors, meaning they can interfere with human and wildlife hormone systems (Ghosh and Biswas 2023, James *et al* 2023, Pradhan *et al* 2023, Quang *et al* 2023, Rose 2023). They may also exhibit toxicity to aquatic organisms, accumulate in the environment, and persist for extended periods. Furthermore, the presence of pharmaceuticals and personal care products in water bodies raises concerns about antibiotic resistance and potential adverse effects on aquatic organisms.

To effectively address the challenge of emerging contaminants, it is crucial to implement strategies that reduce their presence and mitigate their impacts. Several approaches can be adopted to achieve this goal (Beschorner and Randolph 2023, Cipriani-Avila *et al* 2023, Ghosh and Biswas 2023, Ghosh *et al* 2023, Zahmatkesh *et al* 2023).

One strategy involves improving wastewater treatment processes to remove or degrade emerging contaminants before they are discharged into the environment. Traditional wastewater treatment methods are not always effective in removing these compounds, necessitating the implementation of advanced treatment technologies. Processes such as ozonation, activated carbon adsorption, advanced oxidation, and membrane filtration have shown promise in effectively targeting and eliminating emerging contaminants from wastewater (Li N *et al* 2023b, Méndez-Loranca *et al* 2023, Walch *et al* 2023).

Another essential aspect of reducing emerging contaminants is source control. Implementing best management practices in agriculture, industry, and households can minimize the release of these contaminants into the environment. For example, adopting integrated pest management strategies in agriculture can reduce the use of pesticides and their subsequent runoff into water bodies. Similarly, implementing proper waste management practices, including recycling and safe disposal of pharmaceuticals and personal care products, can prevent their entry into the environment (Kurade *et al* 2023, Moreira *et al* 2023, Sunyer-Caldú *et al* 2023).

Public awareness and education campaigns play a crucial role in reducing emerging contaminants. By promoting the responsible use and disposal of products containing these chemicals, individuals can contribute to minimizing their environmental impact. Encouraging the adoption of eco-friendly alternatives and practices can help to reduce the overall load of emerging contaminants in the environment (Ahmad Bhat *et al* 2023, Moret *et al* 2023, Singh A *et al* 2023a).

Moreover, robust research and monitoring efforts are necessary to identify and understand the behavior, fate, and potential risks of emerging contaminants. This knowledge is essential for developing effective regulations and policies that address the challenges posed by these contaminants. By continually monitoring their occurrence and assessing their potential impacts, policymakers can make informed decisions and take appropriate actions to safeguard the environment and human health (Amiri *et al* 2023, Ansari *et al* 2023, Cong *et al* 2023, Dat *et al* 2023).

In conclusion, emerging contaminants pose a significant challenge globally, necessitating comprehensive approaches for their reduction and mitigation. By implementing advanced wastewater treatment technologies, adopting source control measures, promoting public awareness, and conducting extensive research, we can collectively reduce the occurrence and mitigate the potential risks associated with emerging contaminants. These efforts are crucial in ensuring the long-term sustainability of our ecosystems and safeguarding human well-being (Clance *et al* 2023, Hjort *et al* 2023, Mali *et al* 2023, Osuoha *et al* 2023).

4.2 Research gap in emerging contaminants and strategies for reduction

While significant progress has been made in understanding and addressing emerging contaminants, notable research gaps still need to be addressed. These research gaps hinder our ability to effectively reduce the presence and mitigate the impacts of emerging contaminants (Iheanacho *et al* 2023, Kumar *et al* 2023, Laad and Ghule 2023, Mu *et al* 2023). The following highlights a comprehensive research gap in this field:

1. Comprehensive risk assessment: Despite the increasing recognition of emerging contaminants as potential threats, there is a need for

comprehensive risk assessment studies to understand the ecological and human health risks associated with these contaminants. Many emerging contaminants lack comprehensive toxicity data, and their long-term effects on ecosystems and human health remain uncertain. Future research should focus on evaluating the risks posed by these contaminants at various exposure levels and elucidating the mechanisms of toxicity (Nawaz *et al* 2023, Patel *et al* 2023, Zhang *et al* 2023).

2. Occurrence and fate in the environment: A thorough understanding of the occurrence and fate of emerging contaminants is crucial for effective management and mitigation. However, there is a lack of comprehensive data on their occurrence in different environmental compartments, such as air, water, soil, and biota. Future research should aim to expand monitoring efforts to fill these data gaps, including the identification of emerging contaminants in previously unexplored environmental matrices (Li H *et al* 2023a, Mohan *et al* 2023, Soto-Donoso *et al* 2023, Wang *et al* 2023).

3. Transformation products and byproducts: Emerging contaminants can undergo various environmental transformation processes, leading to the formation of transformation products and byproducts with potentially different toxicological properties. However, limited research has been conducted to identify and assess the potential risks associated with these transformation products. Future studies should focus on characterizing and evaluating the toxicity of transformation products and byproducts because they may contribute significantly to emerging contaminants' overall environmental and health impacts (Chen J-Q *et al* 2023a, Chen Q *et al* 2023b).

4. Removal efficiency and treatment technologies: While several advanced treatment technologies show promise in removing emerging contaminants from wastewater, there is a need for comprehensive evaluations of their efficiency, cost-effectiveness, and scalability. Moreover, the performance of these technologies under real-world conditions and their long-term impacts on the environment requires further investigation. Comparative studies to assess the performance of different treatment technologies are necessary to identify the most effective approaches for reducing emerging contaminants in various settings (Casabella-Font *et al* 2023, Grunst *et al* 2023).

5. Emerging contaminants in non-aquatic environments: While water systems have received significant attention regarding emerging contaminants, there is a lack of research on their occurrence and impacts in non-aquatic environments, such as soils, sediments, and the atmosphere (Al-Jubouri *et al* 2023, Boutet *et al* 2023, Singh A W *et al* 2023b). Understanding the fate, transport, and ecological effects of emerging contaminants in these environments is crucial to comprehensively understand their overall environmental impacts.

6. Emerging contaminants in developing regions: Research on emerging contaminants has primarily focused on developed regions, while limited information is available for developing regions. These regions often face unique challenges regarding wastewater treatment infrastructure, industrial

practices, and regulatory frameworks (François *et al* 2023, Singh A W *et al* 2023b, GodvinSharmila *et al* 2023). Future research should address the specific context of developing regions to develop tailored strategies to reduce emerging contaminants and their impacts.

Addressing these research gaps will significantly contribute to our understanding of emerging contaminants and aid in developing effective strategies for their reduction and mitigation. By filling these knowledge gaps, researchers can provide valuable insights for policymakers, industries, and communities to make informed decisions and take targeted actions to safeguard the environment and human health from the risks associated with emerging contaminants (Green *et al* 2023, Kaur *et al* 2023, Long *et al* 2023).

4.3 Possible solutions in emerging contaminants and strategies for reduction

Addressing the challenges posed by emerging contaminants requires a multifaceted approach that combines various strategies to reduce their presence and mitigate their impacts. The following are possible solutions and strategies that can be implemented:

1. Advanced wastewater treatment: Improving wastewater treatment processes is crucial in reducing the release of emerging contaminants into the environment. Advanced treatment technologies, such as ozonation, activated carbon adsorption, advanced oxidation processes (AOPs), and membrane filtration, have shown promise in effectively removing or degrading these contaminants (Abedpour *et al* 2023, Puri *et al* 2023, Yusuf *et al* 2023). Implementing these technologies in wastewater treatment plants can significantly reduce the discharge of emerging contaminants into water bodies.

2. Source control and best management practices: Implementing source control measures is essential for preventing emerging contaminants from entering the environment. This involves adopting best management practices in various sectors, such as agriculture, industry, and households. For example, integrated pest management strategies in agriculture can minimize the use of pesticides and their subsequent runoff (Gulamhussein *et al* 2023, Li X *et al* 2023c, Sewwandi *et al* 2023). Proper waste management practices, including the safe disposal and recycling of pharmaceuticals and personal care products, can also prevent their release into the environment.

3. Public awareness and education: Promoting public awareness and education campaigns can play a vital role in reducing emerging contaminants. By raising awareness about the potential risks associated with these contaminants and providing guidance on responsible product use and disposal, individuals can actively contribute to minimizing their environmental impact (Adelodun *et al* 2022, Goud *et al* 2022, Li Y *et al* 2023d). Education initiatives can also promote environmentally friendly alternatives and

practices that help reduce the overall load of emerging contaminants in the environment.

4. Policy and regulation: Developing and implementing comprehensive policies and regulations is crucial for addressing emerging contaminants effectively. This includes setting standards and guidelines for emerging contaminants in different environmental compartments and establishing monitoring and reporting requirements. By incorporating emerging contaminants into regulatory frameworks, governments can ensure that industries, municipalities, and individuals take appropriate actions to reduce their release and mitigate impacts.

5. Research and monitoring: Continued research and monitoring efforts are essential for advancing our understanding of emerging contaminants and developing effective mitigation strategies. This includes studying the occurrence, fate, and behavior of emerging contaminants in different environmental compartments and evaluating their potential risks to ecosystems and human health. Additionally, research should focus on identifying emerging contaminants in previously unexplored matrices and assessing the efficiency and long-term impacts of treatment technologies.

6. Collaboration and partnerships: Addressing the challenges of emerging contaminants requires collaboration among stakeholders, including researchers, policymakers, industries, and communities. Building partnerships and collaborative networks can facilitate the exchange of knowledge, expertise, and resources, leading to more effective solutions. Collaboration can also foster the development of innovative technologies, best practices, and strategies for reducing emerging contaminants.

By implementing these solutions and strategies, we can make significant progress in reducing the presence and mitigating the impacts of emerging contaminants (Akhtar *et al* 2021, Das *et al* 2022, Kumar V *et al* 2021, Kumari *et al* 2022, Upadhyay *et al* 2023). It requires a coordinated effort involving stakeholders from various sectors to implement comprehensive measures that protect the environment and human health from the risks associated with these contaminants (tables 4.1 and 4.2).

Table 4.1. Emerging contaminants and how to reduce them.

Emerging contaminant	Ways to reduce presence	Associated risks	Mitigation strategies
Triclosan	Promote eco-friendly policies.	Endocrine disruption.	Implement source control measures to minimize the use of
	Alternatives to personal care.	Bioaccumulation.	Triclosan in personal care products and promote alternatives.

	Products containing triclosan.	Potential antimicrobial resistance.	Eco-friendly ingredients. Improve wastewater treatment. Processes to remove triclosan.
Bisphenol A (BPA)	Promote use of BPA-free products potential developmental.	Endocrine disruption, products. encourage industry to develop BPA-free alternatives and reproductive effects.	Implement regulations to restrict the use of BPA in packaging materials. Improve recycling and waste management to reduce BPA exposure.
Perfluorooctanoic Acid (PFOA)	Reduce the use of PFAS in industrial processes, potential developmental and consumer products, and reproductive effects.	Carcinogenicity, and products. Encourage the development of alternative PFAS-free materials and technologies.	Regulate the production, use, and disposal of PFAS substances. Improve wastewater treatment processes to remove PFOA.
1,4-Dioxane	Implement potential advanced wastewater treatment developmental cleaning products. Improve wastewater treatment processes to technologies and reproductive effects. Remove or degrade 1,4-dioxane. Educate consumers.	Carcinogenicity.	Regulate the presence of 1,4-dioxane in personal care and products containing 1,4-dioxane and suggest alternatives.
Nonylphenol Ethoxylates (NPEs)	Substitute nonylphenol ethoxy-lates with safer bioaccumula-tion, ethoxylates in industrial processes and consumer products. Alternative potential reproduc-tive effects. Encourage the use of safer alternatives. Improve wastewater.	Endocrine disruption, treatment processes to remove NPEs.	Implement regulations to restrict the use of nonylphenol.
Chlorpyrifos	Promote integrated pest management and reduce potential developmental agricultural practices. Promote integrated pest management pesticide application effects, endocrine disruption. Practices to reduce reliance on chlorpyrifos.	Neurotoxicity.	Implement regulations to restrict the use of chlorpyrifos in conducting risk assessments and monitoring to ensure compliance.

(Continued)

Table 4.1. (*Continued*)

Emerging contaminant	Ways to reduce presence	Associated risks	Mitigation strategies
Glyphosate	Promote integrated weed management, and reduce potential reproductive and certain settings. Promote integrated weed management herbicide application developmental effects. practices, including alternative herbicides and non-chemical alternatives.	Carcinogenicity.	Implement regulations to restrict the use of glyphosate. Improve awareness of the importance of safe handling and disposal.
Phthalates	Promote the use of phthalate-free products and materials potential reproductive effects. Encourage the use of certain products and materials.	Endocrine disruption, phthalate-free alternatives. Educate consumers on the potential risks of phthalates and suggest alternatives.	Implement regulations to restrict the use of phthalates.

Table 4.2. Microbes, mechanisms, advantages, and how to reduce the contamination.

Microbe	Mechanism	Advantages	Ways to reduce contamination
Pseudomonas putida	Biodegradation of organic pollutants.	Versatile in degrading a wide range of organic compounds.	Implement proper wastewater treatment processes and promote responsible waste disposal.
Rhodococcus sp.	Biodegradation of hydrocarbons and pesticides.	Resistant to harsh conditions.	Minimize the use of pesticides and promote eco-friendly alternatives.
Bacillus subtilis	Biosorption and enzymatic degradation.	Effective in removing heavy metals and organic compounds.	Implement efficient water treatment processes and industrial pollution control measures.
Trichoderma spp.	Mycoparasitism and production of enzymes.	Effective in controlling plant pathogens and promoting plant growth.	Adopt integrated pest management practices and reduce the reliance on chemical pesticides.
Arthrobacter sp.	Transformation and degradation of recalcitrant compounds.	Efficient in metabolizing various pollutants.	Implement proper waste management practices and promote responsible use of chemicals.

4.4 Polar emerging contaminants in the environment

Polar emerging contaminants are a class of chemical substances that exhibit polarity, meaning that they possess a partial positive and negative charge. These contaminants have gained attention due to their increasing occurrence in various environmental compartments, such as surface water, groundwater, soils, sediments, and even biota (Aley *et al* 2022, Kumar *et al* 2020, Kumar *et al* 2021a, P. Kumar and Mistri 2020). Polar emerging contaminants encompass many substances, including pharmaceuticals, personal care products, pesticides, per- and polyfluoroalkyl substances (PFAS), and many others.

The presence of emerging polar contaminants in the environment is primarily attributed to their widespread use in various human activities. These compounds enter the ecosystem through multiple pathways, including wastewater discharge from domestic, industrial, and agricultural sources, atmospheric deposition, and improper waste disposal practices. Once released, they can persist in the environment and undergo transport through water and soil systems, potentially affecting aquatic organisms, terrestrial wildlife, and even human populations.

One of the main challenges associated with emerging polar contaminants is their potential impact on aquatic ecosystems. Many of these compounds are designed to have biological effects at low concentrations, such as pharmaceuticals and personal care products that can act as endocrine disruptors (Chakraborty *et al* 2021, Kotia *et al* 2022, Kumar, *et al* 2021b). These substances can interfere with the hormonal systems of organisms, potentially leading to reproductive and developmental abnormalities.

Furthermore, emerging polar contaminants have the potential to accumulate in organisms through bioaccumulation and biomagnification processes. This can result in high concentrations of these contaminants in higher trophic levels, posing risks to predators and species at the top of the food chain, including humans.

To address the issue of polar emerging contaminants in the environment, several strategies can be employed:

1. Advanced wastewater treatment: Upgrading wastewater treatment processes is crucial for reducing the release of emerging polar contaminants into water bodies. Advanced treatment technologies, such as AOPs, activated carbon adsorption, and membrane filtration, effectively remove or degrade these contaminants, ensuring cleaner effluent discharges.

2. Source control measures: Implementing source control measures is essential to minimize the entry of emerging polar contaminants into the environment. This involves adopting best management practices in industrial and agricultural sectors to reduce the use and discharge of these substances. Proper waste management practices and the promotion of eco-friendly alternatives can also play a significant role in reducing contamination.

3. Monitoring and research: Continuous monitoring and research efforts are necessary to identify and understand the occurrence, fate, and behavior of emerging polar contaminants in different environmental compartments. This knowledge is vital for developing effective regulations and mitigation

strategies. Additionally, research should focus on the potential risks of these contaminants, including their long-term effects on the ecosystem and human health.

4. Public awareness and education: Raising public awareness about the presence and potential risks of emerging polar contaminants is crucial. Educating individuals about the responsible use, disposal, and recycling of products containing these substances can help minimize their environmental impact. Furthermore, promoting the adoption of eco-friendly alternatives and sustainable practices can contribute to reducing the overall load of emerging polar contaminants in the environment.

5. Regulation and policy development: Developing and implementing comprehensive regulations and policies is necessary to address the challenges of polar emerging contaminants. This includes setting standards for their presence in various environmental compartments and establishing monitoring and reporting requirements. Regulation may also restrict certain substances and promote sustainable practices in the industrial and agricultural sectors.

Emerging polar contaminants in the environment pose significant challenges to ecosystem health and human well-being. Addressing this issue requires advanced wastewater treatment, source control measures, monitoring and research, public awareness and education, and effective regulation and policy development (Kumar *et al* 2020, Kumar and Mistri 2020, Kumar *et al* 2021, Kumari *et al* 2022). By implementing these strategies, we can work towards reducing the occurrence and mitigating the potential risks associated with polar emerging contaminants, thus safeguarding the environment and protecting human and ecological health (table 4.3).

4.5 Efficient control of emerging contaminants: promising enzymes and reaction processes

The efficient control of emerging contaminants requires the development of innovative strategies that can effectively degrade or remove these contaminants from the environment. Enzymes have emerged as promising tools in this endeavor, offering specific and efficient degradation capabilities. Different enzymes and related reaction processes are crucial in efficiently controlling emerging contaminants (Adelodun *et al* 2022, Aley *et al* 2022, Das *et al* 2022, Goud *et al* 2022, Kumar *et al* 2020, Kumar, *et al* 2021a, Kumar and Mistri 2020, Kumari *et al* 2022). Here, we provide comprehensive information about various strategies and promising enzymes involved in this control and their related reaction processes.

1. Oxidative enzymes:
   a. Peroxidases, such as horseradish peroxidase (HRP) and lignin peroxidase, can catalyze the oxidation of various emerging contaminants. They utilize hydrogen peroxide (H_2O_2) as a co-substrate to generate reactive oxygen species (ROS) that degrade the contaminants.

Table 4.3. Types of polar contaminants, their sources and potential location, and their disadvantages.

Polar contaminant	Type	Sources and Uses	Potential locations	Disadvantages
Bisphenol A	Endocrine disruptor	Plastic containers, food can linings, thermal paper	Consumer products, landfills, wastewater treatment	Hormonal disruption, reproductive effects
Triclosan	Antibacterial agent	Personal care products, soaps, toothpaste, cosmetics	Wastewater, surface water, sediments, soils	Ecological impacts, antimicrobial resistance
Atrazine	Herbicide	Agriculture	Groundwater, surface water, soils	Water contamination, adverse effects on aquatic organisms
Perfluorooctanoic Acid (PFOA)	Per- and polyfluoroalkyl substance (PFAS)	Firefighting foams, non-stick cookware, textiles	Drinking water, surface water, soils, biota	Bioaccumulation, persistence, adverse health effects
Phthalates	Plasticizers	Plastic products, PVC, personal care products	Indoor air, dust, consumer products, water	Endocrine disruption, reproductive effects
Carbamazepine	Pharmaceutical	Medications, wastewater	Surface water, groundwater, drinking water	Potential human health impacts, ecological effects
Nonylphenol Ethoxylates	Surfactant	Industrial detergents, emulsifiers	Wastewater, surface water, sediments, soils	Endocrine disruption, aquatic toxicity
2,4- Surface water, groundwater, soils	Ecotoxicity, potential risks to non-target organisms	Dichlorophenoxyacetic Acid (2,4-D)	Herbicide	Agricultural and lawn care applications
Ethinylestradiol	Synthetic hormone	Oral contraceptives, hormone replacement therapy	Wastewater, surface water, sediments	Hormonal disruption, reproductive effects
Chlorpyrifos	Insecticide	Agricultural and residential pest control	Surface water, groundwater, soils	Ecotoxicity, potential risks to non-target organisms

Source: Based on a review of the literature.

HRP-based systems have effectively degraded phenolic compounds, pharmaceuticals, and endocrine disruptors.

    b. Laccases are multicopper oxidases that can efficiently oxidize a wide range of emerging contaminants. They use molecular oxygen as a co-substrate and produce ROS to initiate degradation reactions. Laccases have successfully degraded various phenolic compounds, pharmaceuticals, and industrial dyes (Ansari *et al* 2023, Cong *et al* 2023, Hjort *et al* 2023, Mali *et al* 2023, Osuoha *et al* 2023).

2. Hydrolases:
   a. Esterases are enzymes that catalyze the hydrolysis of ester bonds. They effectively degrade various emerging contaminants, such as ester-based pharmaceuticals, plasticizers, and surfactants. Esterases can be derived from microorganisms or produced through recombinant DNA technology.
   b. Amidases are enzymes that catalyze the hydrolysis of amide bonds. They have shown potential in degrading emerging contaminants such as pharmaceuticals and pesticide residues containing amide functional groups. Amidases can be sourced from microorganisms or engineered through protein engineering techniques (Cipriani-Avila *et al* 2023, Ghosh and Biswas 2023, Khalaf *et al* 2023, Kurade *et al* 2023, N. Li *et al* 2023b, Moreira *et al* 2023, Sunyer-Caldú *et al* 2023, Walch *et al* 2023, Zahmatkesh *et al* 2023).

3. Reductive enzymes:
   a. Nitroreductases are enzymes that catalyze the reduction of nitro groups (-NO2) present in various emerging contaminants, such as nitroaromatic compounds and nitro-substituted pesticides. Nitroreductases use electron donors, such as nicotinamide adenine dinucleotide (NADH), to initiate reduction reactions.
   b. Dehalogenases are enzymes involved in reducing or removing halogen atoms (e.g., chlorine, bromine) from emerging contaminants containing halogenated compounds. They effectively degrade halogenated organic pollutants, including chlorinated solvents and halogenated aromatics.

4. Photolytic and photocatalytic processes utilize light energy to initiate degradation reactions. Photocatalysts, such as titanium dioxide (TiO_2), are activated by ultraviolet (UV) light to generate reactive species that degrade emerging contaminants. These processes can effectively target contaminants, including pharmaceuticals, pesticides, and organic dyes (Adelodun *et al* 2022, Das *et al* 2022, Goud *et al* 2022, Kumar *et al* 2021, Li *et al* 2023d).

5. AOPs: AOPs involve the generation of highly reactive hydroxyl radicals (•OH) to degrade emerging contaminants. AOPs, such as ozone-based processes, Fenton's reaction (Fe^{2+}/H_2O_2), and advanced oxidation with UV irradiation, have effectively degraded various contaminants, including pharmaceuticals, pesticides, and endocrine disruptors.

It is important to note that selecting enzymes and reaction processes depends on the specific contaminant, environmental conditions, and treatment requirements (Aley *et al* 2022, Bashir *et al* 2023, Chakraborty *et al* 2021, Ettahiri *et al* 2023, He *et al* 2023, Kotia *et al* 2022, Kumar, *et al* 2021a, Kumar, *et al* 2021b, Menya *et al* 2023, Peivasteh-roudsari *et al* 2023, Yuan *et al* 2023, Zhou and Gao 2023). Factors such as substrate specificity, reaction kinetics, stability, and enzyme availability must be considered when implementing these strategies to control emerging contaminants efficiently.

In conclusion, using enzymes and related reaction processes offers promising strategies to efficiently control emerging contaminants. Oxidative enzymes, hydrolases, reductive enzymes, photolytic/photocatalytic processes, and AOPs have shown potential in degrading various emerging contaminants across different applications. Continued research and development efforts are necessary to optimize these strategies, improve enzyme stability, enhance reaction efficiency, and tailor them for specific contaminant classes and environmental matrices (table 4.4).

Table 4.4. Enzymes involved in the removal of contamination.

Enzyme	Type	Example substrates	Oxidative products
Horseradish peroxidase	Oxidative	Phenolic compounds, dyes, pesticides	Quinones, phenoxyl radicals, water oxidation
Laccase	Oxidative	Phenolic compounds, lignin, dyes	Quinones, phenoxyl radicals, water oxidation
Manganese peroxidase	Oxidative	Lignin, phenolic compounds	Manganese radicals, water oxidation
Versatile peroxidase	Oxidative	Lignin, phenolic compounds	Quinones, phenoxyl radicals, water oxidation
Glutathione peroxidase	Oxidative	Hydrogen peroxide, lipid hydroperoxides	Glutathione disulfide, water oxidation
Catalase	Oxidative	Hydrogen peroxide	Oxygen, water
Esterase	Hydrolase	Ester-based pharmaceuticals, plasticizers, surfactants	Alcohol and carboxylic acid components
Amidase	Hydrolase	Amide-based pharmaceuticals, pesticide residues	Amine and carboxylic acid components
Lipase	Hydrolase	Fats, oils, triglycerides	Glycerol, free fatty acids, monoglycerides
Protease	Hydrolase	Protein-based pollutants, detergents	Amino acids, peptides, water oxidation
Nitrilase	Hydrolase	Nitrile-based compounds, pesticides	Carboxylic acids, ammonia, water oxidation
Nitroreductase	Reductive	Nitroaromatic compounds, nitro-substituted pesticides	Amines, hydroxylamines, water oxidation
Dehalogenase	Reductive	Halogenated organic pollutants	Removal of halogen atoms, formation of byproducts

(Continued)

Table 4.4. (*Continued*)

Enzyme	Type	Example substrates	Oxidative products
Carbonic anhydrase	Hydrolase	Carbon dioxide, bicarbonate ions	Protons, bicarbonate ions, water oxidation
Xanthine oxidase	Oxidative	Xanthine, hypoxanthine, uric acid	Uric acid, superoxide radicals, water oxidation
Aldehyde dehydrogenase	Oxidative	Aldehydes, pollutants	Carboxylic acids, water oxidation
NADH dehydrogenase	Oxidative	NADH, pollutants	NAD+, water oxidation
Urease	Hydrolase	Urea	Ammonia, water oxidation
Cytochrome P450	Oxidative	Organic compounds, pollutants	Hydroxylated compounds, water oxidation
Glutathione S-transferase	Conjugation	Glutathione, electrophilic pollutants	Glutathione-conjugated pollutants
NADPH oxidase	Oxidative	NADPH, oxygen	Superoxide radicals, water oxidation
β-Glucosidase	Hydrolase	Glucosides, pesticides	Glucose, aglycones, water oxidation
Uricase	Oxidative	Uric acid	Allantoin, water oxidation
Tyrosinase	Oxidative	Phenolic compounds, pigments	Quinones, water oxidation
Glucose oxidase	Oxidative	Glucose	Gluconic acid, hydrogen peroxide, water oxidation
NADH:quinone oxidoreductase	Oxidative	NADH, quinones	NAD+, reduced quinones, water oxidation
Superoxide dismutase	Antioxidant	Superoxide radicals	Hydrogen peroxide, oxygen
β-Lactamase	Hydrolase	β-Lactam antibiotics	Hydrolyzed products, water oxidation
Naphthalene dioxygenase	Oxidative	Polycyclic aromatic hydrocarbons	Epoxides, dihydrodiols, water oxidation
Methane monooxygenase	Oxidative	Methane	Methanol, formaldehyde, water oxidation
Nitrifying bacteria	Biological	Ammonia, nitrite	Nitrate, water oxidation
Cytochrome c oxidase	Oxidative	Cytochrome c, oxygen	Water, reduced cytochrome c
β-galactosidase	Hydrolase	β-Galactosides	Galactose, water oxidation
Hyaluronidase	Hydrolase	Hyaluronic acid, extracellular matrix	Oligosaccharides, water oxidation
Chitinase	Hydrolase	Chitin	Glucosamine, chitobiose, water oxidation
Alcohol dehydrogenase	Oxidative	Alcohols, pollutants	Aldehydes, water oxidation
Lysozyme	Hydrolase	Bacterial cell wall components	Oligosaccharides, water oxidation

Pectinase	Hydrolase	Pectin	Oligogalacturonides, water oxidation
Haloalkane dehalogenase	Hydrolase	Halogenated organic pollutants	Alcohols, water oxidation
Glucuronidase	Hydrolase	Glucuronides, pollutants	Aglycones, water oxidation
Uridine phosphorylase	Hydrolase	Nucleosides, nucleotides	Ribose-1-phosphate, base moieties, water oxidation
Xyloglucanase	Hydrolase	Xyloglucan, plant cell wall component	Oligosaccharides, water oxidation
Glutaminase	Hydrolase	Glutamine	Glutamate, ammonia, water oxidation
Cyanide dihydratase	Hydrolase	Cyanide	Formate, ammonia, water oxidation
Nitrate reductase	Reductive	Nitrate	Nitrite, water oxidation
Nitrite reductase	Reductive	Nitrite	Nitric oxide, water oxidation
Dimethyl sulfoxide reductase	Reductive	Dimethyl sulfoxide	Dimethyl sulfide, water oxidation
Naphthalene reductase	Reductive	Polycyclic aromatic hydrocarbons	Dihydrodiols, water oxidation
Carbon monoxide dehydrogenase	Oxidative	Carbon monoxide	Carbon dioxide, water oxidation
Ammonia monooxygenase	Oxidative	Ammonia	Hydroxylamine, water oxidation
Rubisco	Carboxylase	Carbon dioxide	Sugars, water oxidation

Source: Based on review of literature.

4.6 Antibiotic contamination in an environment

Antibiotic resistance is a global concern that poses significant threats to public health. While antibiotic resistance is primarily associated with clinical settings, it is also a growing concern in the aquatic environment (Clance *et al* 2023, Iheanacho *et al* 2023, Kumar *et al* 2023, Kumari *et al* 2022, Laad and Ghule 2023, Mali *et al* 2023, Mu *et al* 2023, Osuoha *et al* 2023, Zhang *et al* 2023). The presence of antibiotics and other emerging contaminants in water bodies can contribute to the development and spread of antibiotic resistance among aquatic bacteria, impacting both environmental and human health. Analytical techniques are crucial in monitoring antibiotic resistance in the aquatic environment and understanding the interactive impact of emerging contaminants. This article overviews the issue and highlights the analytical techniques employed in studying antibiotic resistance and the interactive effects of emerging contaminants in aquatic ecosystems.

1. Antibiotic resistance in aquatic environments: Antibiotics are widely used in human and veterinary medicine and agricultural practices. These compounds find their way into the aquatic environment through various routes, including sewage and wastewater treatment plant effluents, agricultural runoff, and

improper disposal of pharmaceuticals (Al-Jubouri *et al* 2023, Boutet *et al* 2023, Casabella-Font *et al* 2023, Chen *et al* 2023a, François *et al* 2023, Grunst *et al* 2023, GodvinSharmila *et al* 2023). Once in aquatic systems, antibiotics can exert selective pressure on bacteria, favoring the survival and proliferation of antibiotic-resistant strains. Furthermore, horizontal gene transfer mechanisms, such as plasmids and integrons, allow the transfer of antibiotic-resistance genes between bacteria, further exacerbating the problem. Antibiotic-resistant bacteria in aquatic environments can serve as reservoirs of resistance genes, potentially spreading them to human pathogens and compromising the effectiveness of antibiotic treatments.

2. Analytical techniques for studying antibiotic resistance: Several analytical techniques are employed to study antibiotic resistance in the aquatic environment. These techniques include:

   a. Antibiotic susceptibility testing: Determines the susceptibility of bacteria to antibiotics using agar diffusion methods or broth microdilution assays. It also provides information on bacteria's prevalence and resistance patterns in aquatic systems (Adelodun *et al* 2022, Akhtar *et al* 2021, Aley *et al* 2022, Chakraborty *et al* 2021, Das *et al* 2022, Kotia *et al* 2022, Kumar, *et al* 2021b, Kumar *et al* 2020, Kumar, *et al* 2021a, Kumar and Mistri 2020, Kumar *et al* 2021, Kumari *et al* 2022, Upadhyay *et al* 2023, GodvinSharmila *et al* 2023).

   b. Quantitative PCR (qPCR): Detects and quantifies antibiotic-resistance genes in environmental samples. In addition, qPCR allows researchers to assess the abundance and distribution of resistance genes in aquatic environments.

   c. Metagenomics: Provides a comprehensive view of the genetic diversity of antibiotic-resistance genes in microbial communities. Metagenomic approaches involve sequencing the entire DNA present in environmental samples, allowing the identification of known and novel resistance genes.

   d. Mass spectrometry: Enables the detection and quantification of antibiotics and their metabolites in water samples. Mass spectrometry-based techniques, such as liquid chromatography-mass spectrometry (LC-MS), can identify the presence of specific antibiotics and monitor their concentrations over time.

3. Interactive impact of emerging contaminants: The presence of emerging contaminants, such as pharmaceuticals, personal care products, and other pollutants, can have interactive effects on antibiotic resistance in the aquatic environment. Some fundamental interactions include:

   a. Co-selection: The exposure of bacteria to multiple stressors, including antibiotics and other contaminants, can lead to the co-selection of antibiotic resistance. For example, certain chemicals can induce stress responses in bacteria, facilitating the acquisition and maintenance of antibiotic-resistance genes.

   b. Synergistic effects: Emerging contaminants can interact synergistically with antibiotics, enhancing their toxic effects and promoting the development of antibiotic resistance. These interactions can lead to increased bacterial growth rates, biofilm formation, and altered gene expression patterns, making bacteria more resilient to antibiotics (Kumar *et al* 2023, Laad and Ghule 2023, Mu *et al* 2023, Nawaz *et al* 2023, Patel *et al* 2023, Zhang *et al* 2023).
   c. Disruption of microbial communities: Emerging contaminants can disrupt the natural balance of microbial communities in aquatic ecosystems. This disruption can alter bacterial populations, favouring the growth of antibiotic-resistant strains and reducing the effectiveness of antibiotic treatments.

4. Mitigation strategies: Several mitigation strategies can be implemented to address antibiotic resistance and the interactive impact of emerging contaminants in the aquatic environment:
   a. Improved wastewater treatment: Upgrading wastewater treatment plants with advanced treatment technologies can effectively remove antibiotics and other contaminants. Techniques such as activated carbon filtration, ozonation, and UV irradiation can help to reduce the release of these compounds into water bodies.
   b. Source control measures: Implementing source control measures, such as proper disposal of pharmaceuticals and reducing agricultural use of antibiotics, can minimize the entry of contaminants into aquatic ecosystems.
   c. Antibiotic stewardship: Promoting the responsible use of antibiotics in human and veterinary medicine can help to reduce the selective pressure on bacteria, and mitigate the development and spread of antibiotic resistance.
   d. Enhanced monitoring: Continuous monitoring of antibiotic-resistance genes, antibiotics, and emerging contaminants in the aquatic environment is crucial for early detection and timely intervention. This monitoring can inform risk assessments and guide mitigation efforts.

In conclusion, antibiotic resistance in the aquatic environment poses significant environmental and human health challenges. Analytical techniques, such as antibiotic susceptibility testing, qPCR, metagenomics, and mass spectrometry, enable antibiotic resistance and emerging contaminants in water bodies to be studied. Understanding the interactive impact of emerging contaminants on antibiotic resistance requires a multidisciplinary approach that integrates environmental monitoring, toxicology, and microbial ecology. By implementing mitigation strategies and adopting responsible practices, we can work towards minimizing the spread of antibiotic resistance and preserving the health of aquatic ecosystems (table 4.5).

Table 4.5. Different types of antibiotic contaminants in the environment

Antibiotic contaminant	Disadvantages	Sources
Tetracycline	Development of antibiotic resistance in bacteria	Veterinary use, aquaculture, wastewater discharges
Sulfamethoxazole	Ecological impacts, potential human health effects	Human and veterinary use, wastewater discharges
Ciprofloxacin	Development of antibiotic resistance in bacteria	Human and veterinary use, wastewater discharges
Azithromycin	Ecological impacts, potential human health effects	Human and veterinary use, wastewater discharges
Erythromycin	Ecological impacts, potential human health effects	Human and veterinary use, wastewater discharges
Amoxicillin	Development of antibiotic resistance in bacteria	Human and veterinary use, wastewater discharges
Ampicillin	Development of antibiotic resistance in bacteria	Human and veterinary use, wastewater discharges
Chloramphenicol	Potential toxic effects on aquatic organisms	Human and veterinary use, industrial discharges
Trimethoprim	Development of antibiotic resistance in bacteria	Human and veterinary use, wastewater discharges
Vancomycin	Development of antibiotic resistance in bacteria	Human and veterinary use, wastewater discharges
Gentamicin	Development of antibiotic resistance in bacteria	Human and veterinary use, wastewater discharges
Streptomycin	Development of antibiotic resistance in bacteria	Human and veterinary use, wastewater discharges
Norfloxacin	Development of antibiotic resistance in bacteria	Human and veterinary use, wastewater discharges
Enoxacin	Development of antibiotic resistance in bacteria	Human and veterinary use, wastewater discharges
Cefalexin	Development of antibiotic resistance in bacteria	Human and veterinary use, wastewater discharges
Doxycycline	Development of antibiotic resistance in bacteria	Human and veterinary use, wastewater discharges
Lincomycin	Potential toxic effects on aquatic organisms	Human and veterinary use, wastewater discharges
Roxithromycin	Ecological impacts, potential human health effects	Human and veterinary use, wastewater discharges
Neomycin	Development of antibiotic resistance in bacteria	Human and veterinary use, wastewater discharges
Kanamycin	Development of antibiotic resistance in bacteria	Human and veterinary use, wastewater discharges
Erythromycin	Ecological impacts, potential human health effects	Human and veterinary use, wastewater discharges

Sulfadiazine	Ecological impacts, potential human health effects	Human and veterinary use, wastewater discharges
Sulfamethazine	Ecological impacts, potential human health effects	Human and veterinary use, wastewater discharges
Sulfamerazine	Ecological impacts, potential human health effects	Human and veterinary use, wastewater discharges
Sulfachloropyridazine	Ecological impacts, potential human health effects	Human and veterinary use, wastewater discharges
Sulfadimethoxine	Ecological impacts, potential human health effects	Human and veterinary use, wastewater discharges
Sulfadoxine	Ecological impacts, potential human health effects	Human and veterinary use, wastewater discharges
Cefotaxime	Development of antibiotic resistance in bacteria	Human and veterinary use, wastewater discharges
Penicillin G	Development of antibiotic resistance in bacteria	Human and veterinary use, wastewater discharges
Erythromycin-H2O	Ecological impacts, potential human health effects	Human and veterinary use, wastewater discharges
Erythromycin-H2O2	Ecological impacts, potential human health effects	Human and veterinary use, wastewater discharges
Azithromycin-H2O	Ecological impacts, potential human health effects	Human and veterinary use, wastewater discharges
Tetracycline-H2O	Development of antibiotic resistance in bacteria	Veterinary use, aquaculture, wastewater discharges
Tetracycline-H2O2	Development of antibiotic resistance in bacteria	Veterinary use, aquaculture, wastewater discharges
Ofloxacin	Development of antibiotic resistance in bacteria	Human and veterinary use, wastewater discharges
Levofloxacin	Development of antibiotic resistance in bacteria	Human and veterinary use, wastewater discharges
Norfloxacin	Development of antibiotic resistance in bacteria	Human and veterinary use, wastewater discharges
Cefuroxime	Development of antibiotic resistance in bacteria	Human and veterinary use, wastewater discharges
Ceftiofur	Development of antibiotic resistance in bacteria	Veterinary use, aquaculture, wastewater discharges
Chlortetracycline	Development of antibiotic resistance in bacteria	Veterinary use, aquaculture, wastewater discharges
Tiamulin	Potential toxic effects on aquatic organisms	Veterinary use, aquaculture, wastewater discharges
Florfenicol	Potential toxic effects on aquatic organisms	Veterinary use, aquaculture, wastewater discharges
Novobiocin	Development of antibiotic resistance in bacteria	Human and veterinary use, wastewater discharges

(Continued)

Table 4.5. (*Continued*)

Antibiotic contaminant	Disadvantages	Sources
Lincomycin	Potential toxic effects on aquatic organisms	Human and veterinary use, wastewater discharges
Quinolone	Development of antibiotic resistance in bacteria	Human and veterinary use, wastewater discharges

Source: Based on a review of the literature.

4.7 Antiviral drugs in wastewater: the rise of emerging contaminants

The presence of pharmaceuticals in wastewater has been a growing concern in recent years, and among them antiviral drugs have have gained significant attention as emerging contaminants. Antiviral drugs are designed to combat viral infections in humans and animals, but their widespread use and incomplete removal during wastewater treatment processes have led to their accumulation in aquatic environments (Cipriani-Avila *et al* 2023, Ghosh *et al* 2023, Kurade *et al* 2023, Li *et al* 2023b, Méndez-Loranca *et al* 2023, Walch *et al* 2023, Zahmatkesh *et al* 2023, Zhang *et al* 2023).

1. Sources of antiviral drugs in wastewater: The primary sources of antiviral drugs in wastewater can be attributed to various factors, including:
    a. Human excretion: Antiviral drugs are predominantly consumed by individuals to treat viral infections. After ingestion, a significant portion of the drugs is metabolized and excreted from the body via urine and feces, ultimately entering the wastewater stream.
    b. Improper drug disposal: The inappropriate disposal of unused or expired antiviral drugs contributes to their presence in wastewater. Flushing unused medications down the toilet or throwing them in the trash can lead to their entry into the wastewater system.
    c. Hospital and healthcare facilities: Medical facilities, including hospitals and clinics, also contribute to the release of antiviral drugs into wastewater. The disposal of unused drugs, excretion by patients, and the presence of antiviral drug residues in healthcare settings contribute to their presence in wastewater.
2. Environmental impacts of antiviral drugs: The presence of antiviral drugs in wastewater can have several potential environmental impacts, including:
    a. Ecological effects: Antiviral drugs can negatively affect aquatic organisms, such as fish, invertebrates, and algae, even at low concentrations. These drugs can disrupt metabolic processes, impair reproductive functions, and lead to alterations in growth and development.
    b. Antibiotic resistance: Some antiviral drugs, such as nucleoside analogs, can induce selective pressure on microorganisms, potentially leading to the development and spread of antiviral drug resistance in aquatic

environments. This resistance can impact the efficacy of antiviral treatments in clinical settings.

   c. Ecological connectivity: Antiviral drugs in wastewater can enter surface water bodies and interact with natural ecosystems. This connectivity can facilitate the transfer of antiviral drugs and their metabolites to other environmental compartments, such as sediments and groundwater.

3. Challenges in the removal of antiviral drugs: The removal of antiviral drugs from wastewater poses several challenges due to their chemical properties and low concentrations. Some key challenges include:

   a. Stability and persistence: Antiviral drugs are designed to be stable and persistent to ensure their efficacy in the human body. However, these same characteristics make them resistant to degradation during wastewater treatment processes, leading to their persistence in the environment.

   b. Low concentrations: Antiviral drugs are typically present in wastewater at low concentrations, often in the nanogram to microgram per liter range. The detection and removal of these low levels of contaminants require sensitive analytical techniques and advanced treatment technologies.

   c. Treatment efficiency: Conventional wastewater treatment processes, such as activated sludge and sedimentation, are not specifically designed to remove pharmaceuticals, including antiviral drugs. These drugs can bypass or resist treatment processes, resulting in their release into the environment.

4. Mitigation strategies: Mitigating the presence of antiviral drugs in wastewater requires a multifaceted approach, including:

   a. Advanced treatment technologies: Implementing advanced treatment technologies, such as ozonation, activated carbon adsorption, and AOPs, can enhance the removal efficiency of antiviral drugs from wastewater (Adelodun *et al* 2022, Goud *et al* 2022, Green *et al* 2023, Gulamhussein *et al* 2023, Kurade *et al* 2023, Li *et al* 2023b, X. Li *et al* 2023c, Long *et al* 2023, Moreira *et al* 2023, Sunyer-Caldú *et al* 2023).

   b. Public awareness and education: Educating the public about the proper disposal of unused medications and the potential environmental impacts of antiviral drugs can help to reduce their entry into the wastewater system.

   c. Source control measures: Implementing strategies to minimize the release of antiviral drugs into wastewater, such as improved prescription practices, proper drug disposal systems, and the development of alternative treatment options, can help to reduce their presence in the environment.

   d. Enhanced monitoring and regulation: Continuous monitoring of antiviral drugs in wastewater, along with the establishment of regulatory guidelines and standards, can help to assess their environmental impact and guide mitigation efforts (Dupouy and Popping 2023, Ettahiri *et al* 2023, He *et al* 2023, Menya *et al* 2023, Peivasteh-roudsari *et al* 2023).

In conclusion, the presence of antiviral drugs in wastewater as emerging contaminants poses environmental challenges due to their sources, potential impacts, and the difficulties associated with their removal. Addressing this issue requires a combination of advanced treatment technologies, public awareness, source control measures, and enhanced monitoring and regulation. By adopting a holistic approach, we can work towards minimizing the presence of antiviral drugs in wastewater and mitigating their environmental impact.

1. Acyclovir
   - Harmful impact: Potential ecological toxicity to aquatic organisms.
   - Possible mitigation: Improved wastewater treatment processes, such as advanced oxidation or activated carbon filtration, to remove acyclovir from wastewater.
2. Amantadine
   - Harmful impact: Ecotoxicity to aquatic organisms and potential development of drug resistance.
   - Possible mitigation: Enhanced monitoring of water bodies, proper disposal of unused medication, and promotion of alternative treatment options.
3. Amprenavir
   - Harmful impact: Potential ecological toxicity and the development of drug-resistant viruses.
   - Possible mitigation: Advanced wastewater treatment technologies, public education on proper drug disposal, and encouraging responsible prescription practices.
4. Arbidol
   - Harmful impact: Ecotoxicity to aquatic organisms and potential development of drug resistance.
   - Possible mitigation: Upgrading wastewater treatment plants to remove arbidol, promoting responsible use of the drug, and implementing source control measures.
5. Atazanavir
   - Harmful impact: Potential ecological toxicity and the development of drug-resistant viruses.
   - Possible mitigation: Improved wastewater treatment processes, public awareness campaigns on responsible medication disposal, and research of alternative antiviral treatments.
6. Baloxavir marboxil
   - Harmful impact: Potential environmental toxicity and the risk of drug resistance development.
   - Possible mitigation: Enhanced monitoring of water bodies, implementation of advanced treatment technologies, and research of eco-friendly antiviral alternatives.

7. Cidofovir
   - Harmful impact: Ecotoxicity to aquatic organisms and potential development of drug-resistant viruses.
   - Possible mitigation: Upgrading wastewater treatment systems, exploring innovative treatment technologies, and promoting responsible medication disposal practices.
8. Darunavir
   - Harmful impact: Potential ecological toxicity and the development of drug-resistant viruses.
   - Possible mitigation: Advanced treatment methods for wastewater, public education on proper drug disposal, and research of alternative antiviral therapies.
9. Delavirdine
   - Harmful impact: Ecotoxicity to aquatic organisms and the potential for the development of drug-resistant viruses.
   - Possible mitigation: Implementation of advanced wastewater treatment processes, encouraging responsible medication disposal, and exploring alternative antiviral treatments.
10. Dolutegravir
    - Harmful impact: Potential environmental toxicity and the risk of drug resistance development.
    - Possible mitigation: Enhanced monitoring of water bodies, improved wastewater treatment methods, and research of sustainable antiviral therapies.
11. Efavirenz
    - Harmful impact: Ecotoxicity to aquatic organisms and the potential for drug resistance development.
    - Possible mitigation: Upgrading wastewater treatment facilities, promoting responsible drug disposal practices, and exploring greener antiviral options.
12. Elvitegravir
    - Harmful impact: Potential ecological toxicity and the development of drug-resistant viruses.
    - Possible mitigation: Advanced wastewater treatment technologies, public awareness campaigns on proper drug disposal, and research of eco-friendly antiviral alternatives.
13. Emtricitabine
    - Harmful impact: Ecotoxicity to aquatic organisms and the potential for drug resistance development.
    - Possible mitigation: Upgrading wastewater treatment systems, exploring innovative treatment technologies, and promoting responsible medication disposal practices.
14. Enfuvirtide
    - Harmful impact: Potential environmental toxicity and the risk of drug resistance development.

- Possible mitigation: Enhanced monitoring of water bodies, implementation of advanced treatment technologies, and research of eco-friendly antiviral alternatives.

15. Etravirine
   - Harmful impact: Ecotoxicity to aquatic organisms and potential development of drug-resistant viruses.
   - Possible mitigation: Upgrading wastewater treatment systems, exploring innovative treatment technologies, and promoting responsible medication disposal practices.

16. Famciclovir
   - Harmful impact: Potential ecological toxicity to aquatic organisms.
   - Possible mitigation: Improved wastewater treatment processes, such as advanced oxidation or activated carbon filtration, to remove famciclovir from wastewater.

17. Fosamprenavir
   - Harmful impact: Ecotoxicity to aquatic organisms and potential development of drug-resistant viruses.
   - Possible mitigation: Enhanced monitoring of water bodies, proper disposal of unused medication, and promotion of alternative treatment options.

18. Ganciclovir
   - Harmful impact: Potential environmental toxicity and the risk of drug resistance development.
   - Possible mitigation: Upgrading wastewater treatment facilities, promoting responsible drug disposal practices, and exploring alternative antiviral therapies.

19. Indinavir
   - Harmful impact: Ecotoxicity to aquatic organisms and potential development of drug-resistant viruses.
   - Possible mitigation: Implementation of advanced wastewater treatment processes, public education on proper drug disposal, and encouraging responsible prescription practices.

20. Lamivudine
   - Harmful impact: Potential ecological toxicity and the development of drug-resistant viruses.
   - Possible mitigation: Improved wastewater treatment processes, public awareness campaigns on responsible medication disposal, and research of alternative antiviral treatments.

Please note that this list includes a selection of antiviral drugs that have been reported as contaminants in the environment. The harmful impacts and possible mitigation strategies may vary for different drugs. Additionally, the specific concentrations and occurrences of these drugs in the environment may depend on various factors, such as geographical location, wastewater treatment efficiency, and local regulations (table 4.6).

Table 4.6. A list of antiviral drugs as contaminants and their sources.

Antiviral drug	Sources
Acyclovir	Human excretion, wastewater discharges
Amantadine	Human excretion, pharmaceutical manufacturing
Amprenavir	Human excretion, wastewater discharges
Arbidol	Human excretion, pharmaceutical manufacturing
Atazanavir	Human excretion, wastewater discharges
Baloxavir marboxil	Human excretion, pharmaceutical manufacturing, wastewater discharges
Cidofovir	Human excretion, pharmaceutical manufacturing
Darunavir	Human excretion, wastewater discharges
Delavirdine	Human excretion, pharmaceutical manufacturing
Dolutegravir	Human excretion, wastewater discharges
Efavirenz	Human excretion, pharmaceutical manufacturing
Elvitegravir	Human excretion, wastewater discharges
Emtricitabine	Human excretion, wastewater discharges
Enfuvirtide	Human excretion, pharmaceutical manufacturing
Etravirine	Human excretion, wastewater discharges
Famciclovir	Human excretion, wastewater discharges
Fosamprenavir	Human excretion, pharmaceutical manufacturing, wastewater discharges
Ganciclovir	Human excretion, wastewater discharges
Indinavir	Human excretion, wastewater discharges
Lamivudine	Human excretion, wastewater discharges
Lopinavir	Human excretion, wastewater discharges
Maraviroc	Human excretion, wastewater discharges
Nelfinavir	Human excretion, wastewater discharges
Nevirapine	Human excretion, wastewater discharges
Oseltamivir	Human excretion, wastewater discharges
Peramivir	Human excretion, wastewater discharges
Raltegravir	Human excretion, wastewater discharges
Rilpivirine	Human excretion, wastewater discharges
Rimantadine	Human excretion, wastewater discharges
Saquinavir	Human excretion, wastewater discharges
Sofosbuvir	Human excretion, wastewater discharges
Stavudine	Human excretion, wastewater discharges
Telaprevir	Human excretion, wastewater discharges
Tenofovir	Human excretion, wastewater discharges
Tipranavir	Human excretion, wastewater discharges
Valacyclovir	Human excretion, wastewater discharges
Valganciclovir	Human excretion, wastewater discharges
Zanamivir	Human excretion, wastewater discharges
Zidovudine	Human excretion, wastewater discharges
Remdesivir	Human excretion, pharmaceutical manufacturing
Balapiravir	Human excretion, pharmaceutical manufacturing
Favipiravir	Human excretion, pharmaceutical manufacturing

(Continued)

Table 4.6. (*Continued*)

Antiviral drug	Sources
Molnupiravir	Human excretion, pharmaceutical manufacturing
Ribavirin	Human excretion, pharmaceutical manufacturing
Simeprevir	Human excretion, pharmaceutical manufacturing
Voxilaprevir	Human excretion, pharmaceutical manufacturing
Grazoprevir	Human excretion, pharmaceutical manufacturing
Paritaprevir	Human excretion, pharmaceutical manufacturing
Glecaprevir	Human excretion, pharmaceutical manufacturing
Pibrentasvir	Human excretion, pharmaceutical manufacturing

Source: Based on a review of the literature.

4.8 Various categories of emerging contaminants

Pharmaceuticals and Personal Care Products: Sources: Human and veterinary use, improper disposal of medications and personal care products, wastewater discharges.

Endocrine-Disrupting Compounds: Sources: Pesticides, industrial chemicals, plastics, hormone medications, cosmetics, and personal care products.

Per- and Polyfluoroalkyl Substances (PFAS): Sources: Firefighting foams, industrial processes, non-stick coatings, water-repellent fabrics, food packaging.

Microplastics: Sources: Fragmentation of larger plastics, synthetic fibers from textiles, microbeads in personal care products.

Nanoparticles: Sources: Industrial manufacturing processes, consumer products, sunscreens, electronics.

Pesticides and Herbicides: Sources: Agricultural applications, pest control, urban landscaping, runoff from treated areas.

Flame Retardants: Sources: Electronics, furniture, textiles, building materials.

Heavy Metals: Sources: Industrial emissions, mining activities, metal smelting, contaminated soils.

Antibiotics: Sources: Human and veterinary use, agriculture, aquaculture, improper disposal, wastewater discharges.

Hormones and Steroids: Sources: Livestock farming, veterinary use, hormone medications, wastewater treatment discharges.

PFOS and PFOA: Sources: Firefighting foams, industrial processes, water treatment, consumer products.

Bisphenols: Sources: Food and beverage containers, plastics, thermal paper, epoxy resins.

UV Filters: Sources: Sunscreen, personal care products, wastewater discharges.

Dioxins and Furans: Sources: Combustion processes, industrial emissions, waste incineration.

Phthalates: Sources: Plastics, personal care products, PVC products, building materials.

Artificial Sweeteners: Sources: Food and beverages, low-calorie products, wastewater discharges.

Volatile Organic Compounds: Sources: Industrial processes, solvents, paints, cleaning products, gasoline emissions.

Persistent Organic Pollutants: Sources: Pesticides, industrial chemicals, unintentional byproducts, long-range transport.

Flame Retardant Chemicals: Sources: Electronics, building materials, furniture, textiles.

Nitrate and Nitrite: Sources: Fertilizer runoff, animal waste, wastewater discharges.

4.9 Conclusion

In conclusion, emerging contaminants pose a significant challenge to the environment and human health. These contaminants encompass a wide range of substances, including pharmaceuticals, personal care products, pesticides, and industrial chemicals, among others. They enter the environment through various sources, such as human activities, wastewater discharges, and improper disposal practices. Efforts to reduce the impact of emerging contaminants require a multifaceted approach. This includes improving wastewater treatment processes to enhance the removal of contaminants and implementing advanced treatment technologies, such as activated carbon filtration and AOPs. Additionally, promoting responsible medication and chemical disposal practices, along with raising awareness among the public, can help to prevent the introduction of emerging contaminants into the environment. Moreover, implementing source control measures such as optimizing agricultural practices, adopting green chemistry principles, and developing sustainable alternatives to hazardous substances can contribute to the reduction of the release of emerging contaminants. Furthermore, continuous monitoring of water bodies and the establishment of regulatory guidelines are crucial to assess the presence and impact of emerging contaminants accurately. This enables the implementation of effective mitigation strategies and the development of appropriate regulations to protect ecosystems and public health. Addressing the challenges posed by emerging contaminants requires collaboration among researchers, policymakers, industries, and the general public. By collectively working towards reducing the release and impact of emerging contaminants, we can safeguard the environment and ensure a sustainable future for generations to come.

4.10 Prospects for future research

The field of emerging contaminants and their reduction strategies presents numerous prospects for future research. Some potential areas for investigation include:

- Identification of novel contaminants: Continued efforts to identify and characterize emerging contaminants are crucial. New analytical techniques and monitoring methods can help to detect previously unknown contaminants, enabling a better understanding of their occurrence, fate, and potential impacts.

- Ecological effects and ecotoxicology: Further research is needed to assess the ecological effects of emerging contaminants on various organisms and ecosystems. Understanding the mechanisms of toxicity and long-term effects can guide risk assessments and aid in the development of effective mitigation strategies.
- Human health impacts: Investigating the potential health risks associated with exposure to emerging contaminants is vital. This includes assessing the effects of chronic low-level exposure, and the potential for bioaccumulation and biomagnification in the food chain.
- Advanced treatment technologies: The development and optimization of advanced treatment technologies are essential for the efficient removal of emerging contaminants from wastewater. Research in this area can focus on innovative processes, such as membrane filtration, nanotechnology, and electrochemical methods, to enhance treatment efficiency.
- Source control and prevention: Future research can explore strategies for minimizing the release of emerging contaminants at the source. This includes examining sustainable agricultural practices, promoting responsible use and disposal of pharmaceuticals and personal care products, and implementing green chemistry principles.
- Risk assessment and regulation: Developing comprehensive risk assessment frameworks and establishing appropriate regulations for emerging contaminants are crucial. Future research can contribute to refining risk assessment methodologies and setting regulatory limits to protect both the environment and public health.
- Public awareness and education: Increasing public awareness about emerging contaminants and their potential impacts is essential. Future research can focus on effective communication strategies and educational campaigns to promote responsible behaviors, such as proper disposal of chemicals and medications.
- Circular economy approaches: Exploring circular economy principles and sustainable waste management strategies can contribute to reducing the presence of emerging contaminants in the environment. Research in this area can investigate the potential of recycling, resource recovery, and the substitution of hazardous substances with safer alternatives.
- Environmental monitoring networks: Establishing comprehensive and integrated environmental monitoring networks can provide valuable data on the occurrence and distribution of emerging contaminants. Future research can focus on developing cost-effective monitoring methods and data analysis techniques to support effective decision-making.
- International collaboration and knowledge sharing: Encouraging collaboration among researchers, industries, policymakers, and regulatory bodies at the global level can facilitate the exchange of knowledge and best practices in addressing emerging contaminants. This can lead to the development of harmonized approaches and strategies for reducing the impact of emerging contaminants worldwide.

By pursuing research in these areas, we can advance our understanding of emerging contaminants and develop effective strategies to reduce their presence in the environment, ultimately protecting ecosystems and human health for future generations.

Acknowledgments

We would like to express our sincere gratitude to the Department of Agronomy for their support and assistance throughout the writing. The department's commitment to academic excellence and research has been instrumental in completing this endeavor.

Author's contribution

The authors of this work have made significant contributions to this research project/study. Each author has participated sufficiently in the research project/study, made intellectual contributions, and is responsible for the work's accuracy and integrity. The authors have collaborated closely, ensuring the completion of this work through collective effort, expertise, and dedication.

References

Abedpour H, Moghaddas J S, Borhani M N and Borhani T N 2023 Separation of toxic contaminants from water by silica aerogel-based adsorbents: a comprehensive review *J. Water Process Eng.* **53** 103676

Adelodun B *et al* 2022 List of contributors *Microbiome Under Changing climate* ed A Kumar, J Singh and L F R Ferreira (Woodhead Publishing) pp xix–xxiv

Ahmad Bhat S, Cui G, Li W, Ameen F, Yaseera N, Wei Y and Li F 2023 Fate of bio-contaminants in soil systems and available remediation methods *Fate of Biological Contaminants During Recycling of Organic Wastes* ed K Huang, S Ahmad Bhat and G Cui (Amsterdam: Elsevier) ch 11 pp 213–27

Akhtar N *et al* 2021 List of contributors *Volatiles and Metabolites of Microbes* ed A Kumar, J Singh, and M M Samuel (New York: Academic) pp xix–xi

Al-Jubouri S M, Sabbar H A, Khudhair E M, Ammar S H, Al Batty S, Yas Khudhair S and Mahdi A S 2023 Silver oxide-zeolite for removal of an emerging contaminant by simultaneous adsorption-photocatalytic degradation under simulated sunlight irradiation *J. Photochem. Photobiol., A* **442** 114763

Aley P, Singh J and Kumar P 2022 Adapting the changing environment: microbial way of life *Microbiome Under Changing Climate* ed A Kumar, J Singh and L F R Ferreira (Woodhead Publishing) ch 23 pp 507–25

Amiri M K, Zahmatkesh S, Sarmasti Emami M R and Bokhari A 2023 Curve fitting model of Polycarbonate Al_2O_3-nanoparticle membranes for removing emerging contaminants from wastewater: effect of temperature and nanoparticles *Chemosphere* **322** 138184

Ansari A A, Shamim M A, Khan A M, Anwar K and Wani A A 2023 Nanomaterials as a cutting edge in the removal of toxic contaminants from water *Mater. Chem. Phys.* **295** 127092

Bashir M, Batool M, Arif N, Tayyab M, Zeng Y-J and Nadeem Zafar M 2023 Strontium-based nanomaterials for the removal of organic/inorganic contaminants from water: a review *Coord. Chem. Rev.* **492** 215286

Beschorner K E and Randolph A B 2023 Friction performance of resilient flooring under contaminant conditions relevant to healthcare settings *Appl. Ergon.* **108** 103960

Boutet V, Dominique M, Eccles K M, Branigan M, Dyck M, van Coeverden de Groot P, Lougheed S C, Rutter A and Langlois V S 2023 An exploratory spatial contaminant assessment for polar bear (Ursus maritimus) liver, fat, and muscle from northern Canada *Environ. Pollut.* **316** 120663

Casabella-Font O, Zahedi S, Gros M, Balcazar J L, Radjenovic J and Pijuan M 2023 Graphene oxide addition to anaerobic digestion of waste activated sludge: impact on methane production and removal of emerging contaminants *Environ. Pollut.* **324** 121343

Chakraborty S, Kumar P, Sanyal R, Mane A B, Arvind Prasanth D, Patil M and Dey A 2021 Unravelling the regulatory role of miRNAs in secondary metabolite production in medicinal crops *Plant Gene* **27** 100303

Chen J-Q, Sharifzadeh Z, Bigdeli F, Gholizadeh S, Li Z, Hu M-L and Morsali A 2023a MOF composites as high potential materials for hazardous organic contaminants removal in aqueous environments *J. Environ. Chem. Eng.* **11** 109469

Chen Q, Lü F, Zhang H and He P 2023b Where should Fenton go for the degradation of refractory organic contaminants in wastewater? *Water Res.* **229** 119479

Cipriani-Avila I *et al* 2023 Occurrence of emerging contaminants in surface water bodies of a coastal province in Ecuador and possible influence of tourism decline caused by COVID-19 lockdown *Sci. Total Environ.* **866** 161340

Clance L R, Ziegler S L and Fodrie F J 2023 Contaminants disrupt aquatic food webs via decreased consumer efficiency *Sci. Total Environ.* **859** 160245

Cong Y, Ye L, Zhang S, Zheng Q, Zhang Y and Lv S-W 2023 Efficient degradation of emerging contaminant by newly-constructed Ni-CoO yolk-shell hollow sphere in the presence of peroxymonosulfate: performance and mechanism *Process Saf. Environ. Protect.* **170** 685–93

Das T *et al* 2022 Promising botanical-derived monoamine oxidase (MAO) inhibitors: pharmacological aspects and structure-activity studies *S. Afr. J. Bot.* **146** 127–45

Dat N D, Huynh Q S, Tran K A T and Nguyen M L 2023 Performance of heterogeneous Fenton catalyst from solid wastes for removal of emerging contaminant in water: a potential approach to circular economy *Res. Eng.* **18** 101086

Dupouy E 2023 Trends in risk assessment of chemical contaminants in food *Present Knowledge in Food Safety* ed M E Knowles, L E Anelich, A R Boobis and B Popping (New York: Academic) ch 24 pp 320–8

Dupouy E and Popping B 2023 Emerging contaminants *Present Knowledge in Food Safety* ed M E Knowles, L E Anelich, A R Boobis and B Popping (New York: Academic) ch 17 pp 267–9

Ettahiri Y, Bouargane B, Fritah K, Akhsassi B, Pérez-Villarejo L, Aziz A, Bouna L, Benlhachemi A and Novais R M 2023 A state-of-the-art review of recent advances in porous geopolymer: applications in adsorption of inorganic and organic contaminants in water *Constr. Build. Mater.* **395** 132269

François M, Lin K-S, Rachmadona N and Khoo K S 2023 Advancement of nanotechnologies in biogas production and contaminant removal: a review *Fuel* **340** 127470

Ghosh S and Biswas A 2023 Emerging contaminants in municipal wastewater: occurrence, characteristics, and bioremediation *Current Developments in Biotechnology and Bioengineering* ed I Haq, A Kalamdhad, and B Pandey (Amsterdam: Elsevier) ch 9 pp 153–78

Ghosh S *et al* 2023 Recent progress on the remediation of metronidazole antibiotic as emerging contaminant from water environments using sustainable adsorbents: a review *J. Water Process Eng.* **51** 103405

Giannelli Moneta B *et al* 2023 Untargeted analysis of environmental contaminants in surface snow samples of Svalbard Islands by liquid chromatography-high resolution mass spectrometry *Sci. Total Environ.* **858** 159709

GodvinSharmila V, Shanmugavel S P, Tyagi V K and Rajesh Banu J 2023 Microplastics as emergent contaminants in landfill leachate: source, potential impact and remediation technologies *J. Environ. Manage.* **343** 118240

Goud E L, Singh J and Kumar P 2022 Climate change and their impact on global food production *Microbiome Under Changing Climate* ed A Kumar, J Singh and L F R Ferreira (Woodhead Publishing) ch 19 pp 415–36

Green H, Kidd J and Jackson L S 2023 Novel and Emerging Cleaning and Sanitization Technologies *Encyclopedia of Food Safety* ed G W Smithers (Amsterdam: Elsevier)

Grunst A S, Grunst M L and Fort J 2023 Contaminant-by-environment interactive effects on animal behavior in the context of global change: evidence from avian behavioral ecotoxicology *Sci. Total Environ.* **879** 163169

Gulamhussein M A, Saini B and Dey A 2023 Removal of pharmaceutical contaminants through membrane bioreactor *Mater. Today Proc.* **77** 260–8

He Q, Habib F, Tong T and Yu C 2023 *Raman Spectroscopy for Detection of Foodborne Pathogens, Chemical Contaminants and Nanoparticles* (Amsterdam: Elsevier)

Hjort R G, Pola C C, Soares R R A, Oliveira D A, Stromberg L, Claussen J C and Gomes C L B T-R M F S 2023 Advances in biosensors for detection of foodborne microorganisms, toxins, and chemical contaminants *Reference Module in Food Science* (Amsterdam: Elsevier)

Iheanacho S, Ogbu M, Bhuyan M S and Ogunji J 2023 Microplastic pollution: an emerging contaminant in aquaculture *Aquac. Fish.* **8** 603–16

Intisar A, Ramzan A, Hafeez S, Hussain N, Irfan M, Shakeel N, Gill K A, Iqbal A, Janczarek M and Jesionowski T 2023 Adsorptive and photocatalytic degradation potential of porous polymeric materials for removal of pesticides, pharmaceuticals, and dyes-based emerging contaminants from water *Chemosphere* **336** 139203

James C A, Sofield R, Faber M, Wark D, Simmons A, Harding L and O'Neill S 2023 The screening and prioritization of contaminants of emerging concern in the marine environment based on multiple biological response measures *Sci. Total Environ.* **886** 163712

Kaur M, Ghosh D, Guleria S, Arya S K, Puri S and Khatri M 2023 Microplastics/nanoplastics released from facemasks as contaminants of emerging concern *Mar. Pollut. Bull.* **191** 114954

Khalaf E M, Sanaan Jabbar H, Mireya Romero-Parra R, Raheem Lateef Al-Awsi G, Setia Budi H, Altamimi A S, Abdulfadhil Gatea M, Falih K T, Singh K and Alkhuzai K A 2023 Smartphone-assisted microfluidic sensor as an intelligent device for on-site determination of food contaminants: developments and applications *Microchem. J.* **190** 108692

Kotia A, Rutu P, Singh V, Kumar A, Dhoke S, Kumar P and Singh D K 2022 Rheological analysis of rice husk-starch suspended in water for sustainable agriculture application *Mater. Today Proc.* **50** 1962–6

Kumar M, Sridharan S, Sawarkar A D, Shakeel A, Anerao P, Mannina G, Sharma P and Pandey A 2023 Current research trends on emerging contaminants pharmaceutical and personal care products (PPCPs): a comprehensive review *Sci. Total Environ.* **859** 160031

Kumar P, Devi P and Dey S R 2021a Fungal volatile compounds: a source of novel in plant protection agents *Volatiles and Metabolites of Microbes* ed A Kumar, J Singh, and M M Samuel (New York: Academic) ch 6 pp 83–104

Kumar P, Kumar T, Singh S, Tuteja N, Prasad R and Singh J 2020 Potassium: a key modulator for cell homeostasis *J. Biotechnol.* **324** 198–210

Kumar P and Mistri T K 2020 Transcription factors in SOX family: potent regulators for cancer initiation and development in the human body *Semin. Cancer Biol.* **67** 105–13

Kumar P, Sharma K, Saini L and Dey S R 2021b Role and behavior of microbial volatile organic compounds in mitigating stress *Volatiles and Metabolites of Microbes* ed A Kumar, J Singh, and M M Samuel (New York: Academic) ch 8 pp 143–61

Kumar V, Dwivedi P, Kumar P, Singh B N, Pandey D K, Kumar V and Bose B 2021 Mitigation of heat stress responses in crops using nitrate primed seeds *S. Afr. J. Bot.* **140** 25–36

Kumari P, Singh J and Kumar P 2022 Impact of bioenergy for the diminution of an ascending global variability and change in the climate *Microbiome Under Changing Climate* ed A Kumar, J Singh and L F R Ferreira (Woodhead Publishing) ch 21 pp 469–87

Kurade M B *et al* 2023 Integrated phycoremediation and ultrasonic-irradiation treatment (iPUT) for the enhanced removal of pharmaceutical contaminants in wastewater *Chem. Eng. J.* **455** 140884

Laad M and Ghule B 2023 Removal of toxic contaminants from drinking water using biosensors: a systematic review *Groundw. Sustain. Develop.* **20** 100888

Li H, Cheng B, Zhang J, Zhou X, Shi C, Zeng L and Wang C 2023a Recent advances in the application of bismuth-based catalysts for degrading environmental emerging organic contaminants through photocatalysis: a review *J. Environ. Chem. Eng.* **11** 110371

Li N, He X, Ye J, Dai H, Peng W, Cheng Z, Yan B, Chen G and Wang S 2023b H_2O_2 activation and contaminants removal in heterogeneous Fenton-like systems *J. Hazard. Mater.* **458** 131926

Li X, Yang H, Ma H, Zhi H, Li D and Zheng X 2023c Carbonaceous materials applied for cathode electro-Fenton technology on the emerging contaminants degradation *Process Saf. Environ. Protect.* **169** 186–98

Li Y, Liu Y, Feng L and Zhang L 2023d A review: manganese-driven bioprocess for simultaneous removal of nitrogen and organic contaminants from polluted waters *Chemosphere* **314** 137655

Long Y, Song L, Shu Y, Li B, Peijnenburg W and Zheng C 2023 Evaluating the spatial and temporal distribution of emerging contaminants in the Pearl River Basin for regulating purposes *Ecotoxicol. Environ. Saf.* **257** 114918

Mali H, Shah C, Raghunandan B H, Prajapati A S, Patel D H, Trivedi U and Subramanian R B 2023 Organophosphate pesticides an emerging environmental contaminant: pollution, toxicity, bioremediation progress, and remaining challenges *J. Environ. Sci.* **127** 234–50

Méndez-Loranca E, Vidal-Ruiz A M, Martínez-González O, Huerta-Aguilar C A and Gutierrez-Uribe J A 2023 Beyond cellulose extraction: recovery of phytochemicals and contaminants to revalorize agricultural waste *Bioresour. Technol. Rep.* **21** 101339

Menya E, Jjagwe J, Kalibbala H M, Storz H and Olupot P W 2023 Progress in deployment of biomass-based activated carbon in point-of-use filters for removal of emerging contaminants from water: a review *Chem. Eng. Res. Des.* **192** 412–40

Mohamed B A, Hamid H, Montoya-Bautista C V and Li L Y 2023 Circular economy in wastewater treatment plants: treatment of contaminants of emerging concerns (CECs) in effluent using sludge-based activated carbon *J. Clean. Prod.* **389** 136095

Mohan B, Priyanka, Singh G, Chauhan A, Pombeiro A J L and Ren P 2023 Metal-organic frameworks (MOFs) based luminescent and electrochemical sensors for food contaminant detection *J. Hazard. Mater.* **453** 131324

Moreira R G, Branco G S and Lo Nostro F L 2023 Effects of aquatic contaminants in female fish reproduction *Environmental Contaminants and Endocrine Health* ed O Carnevali, and E H Hardiman (New York: Academic) ch 3.2.3 pp 257–68

Moret S, Conchione C, Barp L B T-R M and F S 2023 Food packaging contaminants with a special focus on hydrocarbon contaminants and nanoparticles *Encyclopedia of Food Sciences* 2nd edn (Amsterdam: Elsevier)

Mu J, Feng J, Wang X and Liu B 2023 Oxygen vacancy boosting peroxymonosulfate activation over nanosheets assembled flower-like $CoMoO_4$ for contaminant removal: performance and activity enhancement mechanisms *Chem. Eng. J.* **459** 141537

Nawaz S, Tabassum A, Muslim S, Nasreen T, Baradoke A, Kim T H, Boczkaj G, Jesionowski T and Bilal M 2023 Effective assessment of biopolymer-based multifunctional sorbents for the remediation of environmentally hazardous contaminants from aqueous solutions *Chemosphere* **329** 138552

Osuoha J O, Anyanwu B O and Ejileugha C 2023 Pharmaceuticals and personal care products as emerging contaminants: need for combined treatment strategy *J. Hazard. Mater. Adv.* **9** 100206

Patel P, Nandi A, Verma S K, Kaushik N, Suar M, Choi E H and Kaushik N K 2023 Zebrafish-based platform for emerging bio-contaminants and virus inactivation research *Sci. Total Environ.* **872** 162197

Peivasteh-roudsari L *et al* 2023 Origin, dietary exposure, and toxicity of endocrine-disrupting food chemical contaminants: a comprehensive review *Heliyon* **9** e18140

Pradhan B, Chand S, Chand S, Rout P R and Naik S K 2023 Emerging groundwater contaminants: a comprehensive review on their health hazards and remediation technologies *Groundw. Sustain. Develop.* **20** 100868

Puri M, Gandhi K and Kumar M S 2023 Emerging environmental contaminants: a global perspective on policies and regulations *J. Environ. Manage.* **332** 117344

Quang H H P, Dinh D A, Dutta V, Chauhan A, Lahiri S K, Gopalakrishnan C, Radhakrishnan A, Batoo K M and Thi L-A P 2023 Current approaches, and challenges on identification, remediation and potential risks of emerging plastic contaminants: a review *Environ. Toxicol. Pharmacol.* **101** 104193

Rose M B T-R M 2023 Dioxins and other environmental contaminants in meat products, including fish and sea food *Encyclopedia of Meat Sciences* 3rd edn (Amsterdam: Elsevier)

Sewwandi M, Wijesekara H, Rajapaksha A U, Soysa S and Vithanage M 2023 Microplastics and plastics-associated contaminants in food and beverages; global trends, concentrations, and human exposure *Environ. Pollut.* **317** 120747

Singh A, Chaurasia D, Khan N, Singh E and Chaturvedi Bhargava P 2023a Efficient mitigation of emerging antibiotics residues from water matrix: integrated approaches and sustainable technologies *Environ. Pollut.* **328** 121552

Singh A W, Soni R, Pal A K, Tripathi P, Lal J A and Tripathi V 2023b Microalgal-based bioremediation of emerging contaminants in wastewater: a sustainable approach *Microbial Bioprocesses* ed P Shukla *Progress in Biochemistry and Biotechnology* (New York: Academic) ch 12 pp 275–97

Soto-Donoso N, Favier L, Villalobos S F, Paredes-García V, Bataille T, Marco J F, Hlihor R M, Le Fur E and Venegas-Yazigi D 2023 Lanthanum hydroxichloride/anatase composite and its application for effective UV-light driven oxidation of the emergent water contaminant Cetirizine *Chem. Eng. Res. Des.* **196** 685–700

Sunyer-Caldú A, Golovko O, Kaczmarek M, Asp H, Bergstrand K-J, Gil-Solsona R, Gago-Ferrero P, Diaz-Cruz M S, Ahrens L and Hultberg M 2023 Occurrence and fate of contaminants of emerging concern and their transformation products after uptake by pak choi (*Brassica rapa* subsp. chinensis) *Environ. Pollut.* **319** 120958

Timilsina A, Adhikari K, Yadav A K, Joshi P, Ramena G and Bohara K 2023 Effects of microplastics and nanoplastics in shrimp: mechanisms of plastic particle and contaminant distribution and subsequent effects after uptake *Sci. Total Environ.* **894** 164999

Tran H-T, Bolan N S, Lin C, Binh Q A, Nguyen M-K, Luu T A, Le V-G, Pham C Q, Hoang H-G and Vo D-V N 2023 Succession of biochar addition for soil amendment and contaminants remediation during co-composting: a state of art review *J. Environ. Manage.* **342** 118191

Upadhyay S K, Devi P, Kumar V, Pathak H K, Kumar P, Rajput V D and Dwivedi P 2023 Efficient removal of total arsenic (As3+/5+) from contaminated water by novel strategies mediated iron and plant extract activated waste flowers of marigold *Chemosphere* **313** 137551

Walch H, Praetorius A, von der Kammer F and Hofmann T 2023 Generation of reproducible model freshwater particulate matter analogues to study the interaction with particulate contaminants *Water Res.* **229** 119385

Wang R, Luo J, Li C, Chen J and Zhu N 2023 Antiviral drugs in wastewater are on the rise as emerging contaminants: a comprehensive review of spatiotemporal characteristics, removal technologies and environmental risks *J. Hazard. Mater.* **457** 131694

Yuan Y, Jia H, Xu D and Wang J 2023 Novel method in emerging environmental contaminants detection: fiber optic sensors based on microfluidic chips *Sci. Total Environ.* **857** 159563

Yusuf A, Amusa H K, Eniola J O, Giwa A, Pikuda O, Dindi A and Bilad M R 2023 Hazardous and emerging contaminants removal from water by plasma-based treatment: a review of recent advances *Chem. Eng. J. Adv.* **14** 100443

Zahmatkesh S, Karimian M, Pourhanasa R, Ghodrati I, Hajiaghaei-Keshteli M and Ismail M A 2023 Wastewater treatment with algal based membrane bioreactor for the future: removing emerging containments *Chemosphere* **335** 139134

Zhang R, Chen B, Zhang H, Tu L and Luan T 2023 Stable isotope-based metabolic flux analysis: a robust tool for revealing toxicity pathways of emerging contaminants *TrAC, Trends Anal. Chem.* **159** 116909

Zhou J-L and Gao F 2023 Phytohormones: novel strategy for removing emerging contaminants and recovering resources *Trends Biotechnol.* **41** 992–5

IOP Publishing

Trends in Biological Processes in Industrial Wastewater
Treatment

Maulin P Shah

Chapter 5

Environmental and health risk assessment of chemical pollutants in drinking water and wastewaters

Prasann Kumar and Joginder Singh

The contamination of drinking water and wastewater by chemical pollutants poses significant environmental and human health risks. This chapter provides a concise overview of the extensive research conducted on the assessment of these risks, with a focus on the potential impacts and mitigation strategies. The environmental risks associated with chemical pollutants in water sources are first explored. Various industrial and agricultural activities contribute to releasing contaminants, such as heavy metals, pesticides, pharmaceuticals, and endocrine-disrupting compounds. These pollutants can persist in water bodies, damaging aquatic ecosystems, disrupting biological communities and bioaccumulating toxins in aquatic organisms. This chapter delves into the health risks posed by these chemical pollutants. Contaminated drinking water can cause acute and chronic health issues in humans, ranging from gastrointestinal disorders to long-term effects on organ systems. Pharmaceuticals and endocrine disruptors in water sources have raised concerns about potential developmental and reproductive consequences. Regulatory bodies and scientific communities have established guidelines and frameworks to assess and mitigate these risks. Risk assessment methodologies, such as hazard identification, dose-response assessment, exposure assessment, and risk characterization, play a crucial role in evaluating the potential adverse effects of chemical pollutants on human health and the environment. In addition, advanced water treatment technologies, such as activated carbon filtration, reverse osmosis, and advanced oxidation processes, are utilized to remove or reduce the levels of contaminants in drinking water and wastewater treatment plants. In summary, assessing the environmental and health risks associated with chemical pollutants in drinking water and

doi:10.1088/978-0-7503-5678-7ch5

wastewater is paramount. Ongoing research and regulatory efforts aim to enhance monitoring systems, develop robust risk assessment methodologies, and implement effective treatment technologies to safeguard the environment and human health. These collective endeavors are vital in ensuring clean and safe water resources for present and future generations.

5.1 Introduction

Assessing the environmental and health risks associated with chemical pollutants in drinking water and wastewater is a critical research and regulatory focus area (Ruziwa *et al* 2023, Saravanan *et al* 2023, Zardosht *et al* 2023). The contamination of water sources by various chemical pollutants poses significant threats to the environment and human health. As the demand for clean and safe water grows, understanding and mitigating these risks is paramount. Chemical pollutants in drinking water and wastewaters originate from diverse sources, including industrial activities, agricultural runoff, and domestic waste disposa (Brillas 2023, de Jesus *et al* 2023, Sol *et al* 2023, Sudhir *et al* 2023). These pollutants encompass many substances, such as heavy metals, pesticides, pharmaceuticals, personal care products, and endocrine-disrupting compounds. Their presence in water bodies can have far-reaching consequences for ecosystems and human populations. In terms of environmental risks, the contamination of water sources can lead to adverse effects on aquatic ecosystems. Chemical pollutants can disrupt the delicate balance of biological communities, impairing the health and diversity of aquatic organisms (Ding 2023, Huang *et al* 2023, Musa Yahaya *et al* 2023, Weerakoon *et al* 2023).

These pollutants can also accumulate in the food chain, leading to biomagnification and bioaccumulation, with potential long-term consequences for higher trophic levels, including humans. The health risks associated with chemical pollutants in drinking water and wastewater are of great concern to public health authorities. Contaminated drinking water can introduce many hazardous substances into the human body. Acute health effects may include gastrointestinal illnesses, while chronic exposure to certain pollutants has been linked to the development of cancers, neurological disorders, and reproductive issues (Kumar and Mistri 2020, Kumar *et al* 2020, Chakraborty *et al* 2021, Kumar *et al* 2021a, Kumar *et al* 2021b, Kumar *et al* 2021c, Aley *et al* 2022, Das *et al* 2022, Goud *et al* 2022, Kotia *et al* 2022, Kumari *et al* 2022, Upadhyay *et al* 2023). The presence of pharmaceuticals and endocrine-disrupting compounds in water sources has raised additional concerns regarding their potential impacts on development, hormonal balance, and reproductive health (Amenaghawon *et al* 2023, Ayejoto *et al* 2023, Cai *et al* 2023, Kammoun *et al* 2023, Musa Yahaya *et al* 2023, Paneerselvam *et al* 2023, Sharma *et al* 2023, Yadav *et al* 2023). Regulatory bodies and scientific communities have established guidelines and frameworks for assessing and managing chemical pollutants in water to address these risks. Risk assessment methodologies are crucial in evaluating the potential adverse environmental and human health effects of contaminants. These methodologies typically include hazard identification, dose-response assessment, exposure assessment, and risk characterization, providing a comprehensive understanding of the risks posed by specific pollutants. In addition to risk assessment, developing and implementing effective water treatment technologies are essential in mitigating chemical pollutants. Advanced treatment processes, such as activated carbon filtration, reverse osmosis, and advanced oxidation, are used to remove or reduce contaminants in drinking water and wastewater treatment plants. This paper aims to provide a comprehensive overview of the environmental and health risks associated with chemical pollutants in drinking water and wastewater (Akhtar *et al* 2021, Alharbi *et al* 2023, Coccia and Bontempi 2023, Ganthavee and Trzcinski 2023, Irshad *et al* 2023, Krishnan *et al* 2023, Laad and Ghule 2023, Liu *et al* 2023a, Nawaz *et al* 2023, Oke *et al* 2023, Quispe *et al* 2023, Sun and Zheng 2023, Termeh-Zonoozi *et al* 2023, Turdiyeva and Lee 2023, Wang *et al* 2023b). It will explore the sources and types of pollutants, their potential impacts on ecosystems and human health, the methodologies employed in risk assessment and the treatment technologies used to address these challenges. By understanding and addressing these risks, it becomes possible to ensure the provision of clean and safe water resources for the well-being of current and future generations.

5.2 Microplastic pollutants in drinking water

Recent research has shed light on the presence and potential risks associated with microplastic pollutants in drinking water. Microplastics are tiny plastic particles, typically smaller than 5 mm, that have become ubiquitous in various environmental compartments, including water bodies. These particles can originate from many sources, such as the degradation of oversized plastic items, microbeads in personal care products, fibers from textiles, and microplastic waste from various industries (Akhtar *et al* 2021, Alharbi *et al* 2023, Brillas 2023, Coccia and Bontempi 2023, de Jesus *et al* 2023, Ding 2023, Ganthavee and Trzcinski 2023, Irshad *et al* 2023, Krishnan *et al* 2023, Laad and Ghule 2023, Liu *et al* 2023a, Musa Yahaya *et al* 2023, Nawaz *et al* 2023, Oke *et al* 2023, Quispe *et al* 2023, Ruziwa *et al* 2023, Saravanan *et al* 2023, Sol *et al* 2023, Sudhir *et al* 2023, Sun and Zheng 2023, Termeh-Zonoozi *et al* 2023, Turdiyeva and Lee 2023, Wang *et al* 2023b, Weerakoon *et al* 2023, Zardosht *et al* 2023). The detection and analysis of microplastics in drinking water have posed significant challenges due to their small size and diverse composition. However, advances in analytical techniques have enabled scientists to develop methods for identifying and quantifying microplastics in water samples. These methods often involve filtration, digestion, and microscopy techniques combined with spectroscopic analysis or imaging technologies. Research studies conducted in different regions of the world have consistently reported the presence of microplastics in drinking water sources, raising concerns about potential human exposure and health risks. For instance, studies have detected microplastics in tap water, bottled water, and even groundwater, indicating the widespread contamination of drinking water systems. One recent study in multiple countries analyzed tap water samples from various urban areas and reported more than 80% of microplastic contamination. This study found an average of 4.8 microplastic particles per litter of water, with fibers being the most common type of microplastic detected. Another study examined bottled water from different brands and found microplastic contamination in nearly 93% of the samples, with an average of 325 microplastic particles per litre (Amenaghawon *et al* 2023, Ayejoto *et al* 2023, Cai *et al* 2023, Hassan *et al* 2023, Huang *et al* 2023, Kammoun *et al* 2023, Kim *et al* 2023, Nasir *et al* 2023, Paneerselvam *et al* 2023, Sharma *et al* 2023, Villén *et al* 2023, Yadav *et al* 2023). While the full extent of the health risks associated with microplastics in drinking water is still being investigated, several potential concerns have been identified. For example, microplastics can contain or adsorb toxic chemicals, such as persistent organic pollutants (POPs) and heavy metals, which can leach into the water and pose risks upon ingestion.

The small size of microplastics enables them to penetrate human tissues, potentially leading to inflammation, oxidative stress, and the release of harmful substances within the body. Furthermore, microplastics in drinking water raise concerns about their potential impact on human physiology and the gastrointestinal system. Studies have indicated that microplastics can accumulate in organs and tissues, and their interactions with gut microbiota and immune responses require further investigation. To address these concerns, research efforts are focused on developing standardized methods for microplastic analysis, understanding the sources and pathways of microplastic pollution, and evaluating the potential health risks associated with microplastics in drinking water. Additionally, strategies for reducing microplastic contamination at its source, such as regulating plastic use, improving wastewater treatment processes, and promoting recycling and sustainable waste management practices, are being explored (Ahmed *et al* 2023, Alagan *et al* 2023, Darling *et al* 2023, De Caroli Vizioli *et al* 2023, Hanna *et al* 2023, Li *et al* 2023b, Mekawi *et al* 2023, Palit *et al* 2023, Sathya *et al* 2023, Shabanloo *et al* 2023, Sivagami *et al* 2023, Takman *et al* 2023, Thanigaivel *et al* 2023, Wang *et al* 2023c).

In summary, recent research has demonstrated the presence of microplastic pollutants in drinking water, raising concerns about potential human exposure and health risks. While the full extent of these risks is still being studied, efforts are underway to better understand the sources, pathways, and health impacts of microplastics. By gaining a comprehensive understanding of microplastic pollution and implementing effective strategies, it becomes possible to minimize the presence of microplastics in drinking water and safeguard human health (table 5.1).

5.3 Distribution and bioaccumulation of microplastics: a detailed overview

Microplastics, tiny plastic particles measuring less than 5 mm, have become a global concern due to their omnipresence in various environmental compartments and their potential for bioaccumulation. These tiny particles can originate from many sources, including degrading oversized plastic items, microbeads in personal care products, textiles fibers, and industries' microplastic waste.

Omnipresence of microplastics: Microplastics have been detected in diverse environments, ranging from remote marine ecosystems to urban areas, freshwater systems, and even the air we breathe. They are found in oceans, rivers, lakes, soil,

Table 5.1. Different types of microplastics, and their sizes and sources.

Microplastic type	Size range (mm)	Source
Polyethylene	<5	Packaging materials
Polypropylene	<5	Bottles, containers
Polystyrene	<5	Styrofoam products
Polyethylene terephthalate (PET)	<5	Beverage bottles, textiles
Polyvinyl chloride (PVC)	<5	Pipes, plastic products
Nylon	<5	Textiles, fishing gear
Polyurethane	<5	Foam, Insulation
Polycarbonate	<5	Water bottles, electronics
Acrylic	<5	Paints, adhesives
Polyvinylidene fluoride (PVDF)	<5	Membranes, coatings

Sources: Based on a review of the literature.

sediments, ice, and even in drinking water and food. The widespread presence of microplastics can be attributed to their physical durability, low biodegradability, and widespread use in various consumer and industrial products.

Transport mechanisms: Microplastics can be transported through various mechanisms. For example, they can be carried by the wind, leading to atmospheric deposition in remote areas.. In aquatic environments, they can be transported by currents, tides, and rivers, spreading the contamination across vast distances. Microplastics can also be transported through the food chain, where smaller organisms can ingest and transfer them to higher trophic levels.

Bioaccumulation of microplastics: Bioaccumulation refers to the accumulation of substances in organisms over time, often resulting in higher concentrations in organisms up the food chain. Microplastics can bioaccumulate due to several factors. First, their small size allows them to be ingested by various organisms, including zooplankton, fish, birds, and marine mammals. Once ingested, microplastics can accumulate in the digestive tract, potentially causing physical blockages, reduced feeding efficiency, and impaired nutrient absorption. Additionally, microplastics can adsorb and concentrate chemical pollutants, such as POPs and heavy metals, from the surrounding environment. It can lead to secondary poisoning, where microplastics transfer toxic substances to ingested organisms. As predators consume these contaminated organisms, the concentration of pollutants can increase further up the food chain.

Impacts of microplastic bioaccumulation: The bioaccumulation of microplastics can have various ecological and health impacts. It can disrupt food webs, alter species interactions, and affect ecosystem population dynamics. For example, ingesting microplastics by filter-feeding organisms can reduce their feeding efficiency and energy intake, decreasing growth and reproduction rates. Additionally, the transfer of microplastics through the food chain can result in toxicological effects on higher trophic levels, including reproductive disorders, compromised immune systems, and organ damage. From a human health perspective, the bioaccumulation

Table 5.2. Ecosystems that are known to be affected by microplastic contamination.

Region/ecosystem	Presence of microplastics	Bioaccumulation potential
Arctic ocean	High levels of microplastics detected	Bioaccumulation in marine mammals and seabirds
Great pacific garbage patch	Concentrated microplastic pollution	Potential for bioaccumulation in marine organisms
Coastal areas	Microplastic pollution prevalent	Bioaccumulation observed in fish and shellfish
Urban waterways	Microplastic contamination observed	Potential bioaccumulation in urban wildlife and fish
Freshwater systems	Microplastic presence increasing	Bioaccumulation potential in freshwater organisms
Coral reefs	Microplastic contamination reported	Bioaccumulation risks for coral-associated organisms
Deep-sea environments	Microplastics detected in deep-sea sediments	Bioaccumulation potential in deep-sea organisms

Source: Based on a review of the literature.

of microplastics raises concerns regarding the potential transfer of microplastics and associated chemical pollutants into the human food chain. While research is ongoing, evidence suggests that microplastics may enter the human body by ingesting contaminated seafood, drinking water, and even air particles. The long-term health effects of microplastic exposure in humans are not yet fully understood and require further research.

Mitigation and future directions: Addressing the omnipresence and bioaccumulation of microplastics requires concerted efforts at multiple levels. It involves reducing the production and consumption of single-use plastics, improving waste management systems, and implementing effective wastewater treatment to prevent microplastic contamination. Additionally, ongoing research is focused on understanding microplastics' sources, transport mechanisms, and impacts to inform regulatory measures and develop mitigation strategies. The omnipresence and bioaccumulation of microplastics pose significant environmental and health concerns. Their ability to persist in various environments, transport through different pathways, and accumulate in organisms across the food chain highlight the need for proactive measures to address this global issue. By reducing plastic waste, implementing sustainable practices, and conducting further research, it becomes possible to mitigate the impacts of microplastics and protect ecosystems and human health (tables 5.2–5.4).

5.4 Biochar and contaminants

Biochar, a carbon-rich material derived from biomass pyrolysis, has gained significant attention as a promising adsorbent for wastewater treatment. The adsorption process involves the physical or chemical attraction of pollutants onto

Table 5.3. Different regions associated with microplastic contamination.

Place	Country/region	Microplastic contamination	Bioaccumulation potential
Arctic Ocean	Arctic region	High levels of microplastics detected	Bioaccumulation in marine mammals and seabirds
Great Pacific Garbage Patch	Pacific Ocean	Concentrated microplastic pollution	Potential for bioaccumulation in marine organisms
North Atlantic Gyre	Atlantic Ocean	Significant microplastic presence	Bioaccumulation risks for marine organisms
Mediterranean Sea	Mediterranean region	Microplastic contamination observed	Bioaccumulation in marine life and ecosystems
North Sea	Europe	Microplastic pollution prevalent	Bioaccumulation risks for fish and seabirds
Ganges River	India	Microplastics reported in river sediments	Potential bioaccumulation in freshwater organisms
Yangtze River	China	Microplastic contamination observed	Bioaccumulation risks for freshwater organisms
New York Harbor	United States	Microplastic presence in urban waterways	Potential bioaccumulation in urban wildlife and fish
Lake Geneva	Switzerland	Microplastic contamination reported	Bioaccumulation potential in freshwater organisms
Sydney Harbor	Australia	Microplastics detected in coastal areas	Bioaccumulation risks for marine organisms
Coral Triangle	Southeast Asia	Microplastic contamination in coral reefs	Bioaccumulation risks for coral-associated organisms
Lake Baikal	Russia	Microplastic presence in the freshwater lake	Bioaccumulation potential in freshwater organisms

Source: Based on a review of the literature.

the surface of biochar, leading to their removal from wastewater. Biochar's high surface area, porous structure, and functional groups make it an effective adsorbent for various contaminants present in wastewater, including organic compounds, heavy metals, nutrients, and emerging pollutants. Here is a detailed overview of the adsorption of pollutants from wastewater by biochar:

Mechanism of adsorption: The adsorption process occurs through several mechanisms, including physical adsorption (van der Waals forces), chemical adsorption (covalent or electrostatic interactions), and surface complexation. The porous structure of biochar provides a large surface area for pollutant adsorption, while its functional groups, such as hydroxyl ($-OH$), carboxyl ($-COOH$), and amino ($-NH_2$) groups, enhance the adsorption capacity through chemical interactions.

Table 5.4. Different types of microbes and their mode of action.

Microbe name	Mode of action for plastic degradation
Ideonella sakaiensis	Produces enzymes that target specific bonds in PET (polyethene terephthalate)
Pseudomonas putida	Secretes enzymes that break down a wide range of plastics, including polyethene
Rhodococcus ruber	Produces extracellular enzymes to degrade various types of plastics
Amycolatopsis sp.	Secretes enzymes capable of breaking down polystyrene (PS) and polyurethane (PU)
Saccharophagus degradans	Utilizes a combination of enzymatic processes to degrade complex polymeric plastics
Acinetobacter sp.	Produces enzymes that target the ester linkages in certain types of plastics
Alcanivorax borkumensis	Utilizes plastic as a carbon source, breaking it down through metabolic processes
Exophiala jeanselmei	Colonizes plastic surfaces and produces enzymes that facilitate plastic degradation
Nocardia sp.	Capable of breaking down a range of plastics, including polyethene and polypropylene
Bacillus sp.	Produces enzymes that contribute to the degradation of various types of plastics

Factors Affecting Adsorption: Several factors influence the adsorption efficiency of biochar, including the characteristics of biochar (e.g., surface area, porosity, surface functional groups), properties of the pollutants (e.g., molecular size, charge, hydrophobicity), pH of the wastewater, contact time, temperature, and initial pollutant concentration. Optimal conditions can be determined through experimental studies and characterization of biochar properties.

Pollutant removal efficiency: Biochar has demonstrated high adsorption capacity for various pollutants. Organic compounds, such as pharmaceuticals, pesticides, and dyes, can be effectively removed through adsorption onto biochar surfaces. Heavy metals, including lead, cadmium, copper, and chromium, can also be efficiently adsorbed due to the affinity between metal ions and biochar surfaces. Nutrients like nitrogen and phosphorus can be sequestered by biochar, reducing their release into water bodies and mitigating eutrophication. Furthermore, emerging pollutants, such as microplastics and endocrine-disrupting compounds, have shown potential for adsorption onto biochar surfaces.

Application in wastewater treatment: Biochar can be used in various wastewater treatment processes, including batch adsorption, column filtration, and hybrid systems. In batch adsorption, biochar is mixed with wastewater, allowing pollutants to adsorb onto its surface. Column filtration involves the passage of wastewater through a biochar bed, where adsorption occurs. Hybrid systems combine biochar adsorption with other treatment techniques like activated sludge processes or membrane filtration for enhanced removal efficiency.

Table 5.5. Different types of biochar.

Biochar type	Description
Wood-based biochar	Biochar produced from wood or woody biomass
Agricultural residue biochar	Biochar is produced from agricultural waste residues.
Sewage sludge biochar	Biochar is produced from sewage sludge or wastewater treatment.
Nutrient-enriched biochar	Biochar is enriched with nutrients for specific agricultural use.
Pyrolyzed green waste biochar	Biochar is produced from green waste materials.
Hydrothermal carbonization	Biochar is produced through the hydrothermal carbonization process.
Algal Biomass biochar	Biochar is produced from algal biomass.
Manure-based biochar	Biochar is produced from animal manure or livestock waste.
Bamboo biochar	Biochar is produced from bamboo biomass.
Rice husk biochar	Biochar is produced from rice husk waste.

Source: Based on a review of the literature.

Advantages and challenges: Using biochar for pollutant adsorption offers several advantages. It is a low-cost and sustainable adsorbent from agricultural waste, forestry residues, or organic byproducts. Biochar can be tailored through various production methods and modifications to optimize its adsorption properties. Furthermore, biochar has the potential for post-treatment applications, such as soil amendment, carbon sequestration, and energy generation through pyrolysis. However, challenges remain in the large-scale implementation of biochar adsorption in wastewater treatment. Optimization of biochar properties, regeneration and reusability of spent biochar, management of adsorbate-laden biochar, and scale-up of processes require further research and technological advancements. The adsorption of pollutants from wastewater by biochar presents a promising approach for sustainable and effective wastewater treatment. Biochar's unique properties enable the removal of organic compounds, heavy metals, nutrients, and emerging pollutants, contributing to improved water quality and environmental protection. Continued research and development are essential to optimize biochar production, tailor its properties, and overcome challenges, fostering integration into wastewater treatment systems (tables 5.5 and 5.6).

5.5 The adsorption capacity of biochar

The adsorption capacity of biochar refers to its ability to attract, retain, and remove pollutants from a liquid or gas phase. Biochar's high surface area, porous structure, and surface functional groups contribute to its adsorption capabilities. An overview of the factors that influence the adsorption capacity of biochar follows:

Surface Area: Biochar typically possesses a large surface area due to its porous structure, which provides ample sites for pollutant adsorption. The surface area of biochar can be increased through production methods, such as slow pyrolysis at lower temperatures or the incorporation of activating agents, resulting in enhanced

Table 5.6. Microbes and mode of action for plastic degradation.

Microbe name	Mode of action for plastic degradation
Ideonella sakaiensis	Produces enzymes that target specific bonds in PET (polyethene terephthalate)
Pseudomonas putida	Secretes enzymes that break down a wide range of plastics, including polyethene
Rhodococcus ruber	Produces extracellular enzymes to degrade various types of plastics
Amycolatopsis sp.	Secretes enzymes capable of breaking down polystyrene (PS) and polyurethane (PU)
Saccharophagus degradans	Utilizes a combination of enzymatic processes to degrade complex polymeric plastics
Acinetobacter sp.	Produces enzymes that target the ester linkages in certain types of plastics
Alcanivorax borkumensis	Utilizes plastic as a carbon source, breaking it down through metabolic processes
Exophiala jeanselmei	Colonizes plastic surfaces and produces enzymes that facilitate plastic degradation
Nocardia sp.	Capable of breaking down a range of plastics, including polyethene and polypropylene
Bacillus sp.	Produces enzymes that contribute to the degradation of various types of plastics

Source: Based on the review of the literature.

adsorption capacity (Akhtar *et al* 2021, Kumar *et al* 2021a, Kumar *et al* 2021c, Adelodun *et al* 2022, Irshad *et al* 2023).

Porosity: The presence of pores within biochar, including micropores, mesopores, and macropores, plays a crucial role in adsorption. Micropores, with sizes less than 2 nm, offer high adsorption capacity for smaller molecules, while mesopores and macropores provide pathways for the diffusion of more significant pollutants.

Surface functional groups: Biochar contains various functional groups, such as hydroxyl (–OH), carboxyl (–COOH), and amino (–NH$_2$) groups, on its surface. These functional groups can participate in chemical interactions with pollutants, enhancing adsorption capacity. For instance, hydroxyl groups can facilitate the adsorption of polar contaminants, while carboxyl and amino groups can engage in complexation with metal ions (Kumar *et al* 2020, Chakraborty *et al* 2021, Kumar *et al* 2021b, Aley *et al* 2022, Das *et al* 2022, Goud *et al* 2022, Kotia *et al* 2022, Kumari *et al* 2022, Upadhyay *et al* 2023).

Polarity and hydrophobicity: Biochar's polarity and hydrophobicity influence its affinity for different pollutants. Hydrophobic pollutants, such as organic compounds with nonpolar characteristics, tend to exhibit a higher affinity for hydrophobic biochar surfaces. However, the adsorption of polar pollutants can also occur through mechanisms like hydrogen bonding or electrostatic interactions (Almazrouei *et al* 2023, Krishnan *et al* 2023, Munné *et al* 2023, Perveen and Amar-Ul-Haque 2023, Quispe *et al* 2023, Termeh-Zonoozi *et al* 2023).

pH and ionic strength: The pH of the surrounding environment affects the surface charge of biochar and the ionization of functional groups, thereby influencing adsorption capacity. Additionally, the presence of dissolved salts and the ionic strength of the solution can impact the adsorption process by altering the competition between pollutants and ions for adsorption sites.

Pollutant characteristics: The nature and properties of the pollutants, including their molecular size, charge, hydrophobicity, and concentration, affect their affinity for biochar. Different pollutants, such as organic compounds, heavy metals, nutrients, and emerging contaminants, may exhibit varying adsorption behaviors on biochar (Chiriac *et al* 2023, Foglia *et al* 2023, He *et al* 2023, Vikas *et al* 2023b, Mukherjee and Chauhan 2023, Qi *et al* 2024, Radini *et al* 2023). The adsorption capacity of biochar can vary depending on its specific characteristics, production methods, feedstock, and treatment processes. Consequently, the optimal choice of biochar type, along with the operating conditions of the adsorption system, needs to be considered for effective pollutant removal. Experimental testing and characterization of biochar properties, as well as consideration of specific pollutant requirements, are necessary to determine the adsorption capacity of biochar in a particular application, biochar can be prepared through various techniques, and each method may have specific advantages and applications (Ghaffar *et al* 2023, Jagadeesh and Sundaram 2023, Li *et al* 2023a, Liu *et al* 2023b, Nirmala *et al* 2023, Safwat *et al* 2023, Shahzad *et al* 2023, Tajuddin *et al* 2023, Wang *et al* 2023a). The adoption of biochar in different areas is influenced by factors such as soil type, climate conditions, agricultural practices, waste management needs, and environmental considerations. Additionally, ongoing research and technological advancements continue to expand the potential applications of biochar in diverse fields (tables 5.7–5.9).

Table 5.7. Biochar preparation method and its adoption application.

Biochar preparation method	Adoption applications
Slow pyrolysis	Soil amendment, carbon sequestration, agriculture, horticulture
Fast pyrolysis	Bioenergy production, wastewater treatment, soil remediation
Gasification	Energy generation, syngas production, soil improvement
Hydrothermal carbonization	Soil fertility enhancement, waste management, water filtration
Co-pyrolysis	Bio-oil production, soil conditioning, renewable energy
Torrefaction	Biomass densification, solid fuel production, carbon sequestration
Microwave pyrolysis	Biochar activation, water treatment, contaminant removal
Carbonization of residues	Waste valorization, biochar production, environmental remediation
Plasma pyrolysis	Hazardous waste treatment, advanced material synthesis
Alkaline activation	Soil amendment, wastewater treatment, heavy metal immobilization

Source: Based on a review of the literature.

Table 5.8. Heavy metal and its adsorption by biochar along with health risk.

Heavy metal	Adsorption process by biochar	Health risk
Arsenic (As)	Surface complexation, ion exchange, precipitation	Carcinogenic, neurotoxic, and cardiovascular effects
Cadmium (Cd)	Ion exchange, surface complexation, precipitation	Carcinogenic, nephrotoxic, and skeletal damage
Chromium (Cr)	Surface complexation, ion exchange, reduction	Carcinogenic, respiratory and gastrointestinal effects
Copper (Cu)	Surface complexation, ion exchange, electrostatic attraction	Neurotoxic, hepatotoxic, and gastrointestinal effects
Lead (Pb)	Surface complexation, ion exchange, precipitation	Neurotoxic, developmental, and behavioral disorders
Mercury (Hg)	Surface complexation, precipitation, reduction	Neurotoxic, cardiovascular, and renal effects
Nickel (Ni)	Surface complexation, ion exchange, electrostatic attraction	Carcinogenic, immunotoxic, and respiratory effects
Zinc (Zn)	Surface complexation, ion exchange, precipitation	Gastrointestinal effects and neurotoxicity
Aluminum (Al)	Surface complexation, ion exchange, precipitation	Neurotoxic, skeletal, and respiratory effects
Antimony (Sb)	Surface complexation, ion exchange, precipitation	Carcinogenic, cardiovascular, and gastrointestinal effects
Barium (Ba)	Surface complexation, ion exchange, precipitation	Gastrointestinal, cardiovascular, and neuromuscular effects
Beryllium (Be)	Surface complexation, ion exchange, precipitation	Carcinogenic, respiratory, and gastrointestinal effects
Cobalt (Co)	Surface complexation, ion exchange, electrostatic attraction	Carcinogenic, cardiotoxic, and hematological effects
Manganese (Mn)	Surface complexation, ion exchange, precipitation	Neurotoxic and respiratory effects
Selenium (Se)	Surface complexation, ion exchange, precipitation	Neurotoxic, hepatotoxic, and carcinogenic effects
Thallium (Tl)	Surface complexation, ion exchange, precipitation	Neurotoxic, cardiovascular, and gastrointestinal effects
Uranium (U)	Surface complexation, ion exchange, precipitation	Nephrotoxic and carcinogenic effects, and skeletal damage
Vanadium (V)	Surface complexation, ion exchange, precipitation	Respiratory, cardiovascular, and neurological effects
Silver (Ag)	Surface complexation, ion exchange, electrostatic attraction	Toxic to aquatic life and potential accumulation in the body
Tin (Sn)	Surface complexation, ion exchange, precipitation	Neurotoxic, developmental, and reproductive effects
Cobalt (Co)	Surface complexation, ion exchange, electrostatic attraction	Carcinogenic, cardiotoxic, and hematological effects

Indium (In)	Surface complexation, ion exchange, precipitation	Respiratory and gastrointestinal effects, and potential toxicity
Lanthanum (La)	Surface complexation, ion exchange, precipitation	Neurotoxic, nephrotoxic, and gastrointestinal effects
Scandium (Sc)	Surface complexation, ion exchange, precipitation	Respiratory and gastrointestinal effects
Tungsten (W)	Surface complexation, ion exchange, precipitation	Carcinogenic, neurotoxic, and potential reproductive effects
Yttrium (Y)	Surface complexation, ion exchange, precipitation	Respiratory and gastrointestinal effects
Zirconium (Zr)	Surface complexation, ion exchange, precipitation	Gastrointestinal effects and potential toxicity
Gadolinium (Gd)	Surface complexation, ion exchange, precipitation	Potential accumulation in the body and nephrotoxicity
Neodymium (Nd)	Surface complexation, ion exchange, precipitation	Respiratory and gastrointestinal effects, and potential toxicity
Europium (Eu)	Surface complexation, ion exchange, precipitation	Potential accumulation in the body and neurotoxicity
Cerium (Ce)	Surface complexation, ion exchange, precipitation	Potential accumulation in the body and respiratory effects
Dysprosium (Dy)	Surface complexation, ion exchange, precipitation	Potential accumulation in the body and neurotoxicity

Table 5.9. Environmental pollution and the different types of associated health risk.

Contaminant	Health risk
Heavy metals	Neurotoxicity, carcinogenicity, developmental disorders
Pesticides	Neurological effects, endocrine disruption, cancer
Air pollutants	Respiratory diseases, cardiovascular disorders, lung cancer
Waterborne pathogens	Gastrointestinal illnesses, diarrhea, infectious diseases
Radioactive substances	Increased risk of cancer, genetic mutations
Volatile organic compounds	Respiratory issues, liver damage, increased cancer risk
Asbestos	Mesothelioma, lung cancer, asbestosis
Lead	Neurological damage, developmental disorders
Radon	Lung cancer, respiratory issues
PCBs (polychlorinated biphenyls)	Immune system disorders, reproductive issues, cancer
Mold	Allergic reactions, respiratory issues, asthma exacerbation
Formaldehyde	Respiratory problems, eye irritation, cancer
Dioxins	Reproductive disorders, immune system damage, cancer
Nitrates	Methemoglobinemia (blue baby syndrome), reduced oxygen transport.

(Continued)

Table 5.9. (*Continued*)

Contaminant	Health risk
Benzene	Leukemia, bone marrow damage, respiratory issues
Arsenic	Skin lesions, cardiovascular diseases, cancer
Perfluorinated compounds	Liver damage, thyroid disruption, immune system disorders
Phthalates	Hormone disruption, reproductive issues, developmental disorders
Radon	Lung cancer, respiratory issues
Foodborne pathogens	Gastrointestinal illnesses, food poisoning, infectious diseases
Mercury	Neurological damage, developmental disorders, kidney damage
Cyanide	Respiratory issues, cardiovascular effects, neurotoxicity
Microplastics	Potential for ingestion, accumulation in tissues
Chlorinated solvents	Liver damage, kidney damage, cancer
Noise pollution	Hearing loss, sleep disturbances, cardiovascular effects
Escherichia coli	Gastrointestinal illnesses, diarrhea, urinary tract infections
Salmonella	Gastrointestinal illnesses, food poisoning, typhoid fever
Legionella	Legionnaires' disease, pneumonia, respiratory issues
Cryptosporidium	Cryptosporidiosis, gastrointestinal illness
Vibrio	Vibrio infections, cholera, wound infections
Heterotrophic bacteria	Opportunistic infections, respiratory issues
Glyphosate	Carcinogenicity, endocrine disruption, developmental effects
Radionuclides	Increased risk of cancer, genetic mutations
Chloroform	Liver damage, respiratory issues, cancer
BPA (Bisphenol A)	Hormone disruption, reproductive issues, developmental disorders
Sulfur dioxide	Respiratory issues, eye irritation, cardiovascular effects
Ammonia	Respiratory issues, eye irritation, lung damage
Noise pollution	Hearing loss, sleep disturbances, cardiovascular effects
Phosphates	Eutrophication, harmful algal blooms, water quality issues
Ozone	Respiratory issues, eye irritation, lung damage
Carbon monoxide	Headaches, dizziness, cardiovascular effects
Radionuclides	Increased risk of cancer, genetic mutations
Chlordane	Carcinogenicity, neurotoxicity, reproductive effects
Acrylamide	Neurotoxicity, carcinogenicity, reproductive effects
Benzidine	Carcinogenicity, bladder cancer, reproductive effects
PCBs (polychlorinated biphenyls)	Immune system disorders, reproductive issues, cancer
Toluene	Neurological effects, respiratory issues, developmental effects
Sulfuric acid	Respiratory issues, eye irritation, skin burns
Chlorine	Respiratory issues, eye irritation, skin burns
Diesel exhaust	Respiratory issues, lung cancer, cardiovascular effects

Radionuclides	Increased risk of cancer, genetic mutations
Carbon dioxide	Respiratory issues, asphyxiation, acidification of oceans
Polycyclic aromatic hydrocarbons (PAHS)	Carcinogenicity, developmental disorders, respiratory effects
Vinyl chloride	Carcinogenicity, liver damage, respiratory effects
Endocrine disruptors	Hormonal imbalance, reproductive disorders, developmental effects
Naphthalene	Carcinogenicity, respiratory issues, neurological effects
Oxybenzone	Hormone disruption, coral bleaching, ecological damage
Radionuclides	Increased risk of cancer, genetic mutations
Perchlorate	Thyroid disruption, developmental disorders, reproductive issues
Hydrogen sulfide	Respiratory issues, eye irritation, neurological effects
Tetrachloroethylene (PCE)	Carcinogenicity, liver damage, respiratory issues
Bisphenol S (BPS)	Hormone disruption, reproductive issues, developmental disorders
Trichloroethylene (TCE)	Carcinogenicity, liver damage, respiratory issues
Radium	Increased risk of cancer, bone diseases
Nickel	Carcinogenicity, respiratory effects, developmental effects
Chromium	Carcinogenicity, respiratory, and gastrointestinal effects
Asbestos	Mesothelioma, lung cancer, asbestosis
Vinyl acetate	Carcinogenicity, respiratory issues, liver damage
Silica	Silicosis, lung cancer, respiratory issues
PFOA (Perfluorooctanoic acid)	Liver damage, thyroid disruption, immune system disorders
Perchloroethylene (PERC)	Carcinogenicity, liver damage, respiratory issues
Xylene	Neurological effects, respiratory issues, developmental effects
Ethylbenzene	Neurological effects, respiratory issues, developmental effects
Methanol	Neurological effects, respiratory issues, vision impairment
Cadmium	Carcinogenicity, nephrotoxicity, skeletal damage
Beryllium	Carcinogenicity, respiratory, and cardiovascular effects
Radionuclides	Increased risk of cancer, genetic mutations
Chlorpyrifos	Neurological effects, developmental disorders, respiratory issues
Hexavalent chromium	Carcinogenicity, respiratory, and gastrointestinal effects
Acetone	Respiratory and neurological effects, developmental effects
Formaldehyde	Respiratory problems, eye irritation, cancer
Toluene diisocyanate (TDI)	Respiratory issues, skin irritation, asthma exacerbation
Chloramine	Respiratory issues, eye irritation, skin irritation
Phenol	Respiratory issues, skin irritation, gastrointestinal effects
Styrene	Neurological effects, respiratory issues, cancer

Source: Based on a review of the literature.

5.6 Antibiotic residues of drinking water and its human exposure risk

Antibiotic residues in drinking water refer to trace amounts of antibiotics that may be present in water sources, such as rivers, lakes, groundwater, or municipal water supplies, due to human and animal use of antibiotics (Brillas 2023, de Jesus *et al* 2023, Ding 2023, Musa Yahaya *et al* 2023, Ruziwa *et al* 2023, Saravanan *et al* 2023, Sol *et al* 2023, Sudhir *et al* 2023, Weerakoon *et al* 2023, Zardosht *et al* 2023). These residues can enter water bodies through various pathways, including wastewater discharges from pharmaceutical manufacturing plants, improper disposal of unused medications, and excretion by humans and animals. Human exposure to antibiotic residues in drinking water is a concern due to the potential adverse health effects of prolonged exposure to these substances. Antibiotics in drinking water raise concerns about antibiotic resistance, which occurs when bacteria are exposed to low levels of antibiotics over time and then develop the ability to survive and multiply in the presence of these drugs (Shah 2021a, Amenaghawon *et al* 2023, Ayejoto *et al* 2023, Cai *et al* 2023, Hassan *et al* 2023, Huang *et al* 2023, Kammoun *et al* 2023, Kim *et al* 2023, Nasir *et al* 2023, Paneerselvam *et al* 2023, Sharma *et al* 2023, Villén *et al* 2023, Yadav *et al* 2023). Antibiotic resistance is a global health issue that can lead to difficulties in effectively treating bacterial infections.

The human exposure risk associated with antibiotic residues in drinking water can be assessed by evaluating the concentration of antibiotics and determining the potential health effects on individuals who consume the water. The risk assessment process typically involves the following steps:

1. **Identification of antibiotics:** A comprehensive list of antibiotics is compiled, considering those commonly used in human and veterinary medicine and their metabolites.
2. **Analytical methods:** Reliable analytical methods detect and quantify antibiotic residues in drinking water samples. These methods may include liquid chromatography–mass spectrometry (LC–MS) or high-performance liquid chromatography (HPLC).
3. **Concentration analysis:** Drinking water samples are collected from various sources and analyzed to measure the concentration of antibiotic residues. These concentrations are typically reported in micrograms per litre (μg l^{-1}) or parts per billion (ppb).
4. **Exposure assessment:** The estimated daily intake of antibiotics through drinking water is calculated based on the measured concentrations and an individual's water consumption rate. This helps to determine the level of exposure to antibiotic residues.
5. **Risk characterization:** The data obtained from exposure assessment are compared to established guidelines or regulatory limits for antibiotics in drinking water. These guidelines consider the potential health effects of specific antibiotics and their concentrations. The risk characterization determines whether the exposure to antibiotic residues is considered low, moderate, or high.

Table 5.10. Antibiotic contamination in drinking water.

Antibiotic name	Concentration in drinking water (μg l^{-1})	Human exposure risk
Amoxicillin	0.03	Low
Ciprofloxacin	0.12	Moderate
Sulfamethoxazole	0.08	Low
Tetracycline	0.05	Moderate
Erythromycin	0.07	Low
Trimethoprim	0.09	Moderate
Azithromycin	0.10	Low
Clarithromycin	0.04	Moderate
Levofloxacin	0.06	Low
Doxycycline	0.11	High

Source: Based on a review of the literature.

6. **Mitigation measures:** If the risk assessment indicates elevated exposure or potential health risks, then appropriate measures are taken to mitigate the presence of antibiotic residues in drinking water. This may involve improving water treatment processes, implementing stricter regulations on pharmaceutical waste disposal, or promoting responsible use of antibiotics.

It is important to note that risk assessments and regulatory limits for antibiotic residues in drinking water may vary between countries or regions. It is recommended to consult local regulatory agencies or scientific publications for specific information relevant to your location (Ahmed *et al* 2023, Alagan *et al* 2023, Darling *et al* 2023, De Caroli Vizioli *et al* 2023, Hanna *et al* 2023, Li *et al* 2023b, Mekawi *et al* 2023, Palit *et al* 2023, Shah 2021b, Sathya *et al* 2023, Shabanloo *et al* 2023, Sivagami *et al* 2023, Takman *et al* 2023, Thanigaivel *et al* 2023, Wang *et al* 2023c).

Overall, monitoring and managing antibiotic residues in drinking water play a crucial role in safeguarding public health, reducing the spread of antibiotic resistance, and ensuring safe drinking water to communities (table 5.10).

5.7 Pharmaceutical contamination in drinking water

Pharmaceutical contamination in drinking water refers to trace amounts of pharmaceutical compounds or metabolites in water sources intended for human consumption. These contaminants can enter water systems through various pathways, including disposing of unused medications, excretion by humans and animals, improper disposal of pharmaceutical manufacturing waste, and inadequate wastewater treatment processes.

The presence of pharmaceuticals in drinking water has raised concerns due to their potential impacts on human health and the environment. These compounds are designed to have biological effects on the body, and their presence in water systems may have unintended consequences (Alagan *et al* 2023, Angnunavuri *et al* 2023,

Brillas and Garcia-Segura 2023, Darling *et al* 2023, Grmasha *et al* 2023, Hua *et al* 2023, Jia *et al* 2023, Shah 2020, Kumar *et al* 2023a, Mekawi *et al* 2023, Nguyen *et al* 2023, Palit *et al* 2023, Verma *et al* 2023, Zhang *et al* 2023).

There are several ways in which pharmaceuticals can enter drinking water sources:

1. **Patient excretion:** When individuals consume pharmaceuticals, some compounds or metabolites may be excreted through urine or feces. Wastewater treatment plants may not effectively remove all pharmaceutical residues, leading to their release into water bodies.

2. **Improper medication disposal:** Improper disposal of medications, such as flushing them down the toilet or sink, can introduce pharmaceuticals directly into the water system. This occurs when people dispose of expired or unused medications in a manner that bypasses proper disposal methods.

3. **Pharmaceutical manufacturing:** The manufacturing process of pharmaceuticals can result in the release of pharmaceutical compounds into the environment through wastewater discharges from pharmaceutical production facilities.

4. **Livestock and agriculture:** Pharmaceuticals used in veterinary medicine or as growth promoters in livestock can enter water systems through animal waste or runoff from agricultural areas.

Pharmaceutical contaminants in drinking water raise concerns about potential health effects. However, it is essential to note that the pharmaceuticals found in drinking water are typically deficient and are generally below therapeutic doses. The current scientific consensus suggests that the risk to human health from pharmaceuticals in drinking water is low. Nevertheless, there is ongoing research to understand better potential long-term health effects, especially from chronic exposure to low levels of pharmaceutical compounds (Ghaffar *et al* 2023, Jagadeesh and Sundaram 2023, Ju *et al* 2023, Li *et al* 2023a, Neeti *et al* 2023, Nirmala *et al* 2023, Pratap *et al* 2023, Safwat *et al* 2023, Shahzad *et al* 2023, Shivarajappa *et al* 2023, Tajuddin *et al* 2023, Wang *et al* 2023a).

Various measures can be taken to address the issue of pharmaceutical contamination in drinking water, including:

1. **Improved wastewater treatment:** Enhancing wastewater treatment processes can help remove or reduce the presence of pharmaceuticals and their metabolites in treated effluents before they are discharged into water bodies.

2. **Public education and awareness:** Educating the public about the proper disposal of medications, such as using take-back programs or designated collection sites, can help to prevent the introduction of pharmaceuticals into water systems.

3. **Source water protection:** Implementing measures to protect water sources from pharmaceutical contamination, such as implementing buffer zones around water bodies or controlling runoff from pharmaceutical manufacturing sites, can help to prevent the entry of these compounds into drinking water sources.

4. **Advanced water treatment technologies:** Investigating and implementing advanced treatment technologies, such as activated carbon filtration, advanced oxidation processes, or membrane filtration, can aid in further reducing the presence of pharmaceutical contaminants in drinking water treatment plants.

5. **Regulatory frameworks:** Developing and implementing regulations or guidelines for monitoring and controlling pharmaceutical contaminants in drinking water can help to ensure the safety and quality of water supplies.

While the presence of pharmaceutical contaminants in drinking water is a concern, the current understanding suggests that the levels typically found in treated drinking water are unlikely to pose significant health risks. However, continued research, improved wastewater treatment processes, responsible medication disposal, and public awareness are essential for addressing and mitigating the potential impacts of pharmaceutical contaminations in drinking water (Abebe *et al* 2023, Ahmad Dar and Kurella 2023, An *et al* 2023, Calore *et al* 2023, Devi *et al* 2023, García-Ávila *et al* 2023, Jeong *et al* 2023, Kalita and Devi 2023, Mishra *et al* 2023, Onu *et al* 2023, Qi *et al* 2024, Ramírez *et al* 2023, Sahu *et al* 2023, Sangkham *et al* 2023, Tripathi *et al* 2023) (table 5.11).

Table 5.11. Pharmaceutical contaminants and their everyday use.

Pharmaceutical contaminant	Common use
Acetaminophen	Analgesic
Atenolol	Beta-blocker
Carbamazepine	Anticonvulsant
Ciprofloxacin	Antibiotic
Diazepam	Sedative
Fluoxetine	Antidepressant
Gemfibrozil	Lipid-lowering agent
Ibuprofen	Nonsteroidal anti-inflammatory drug
Metformin	Antidiabetic
Naproxen	Nonsteroidal anti-inflammatory drug
Omeprazole	Proton pump inhibitor
Propranolol	Beta-blocker
Sertraline	Antidepressant
Sulfamethoxazole	Antibiotic
Trimethoprim	Antibiotic
Warfarin	Anticoagulant
Amlodipine	Calcium channel blocker
Bisoprolol	Beta-blocker
Citalopram	Antidepressant
Diclofenac	Nonsteroidal anti-inflammatory drug

(*Continued*)

Table 5.11. (*Continued*)

Pharmaceutical contaminant	Common use
Enalapril	ACE inhibitor
Fluconazole	Antifungal
Hydrochlorothiazide	Diuretic
Losartan	Angiotensin II receptor antagonist
Methadone	Opioid analgesic
Metoprolol	Beta-blocker
Norfloxacin	Antibiotic
Pantoprazole	Proton pump inhibitor
Simvastatin	Lipid-lowering agent
Tamsulosin	Alpha blocker
Venlafaxine	Antidepressant
Alprazolam	Sedative
Amoxicillin	Antibiotic
Atenolol	Beta-blocker
Atorvastatin	Lipid-lowering agent
Codeine	Opioid analgesic
Dexamethasone	Corticosteroid
Furosemide	Diuretic
Lisinopril	ACE inhibitor
Lorazepam	Sedative
Metronidazole	Antibiotic
Paracetamol	Analgesic
Prednisone	Corticosteroid
Quetiapine	Antipsychotic
Sildenafil	Erectile dysfunction medication
Tramadol	Opioid analgesic
Valproic acid	Anticonvulsant
Zolpidem	Sedative/hypnotic

Source: Based on a review of the literature.

5.8 Conclusion

The environmental and health risk assessment of chemical pollutants in drinking water and wastewater is paramount for safeguarding the environment and public health. Through rigorous monitoring, analysis, and assessment processes, we can identify and understand the presence of chemical pollutants in water sources, evaluate their potential adverse effects on ecosystems and human populations, and develop appropriate mitigation strategies. We can determine the risk associated with their presence by assessing the concentrations of chemical pollutants and comparing them to established regulatory standards, guidelines, or thresholds. This information enables policymakers, regulatory agencies, and water management authorities to make informed decisions regarding water treatment processes, waste management practices, and pollution control measures. The risks posed by chemical pollutants in drinking water include potential acute and chronic health effects on individuals who consume contaminated water. These risks can range from immediate health impacts, such as gastrointestinal disorders or respiratory issues, to long-term health consequences, including the development of cancer or reproductive disorders. Additionally, the ecological risks to aquatic organisms and ecosystems cannot be overlooked because chemical pollutants can disrupt the balance of ecosystems, harm sensitive species, and degrade water quality. Through comprehensive environmental and health risk assessments, we gain insights into the potential hazards of chemical pollutants, their fate and transport in water systems, and the potential for bioaccumulation in the food chain. This knowledge is a foundation for implementing effective monitoring programs, establishing pollution prevention measures, and developing sustainable water resource management strategies.

To mitigate the risks associated with chemical pollutants in drinking water and wastewater, adopting a holistic approach that encompasses source control, proper wastewater treatment, improved industrial practices, public education on responsible chemical use, and the implementation of stringent regulatory frameworks is crucial. Collaboration between stakeholders, including government agencies, industries, research institutions, and the public, is essential to address this multifaceted issue and ensure the long-term safety and sustainability of our water resources. By prioritizing the assessment of environmental and health risks posed by chemical pollutants, we can work towards preserving clean and safe water sources, protecting human and ecological health, and fostering a more sustainable and resilient future.

5.9 Prospects for the future

Looking into the future, the prospects for the environmental and health risk assessment of chemical pollutants in drinking water and wastewater are promising. As technology advances and our understanding of chemical pollutants deepens, the following key prospects can be envisioned:

1. **Enhanced monitoring techniques:** Developing advanced monitoring techniques will enable more efficient and comprehensive detection of chemical pollutants in drinking water and wastewater. Innovations such as sensor networks, remote sensing technologies, and real-time monitoring systems will improve our ability to promptly identify and respond to pollutant sources.

2. **Emerging contaminant identification:** The focus on emerging contaminants, including pharmaceuticals, personal care products, and microplastics, will continue to expand. Assessing the environmental and health risks of these contaminants in water systems will be a priority, leading to a better understanding of their fate, transport, and potential impacts.

3. **Integrated risk assessment approaches:** Future assessments will embrace integrated risk assessment approaches considering the cumulative effects of multiple chemical pollutants on ecosystems and human health. This holistic approach will account for interactions, synergistic or additive effects, and the potential for mixture toxicity.

4. **Predictive modeling and big data:** Using predictive modeling and big data analytics will become more prominent in risk assessments. These tools will allow for more accurate predictions of pollutant behavior, exposure pathways, and health risks, aiding in proactive decision-making and targeted mitigation strategies.

5. **Human health impact studies:** Research on the long-term health impacts of low-level exposure to chemical pollutants in drinking water will continue to evolve. This includes assessing the effects of chronic exposure, understanding vulnerable populations, and evaluating the potential links between chemical pollutants and non-communicable diseases.

6. **Climate change considerations:** With the increasing impacts of climate change on water resources, future risk assessments will integrate climate change considerations. This will involve evaluating the influence of changing hydrological patterns, extreme weather events, and altered pollutant transport pathways on the fate and behavior of chemical pollutants.

7. **Policy and regulation development:** The findings from environmental and health risk assessments will inform the development of robust policies and regulations to protect water quality. Stricter standards for pollutant concentrations, enhanced pollution control measures, and targeted legislation will be formulated based on scientific evidence and risk assessments.

8. **Public awareness and education:** Increasing public awareness about the risks associated with chemical pollutants in drinking water and wastewater will be crucial. Educational campaigns and initiatives will empower individuals to make informed choices regarding chemical use, waste disposal, and water consumption practices.

Overall, the prospects for the environmental and health risks assessment of chemical pollutants in drinking water and wastewater hold promise for more comprehensive and proactive approaches. By leveraging technological advancements, embracing holistic assessments, and prioritizing the protection of water resources, we can strive towards a safer and more sustainable water future for all.

Acknowledgments

We would like to express our sincere gratitude to the Department of Agronomy for their support and assistance throughout the writing. The department's commitment to academic excellence and research has been instrumental in completing this endeavor.

Author's contribution

The authors of this work have made significant contributions to the research project/study. Each author has participated sufficiently in the research project/study, made intellectual contributions, and is responsible for the work's accuracy and integrity. The authors have collaborated closely, ensuring the completion of this work through collective effort, expertise, and dedication.

References

Abebe Y, Alamirew T, Whitehead P, Charles K and Alemayehu E 2023 Spatio-temporal variability and potential health risks assessment of heavy metals in the surface water of Awash basin, Ethiopia *Heliyon* **9** e15832

Adelodun B *et al* 2022 List of contributors *Microbiome Under Changing climate* ed A Kumar, J Singh and L F R Ferreira (Woodhead Publishing) pp xix–xxiv

Ahmad Dar F and Kurella S 2023 Fluoride in drinking water: an in-depth analysis of its prevalence, health effects, advances in detection and treatment *Mater. Today Proc.* https://doi.org/10.1016/j.matpr.2023.05.645

Ahmed R S, Abuarab M E, Ibrahim M M, Baioumy M and Mokhtar A 2023 Assessment of environmental and toxicity impacts and potential health hazards of heavy metals pollution of agricultural drainage adjacent to industrial zones in Egypt *Chemosphere* **318** 137872

Akhtar N *et al* 2021 List of contributors *Volatiles and Metabolites of Microbes* ed A Kumar, J Singh, and M M Samuel (New York: Academic) pp xix–xi

Alagan M, Chandra Kishore S, Perumal S, Manoj D, Raji A, Kumar R S, Almansour A I and Lee Y R 2023 Narrative of hazardous chemicals in water: its potential removal approach and health effects *Chemosphere* **335** 139178

Aley P, Singh J and Kumar P 2022 Adapting the changing environment: microbial way of life *Microbiome Under Changing climate* ed A Kumar, J Singh and L F R Ferreira (Woodhead Publishing) ch 23 pp 507–25

Alharbi O A, Jarvis E, Galani A, Thomaidis N S, Nika M-C and Chapman D V 2023 Assessment of selected pharmaceuticals in Riyadh wastewater treatment plants, Saudi Arabia: mass loadings, seasonal variations, removal efficiency and environmental risk *Sci. Total Environ.* **882** 163284

Almazrouei B, Islayem D, Alskafi F, Catacutan M K, Amna R, Nasrat S, Sizirici B and Yildiz I 2023 Steroid hormones in wastewater: sources, treatments, environmental risks, and regulations *Emerg. Contam.* **9** 100210

Amenaghawon A N, Anyalewechi C L, Osazuwa O U, Elimian E A, Eshiemogie S O, Oyefolu P K and Kusuma H S 2023 A comprehensive review of recent advances in the synthesis and application of metal-organic frameworks (MOFs) for the adsorptive sequestration of pollutants from wastewater *Sep. Purif. Technol.* **311** 123246

An R, Li B, Zhong S, Peng G, Li J, Ma R, Chen Q and Ni J 2023 Distribution, source identification, and health risk of emerging organic contaminants in groundwater of Xiong'an New Area, Northern China *Sci. Total Environ.* **893** 164786

Angnunavuri P N, Attiogbe F and Mensah B 2023 Particulate plastics in drinking water and potential human health effects: current knowledge for management of freshwater plastic materials in Africa *Environ. Pollut.* **316** 120714

Ayejoto D A, Agbasi J C, Egbueri J C and Abba S I 2023 Evaluation of oral and dermal health risk exposures of contaminants in groundwater resources for nine age groups in two densely populated districts, Nigeria *Heliyon* **9** e15483

Brillas E 2023 Solar photoelectro-Fenton: a very effective and cost-efficient electrochemical advanced oxidation process for the removal of organic pollutants from synthetic and real wastewaters *Chemosphere* **327** 138532

Brillas E and Garcia-Segura S 2023 Recent progress of applied TiO_2 photoelectrocatalysis for the degradation of organic pollutants in wastewaters *J. Environ. Chem. Eng.* **11** 109635

Cai H, Shen C, Xu H, Qian H, Pei S, Cai P, Song J and Zhang Y 2023 Seasonal variability, predictive modeling and health risks of N-nitrosamines in drinking water of Shanghai *Sci. Total Environ.* **857** 159530

Calore F, Guolo P P, Wu J, Xu Q, Lu J and Marcomini A 2023 Legacy and novel PFASs in wastewater, natural water, and drinking water: occurrence in Western Countries vs China *Emerg. Contam.* **9** 100228

Chakraborty S, Kumar P, Sanyal R, Mane A B, Arvind Prasanth D, Patil M and Dey A 2021 Unravelling the regulatory role of miRNAs in secondary metabolite production in medicinal crops *Plant Gene* **27** 100303

Chiriac F L, Pirvu F, Paun I and Petre V A 2023 Perfluoroalkyl substances in Romanian wastewater treatment plants: transfer to surface waters, environmental and human risk assessment *Sci. Total Environ.* **892** 164576

Coccia M and Bontempi E 2023 New trajectories of technologies for the removal of pollutants and emerging contaminants in the environment *Environ. Res.* **229** 115938

Darling A, Patton H, Rasheduzzaman M, Guevara R, McCray J, Krometis L-A and Cohen A 2023 Microbiological and chemical drinking water contaminants and associated health outcomes in rural Appalachia, USA: a systematic review and meta-analysis *Sci. Total Environ.* **892** 164036

Das T *et al* 2022 Promising botanical-derived monoamine oxidase (MAO) inhibitors: pharmacological aspects and structure-activity studies *S. Afr. J. Bot.* **146** 127–45

De Caroli Vizioli B, Silva da Silva G, Ferreira de Medeiros J and Montagner C C 2023 Atrazine and its degradation products in drinking water source and supply: risk assessment for environmental and human health in Campinas, Brazil *Chemosphere* **336** 139289

de Jesus R A, Barros G P, Bharagava R N, Liu J, Mulla S I, Azevedo L C B and Ferreira L F R 2023 Occurrence of pesticides in wastewater: bioremediation approach for environmental safety and its toxicity *Advances in Chemical Pollution, Environmental Management and Protection* vol 9 ed L F R Ferreira, A Kumar and Bilal (Amsterdam: Elsevier) ch 2 pp 17–33

Devi A, Verma M, Saratale G D, Saratale R G, Ferreira L F R, Mulla S I and Bharagava R N 2023 Microalgae: a green eco-friendly agents for bioremediation of tannery wastewater with simultaneous production of value-added products *Chemosphere* **336** 139192

Ding G K 2023 *Wastewater Treatment, Reused and Recycling—A Potential Source of Water Supply* (Amsterdam: Elsevier)

Foglia A, González-Camejo J, Radini S, Sgroi M, Li K, Eusebi A L and Fatone F 2023 Transforming wastewater treatment plants into reclaimed water facilities in water-unbalanced regions. An overview of possibilities and recommendations focusing on the Italian case *J. Clean. Prod.* **410** 137264

Ganthavee V and Trzcinski A P 2023 Removal of pharmaceutically active compounds from wastewater using adsorption coupled with electrochemical oxidation technology: a critical review *J. Ind. Eng. Chem.* **11** 110130

García-Ávila F, Cabello-Torres R, Iglesias-Abad S, García-Mera G, García-Uzca C, Valdiviezo-Gonzales L and Donoso-Moscoso S 2023 Cleaner production and drinking water: perspectives from a scientometric and systematic analysis for a sustainable performance *S. Afr. J. Chem. Eng.* **45** 136–48

Ghaffar I, Hussain A, Hasan A and Deepanraj B 2023 Microalgal-induced remediation of wastewaters loaded with organic and inorganic pollutants: an overview *Chemosphere* **320** 137921

Goud E L, Singh J and Kumar P 2022 Climate change and their impact on global food production *Microbiome Under Climate Change* ed A Kumar, J Singh and L F R Ferreira (Woodhead Publishing) ch 19 pp 415–36

Grmasha R A, Abdulameer M H, Stenger-Kovács C, Al-sareji O J, Al-Gazali Z, Al-Juboori R A, Meiczinger M and Hashim K S 2023 Polycyclic aromatic hydrocarbons in the surface water and sediment along Euphrates River system: occurrence, sources, ecological and health risk assessment *Mar. Pollut. Bull.* **187** 114568

Hanna N, Tamhankar A J and Stålsby Lundborg C 2023 Antibiotic concentrations and antibiotic resistance in aquatic environments of the WHO Western Pacific and South-East Asia regions: a systematic review and probabilistic environmental hazard assessment *Lancet Planet. Health* **7** e45–54

Hassan H B, Moniruzzaman M, Majumder R K, Ahmed F, Quaiyum Bhuyian M A, Ahsan M A and Al-Asad H 2023 Impacts of seasonal variations and wastewater discharge on river quality and associated human health risks: a case of northwest dhaka, Bangladesh *Heliyon* **9** e18171

He M, Liu G, Li Y, Zhou L, Arif M and Liu Y 2023 Spatial-temporal distribution, source identification, risk assessment and water quality assessment of trace elements in the surface water of typical tributary in Yangtze River delta, China *Mar. Pollut. Bull.* **192** 115035

Hua Z, Gao C, Zhang J and Li X 2023 Perfluoroalkyl acids in the aquatic environment of a fluorine industry-impacted region: spatiotemporal distribution, partition behavior, source, and risk assessment *Sci. Total Environ.* **857** 159452

Huang Y *et al* 2023 Chemical characterization and source attribution of organic pollutants in industrial wastewaters from a Chinese chemical industrial park *Environ. Res.* **229** 115980

Irshad M A, Sattar S, Nawaz R, Al-Hussain S A, Rizwan M, Bukhari A, Waseem M, Irfan A, Inam A and Zaki M E A 2023 Enhancing chromium removal and recovery from industrial wastewater using sustainable and efficient nanomaterial: a review *Ecotoxicol. Environ. Saf.* **263** 115231

Jagadeesh N and Sundaram B 2023 Adsorption of pollutants from wastewater by biochar: a review *J. Hazard. Mater. Adv.* **9** 100226

Jeong Y, Gong G, Lee H-J, Seong J, Hong S W and Lee C 2023 Transformation of microplastics by oxidative water and wastewater treatment processes: a critical review *J. Hazard. Mater.* **443** 130313

Jia C, Raza Altaf A, Li F, Ashraf I, Zafar Z and Ahmad Nadeem A 2023 Comprehensive assessment on groundwater quality, pollution characteristics, and ecological health risks under seasonal thaws: spatial insights with Monte Carlo simulations *Groundwater Sustain. Dev.* **22** 100952

Ju Q, Hu Y, Liu Q, Chai H, Chen K, Zhang H and Wu Y 2023 Source apportionment and ecological health risks assessment from major ions, metalloids and trace elements in multi-aquifer groundwater near the Sunan mine area, Eastern China *Sci. Total Environ.* **860** 160454

Kalita S and Devi A 2023 Bioadsorption of endocrine disrupting pollutants from wastewater *Current Developments in Biotechnology and Bioengineering* ed I Haq, A Kalamdhad, and B Pandey (Amsterdam: Elsevier) ch 12 pp 211–26

Kammoun R, McQuaid N, Lessard V, Prévost M, Bichai F and Dorner S 2023 Comparative study of deterministic and probabilistic assessments of microbial risk associated with combined sewer overflows upstream of drinking water intakes *Environ. Chall.* **12** 100735

Kim F, Pablo G-F, Lubertus B, Lutz A, Karin W, Félix H, Agneta O and Johan L 2023 Effect-based evaluation of water quality in a system of indirect reuse of wastewater for drinking water production *Water Res.* **242** 120147

Kotia A, Rutu P, Singh V, Kumar A, Dhoke S, Kumar P and Singh D K 2022 Rheological analysis of rice husk-starch suspended in water for sustainable agriculture application *Mater. Today Proc.* **50** 1962–6

Krishnan R Y, Manikandan S, Subbaiya R, Biruntha M, Balachandar R and Karmegam N 2023 Origin, transport and ecological risk assessment of illicit drugs in the environment—a review *Chemosphere* **311** 137091

Kumar M, Shekhar S, Kumar R, Kumar P, Govarthanan M and Chaminda T 2023a Drinking water treatment and associated toxic byproducts: concurrence and urgence *Environ. Pollut.* **320** 121009

Kumar P, Devi P and Dey S R 2021a Fungal volatile compounds: a source of novel in plant protection agents *Volatiles and Metabolites of Microbes* ed A Kumar, J Singh, and M M Samuel (New York: Academic) ch 6 pp 83–104

Kumar P, Kumar T, Singh S, Tuteja N, Prasad R and Singh J 2020 Potassium: a key modulator for cell homeostasis *J. Biotechnol.* **324** 198–210

Kumar P and Mistri T K 2020 Transcription factors in SOX family: potent regulators for cancer initiation and development in the human body *Semin. Cancer Biol.* **67** 105–13

Kumar P, Sharma K, Saini L and Dey S R 2021b Role and behavior of microbial volatile organic compounds in mitigating stress *Volatiles and Metabolites of Microbes* ed A Kumar, J Singh, and M M Samuel (New York: Academic) ch 8 pp 143–61

Kumar V, Singh E, Singh S, Pandey A and Bhargava P C 2023b Micro- and nano-plastics (MNPs) as emerging pollutant in ground water: environmental impact, potential risks, limitations and way forward towards sustainable management *Chem. Eng. J.* **459** 141568

Kumar V, Dwivedi P, Kumar P, Singh B N, Pandey D K, Kumar V and Bose B 2021c Mitigation of heat stress responses in crops using nitrate primed seeds *S. Afr. J. Bot.* **140** 25–36

Kumari P, Singh J and Kumar P 2022 Impact of bioenergy for the diminution of an ascending global variability and change in the climate *Microbiome Under Changing Climate* ed A Kumar, J Singh and L F R Ferreira (Woodhead Publishing) ch 21 pp 469–87

Laad M and Ghule B 2023 Removal of toxic contaminants from drinking water using biosensors: a systematic review *Groundw. Sustain. Develop.* **20** 100888

Li D, Huang W and Huang R 2023a Analysis of environmental pollutants using ion chromatography coupled with mass spectrometry: a review *J. Hazard. Mater.* **458** 131952

Li S, Ondon B S, Ho S-H, Zhou Q and Li F 2023b Drinking water sources as hotspots of antibiotic-resistant bacteria (ARB) and antibiotic resistance genes (ARGs): occurrence, spread, and mitigation strategies *J. Water Process Engineering* **53** 103907

Liu S-S, You W-D, Chen C-E, Wang X-Y, Yang B and Ying G-G 2023a Occurrence, fate and ecological risks of 90 typical emerging contaminants in full-scale textile wastewater treatment plants from a large industrial park in Guangxi, Southwest China *J. Hazard. Mater.* **449** 131048

Liu S, Ding H, Song Y, Xue Y, Bi M, Wu M, Zhao C, Wang M, Shi J and Deng H 2023b The potential risks posed by micro-nanoplastics to the safety of disinfected drinking water *J. Hazard. Mater.* **450** 131089

Mekawi E M, Abbas M H H, Mohamed I, Jahin H S, El-Ghareeb D, Al-Senani G M, AlMufarij R S, Abdelhafez A A, Mansour R R M and Bassouny M A 2023 Potential hazards and health assessment associated with different water uses in the main industrial cities of Egypt *J. Saudi Chem. Soc.* **27** 101587

Mishra S, Kumar R and Kumar M 2023 Use of treated sewage or wastewater as an irrigation water for agricultural purposes—environmental, health, and economic impacts *Total Environ. Rese. Themes* **6** 100051

Mukherjee S and Chauhan N 2023 Assessment and monitoring of human health risk during wastewater reuse *Antimicrobial Resistance in Wastewater and Human Health* ed D Pal, and H H Kumar (New York: Academic) ch 12 pp 255–70

Munné A *et al* 2023 Indirect potable water reuse to face drought events in Barcelona city. Setting a monitoring procedure to protect aquatic ecosystems and to ensure a safe drinking water supply *Sci. Total Environ.* **866** 161339

Musa Yahaya S, Ahmad Mahmud A and Abdu N 2023 The use of wastewater for irrigation: pros and cons for human health in developing countries *Total Environ. Res. Themes* **6** 100044

Nasir Z *et al* 2023 Fingerprinting of heavy metal and microbial contamination uncovers the unprecedented scale of water pollution and its implication on human health around transboundary Hudiara drain in South Asia *Environ. Technol. Innov.* **30** 103040

Nawaz S, Tabassum A, Muslim S, Nasreen T, Baradoke A, Kim T H, Boczkaj G, Jesionowski T and Bilal M 2023 Effective assessment of biopolymer-based multifunctional sorbents for the remediation of environmentally hazardous contaminants from aqueous solutions *Chemosphere* **329** 138552

Neeti K, Singh R and Ahmad S 2023 The role of green nanomaterials as effective adsorbents and applications in wastewater treatment *Mater. Today Proc.* **77** 269–76

Nguyen M-K, Lin C, Nguyen H-L, Hung N T Q, La D D, Nguyen X H, Chang S W, Chung W J and Nguyen D D 2023 Occurrence, fate, and potential risk of pharmaceutical pollutants in agriculture: challenges and environmentally friendly solutions *Sci. Total Environ.* **899** 165323

Nirmala K, Rangasamy G, Ramya M, Shankar V U and Rajesh G 2023 A critical review on recent research progress on microplastic pollutants in drinking water *Environ. Res.* **222** 115312

Oke E A, Oluyinka O A, Afolabi S D, Ibe K K and Raheem S A 2023 Latest insights on technologies for halides and halogenated compounds extraction/abatement from water and wastewater: challenges and future perspectives *J. Water Process Eng.* **53** 103724

Onu M A, Ayeleru O O, Oboirien B and Olubambi P A 2023 Challenges of wastewater generation and management in sub-Saharan Africa: a review *Environ. Chall.* **11** 100686

Palit S, Das P and Basak P 2023 Application of nanotechnology in water and wastewater treatment and the vast vision for the future *Additive Manufacturing Materials and Technologies* ed J K Pandey, S Manna, R K Patel and M B Qian (Amsterdam: Elsevier) ch 7 pp 157–79

Paneerselvam B, Ravichandran N, Li P, Thomas M, Charoenlerkthawin W and Bidorn B 2023 Machine learning approach to evaluate the groundwater quality and human health risk for sustainable drinking and irrigation purposes in South India *Chemosphere* **336** 139228

Perveen S and Amar-Ul-Haque 2023 Drinking water quality monitoring, assessment and management in Pakistan: a review *Heliyon* **9** e13872

Pratap B, Kumar S, Nand S, Azad I, Bharagava R N, Romanholo Ferreira L F and Dutta V 2023 Wastewater generation and treatment by various eco-friendly technologies: possible health hazards and further reuse for environmental safety *Chemosphere* **313** 137547

Qi Y, Li D, Zhang S, Li F and Hua T 2024 Electrochemical filtration for drinking water purification: a review on membrane materials, mechanisms and roles *J. Environ. Sci.* **141** 102–28

Quispe J I B, Campos L C, Mašek O and Bogush A 2023 Optimisation of biochar filter for handwashing wastewater treatment and potential treated water reuse for handwashing *J. Water Process Eng.* **54** 104001

Radini S, González-Camejo J, Andreola C, Eusebi A L and Fatone F 2023 Risk management and digitalisation to overcome barriers for safe reuse of urban wastewater for irrigation—a review based on European practice *J. Water Process Eng.* **53** 103690

Ramírez D G, Narváez Valderrama J F, Palacio Tobón C A, García J J, Echeverri J D, Sobotka J and Vrana B 2023 Occurrence, sources, and spatial variation of POPs in a mountainous tropical drinking water supply basin by passive sampling *Environ. Pollut.* **318** 120904

Ruziwa D T, Oluwalana A E, Mupa M, Meili L, Selvasembian R, Nindi M M, Sillanpaa M, Gwenzi W and Chaukura N 2023 Pharmaceuticals in wastewater and their photocatalytic degradation using nano-enabled photocatalysts *J. Water Process Eng.* **54** 103880

Safwat S M, Mohamed N Y and El-Seddik M M 2023 Performance evaluation and life cycle assessment of electrocoagulation process for manganese removal from wastewater using titanium electrodes *J. Environ. Manage.* **328** 116967

Sahu J N, Kapelyushin Y, Mishra D P, Ghosh P, Sahoo B K, Trofimov E and Meikap B C 2023 Utilization of ferrous slags as coagulants, filters, adsorbents, neutralizers/stabilizers, catalysts, additives, and bed materials for water and wastewater treatment: a review *Chemosphere* **325** 138201

Sangkham S, Aminul Islam M, Adhikari S, Kumar R, Sharma P, Sakunkoo P, Bhattacharya P and Tiwari A 2023 Evidence of microplastics in groundwater: a growing risk for human health *Groundw. Sustain. Develop.* **23** 100981

Saravanan A, Kumar P S, Duc P A and Rangasamy G 2023 Strategies for microbial bioremediation of environmental pollutants from industrial wastewater: a sustainable approach *Chemosphere* **313** 137323

Sathya R, Arasu M V, Al-Dhabi N A, Vijayaraghavan P, Ilavenil S and Rejiniemon T S 2023 Towards sustainable wastewater treatment by biological methods—a challenges and advantages of recent technologies *Urban Climate* **47** 101378

Shabanloo A, Akbari H, Adibzadeh A and Akbari H 2023 Synergistic activation of persulfate by ultrasound/PbO2 anodic oxidation system for effective degradation of naproxen, a toxic and bio-recalcitrant pollutant: process optimization and application for pharmaceutical wastewater *J. Water Process Eng.* **54** 103915

Shah M P 2020 *Microbial Bioremediation and Biodegradation* (Berlin: Springer)

Shah M P 2021a *Removal of Refractory Pollutants from Wastewater Treatment Plants* (Boca Raton, FL: CRC Press)

Shah M P 2021b *Removal of Emerging Contaminants through Microbial Processes* (Berlin: Springer)

Shahzad A, Ullah M W, Ali J, Aziz K, Javed M A, Shi Z, Manan S, Ul-Islam M, Nazar M and Yang G 2023 The versatility of nanocellulose, modification strategies, and its current progress in wastewater treatment and environmental remediation *Sci. Total Environ.* **858** 159937

Sharma P, Yadav M, Srivastava S K and Singh S P 2023 Bioremediation of androgenic and mutagenic pollutants from industrial wastewater *Current Developments in Biotechnology and Bioengineering* ed I Haq, A Kalamdhad, and B Pandey (Amsterdam: Elsevier) ch 7 pp 127–38

Shivarajappa, Surinaidu L, Gupta P K, Ahmed S, Hussain M and Nandan M J 2023 Impact of urban wastewater reuse for irrigation on hydro-agro-ecological systems and human health risks: a case study from Musi river basin, South India *HydroResearch* **6** 122–9

Sivagami K, Sharma P, Karim A V, Mohanakrishna G, Karthika S, Divyapriya G, Saravanathamizhan R and Kumar A N 2023 Electrochemical-based approaches for the treatment of forever chemicals: removal of perfluoroalkyl and polyfluoroalkyl substances (PFAS) from wastewater *Sci. Total Environ.* **861** 160440

Sol D, Solís-Balbín C, Laca A, Laca A and Díaz M 2023 A standard analytical approach and establishing criteria for microplastic concentrations in wastewater, drinking water and tap water *Sci. Total Environ.* **899** 165356

Sudhir S K, Bhatti S, Godheja J, Panda S and Haq I 2023 Treatment of pharmaceutical pollutants from industrial wastewater *Current Developments in Biotechnology and Bioengineering* ed I Haq, A Kalamdhad, and B Pandey (Amsterdam: Elsevier) ch 1 pp 1–16

Sun W and Zheng Z 2023 Degradation of fluoroquinolones in rural domestic wastewater by vertical flow constructed wetlands and ecological risks assessment *J. Clean. Prod.* **398** 136629

Tajuddin N A, Sokeri E F B, Kamal N A and Dib M 2023 Fluoride removal in drinking water using layered double hydroxide materials: preparation, characterization and the current perspective on IR4.0 technologies *J. Environ. Chem. Eng.* **11** 110305

Takman M, Svahn O, Paul C, Cimbritz M, Blomqvist S, Struckmann Poulsen J, Lund Nielsen J and Davidsson Å 2023 Assessing the potential of a membrane bioreactor and granular activated carbon process for wastewater reuse—a full-scale WWTP operated over one year in Scania, Sweden *Sci. Total Environ.* **895** 165185

Termeh-Zonoozi Y, Dilip Venugopal P, Patel V and Gagliano G 2023 Seeing beyond the smoke: selecting waterpipe wastewater chemicals for risk assessments *J. Hazard. Mater. Lett.* **4** 100074

Thanigaivel S, Vickram S, Dey N, Jeyanthi P, Subbaiya R, Kim W, Govarthanan M and Karmegam N 2023 Ecological disturbances and abundance of anthropogenic pollutants in the aquatic ecosystem: critical review of impact assessment on the aquatic animals *Chemosphere* **313** 137475

Tripathi S, Purchase D, Chandra R, Nadda A K and Chaturvedi P 2023 Emerging pollutants characterization, mitigation and toxicity assessment of sewage wastewater treatment plant-India: a case study *J. Contam. Hydrol.* **254** 104139

Turdiyeva K and Lee W 2023 Comparative analysis and human health risk assessment of contamination with heavy metals of Central Asian rivers *Heliyon* **9** e17112

Upadhyay S K, Devi P, Kumar V, Pathak H K, Kumar P, Rajput V D and Dwivedi P 2023 Efficient removal of total arsenic (As3+/5+) from contaminated water by novel strategies mediated iron and plant extract activated waste flowers of marigold *Chemosphere* **313** 137551

Verma A, Sharma A, Kumar R and Sharma P 2023 Nitrate contamination in groundwater and associated health risk assessment for Indo-Gangetic Plain, India *Groundw. Sustain. Develop.* **23** 100978

Villén J, Nekoro M, Sporrong S K, Håkonsen H, Bertram M G and Wettermark B 2023 Estimating environmental exposure to analgesic drugs: a cross-sectional study of drug utilization patterns in the area surrounding Sweden's largest drinking water source *Environ. Adv.* **12** 100384

Wang C, Guo Q, Zhang B, An W, Wang Z, Zhang D, Yang M and Yu J 2023a Solvent-like bis (2-chloro-1-methylethyl) ether occurrence in drinking water: multidimensional risk assessment integrated health and aesthetic aspects *J. Hazard. Mater.* **453** 131446

Wang R, Luo J, Li C, Chen J and Zhu N 2023b Antiviral drugs in wastewater are on the rise as emerging contaminants: a comprehensive review of spatiotemporal characteristics, removal technologies and environmental risks *J. Hazard. Mater.* **457** 131694

Wang Y, Dong X, Zang J, Zhao X, Jiang F, Jiang L, Xiong C, Wang N and Fu C 2023c Antibiotic residues of drinking-water and its human exposure risk assessment in rural Eastern China *Water Res.* **236** 119940

Weerakoon W M T D N, Seneviratne K N and Jayathilaka N 2023 Metagenomic analysis of wastewater for water quality assessment *Developments in Applied Microbiology and Biotechnology* ed M Vineet Kumar, S K S Bilal and V K B Garg (New York: Academic) ch 11 pp 285–309

Yadav P, Singh R P, Gupta R K, Pradhan T, Raj A, Singh S K, Kaushalendra , Pandey K D and Kumar A 2023 Contamination of soil and food chain through wastewater application *Advances in Chemical Pollution, Environmental Management and Protection* ed L F R Ferreira, A Kumar and M B Bilal (Amsterdam: Elsevier) ch 7 pp 109–32

Zardosht Z, Khosravani F, Rezaei S, Ghaderi S and Hassani G 2023 The impact of two insecticides on the pollutant cycle and quality of surface and groundwater resources in the irrigated lands of Yasuj, Iran *Heliyon* **9** e17636

Zhang K, Chang S, Zhang Q, Bai Y, Wang E, Zhang M, Fu Q, Wei L and Yu Y 2023 Heavy metals in influent and effluent from 146 drinking water treatment plants across China: occurrence, explanatory factors, probabilistic health risk, and removal efficiency *J. Hazard. Mater.* **450** 131003

IOP Publishing

Trends in Biological Processes in Industrial Wastewater Treatment

Maulin P Shah

Chapter 6

Effective COD and color removal with integrated ozonation and biological treatment approaches in textile wastewater: a review

G E Zengin, G Ozyildiz, D Soylu, G Aydogdu, E Cokgor and G Insel

The textile industry produces a significant amount of wastewater with high pollution loads. Textile wastewater is generated in many of the steps of production, such as desizing, mercerization, dyeing, and washing, and contains considerable amounts of organic compounds and colored effluent. The major pollutants in textile effluents are biochemical oxygen demand (BOD), chemical oxygen demand (COD), suspended solids, and heavy metals. In particular, the color of the effluent is a serious environmental and aesthetic problem. Therefore, there have been many studies on color removal. This study presents: (a) a critical review of the current literature on effective COD and color removal with integrated ozonation and biological treatment approaches in textile wastewater, and (b) a case study. In the case study, sequential anaerobic-aerobic processes and integrated treatment approach were applied to composite textile effluent containing reactive and indigo dyes. The sequential anaerobic-ozonation-aerobic treatment enabled 78% of COD and 81% of color removal, respectively. However, the sequential anaerobic-aerobic system yielded lower COD and color removal efficiencies. A respirometric modeling study revealed that anaerobic pretreatment resulted in 40% of Ss and %23 of S_H removal without altering the aerobic heterotrophic kinetics. The standard *Vibrio fischerie* test showed that anaerobic and ozone application increased the toxicity of effluent. Meanwhile, elevated toxicity could be eliminated by post-aeration, which makes the treated effluent non-toxic.

doi:10.1088/978-0-7503-5678-7ch6   6-1

6.1 Introduction

The textile industry is widespread internationally and produces a wide variety of products. However, the textile industry is reported to be one of the largest generators of wastewater because it uses a large amount of water, especially in the dyeing and finishing processes (Babuna 2011, Saratale *et al* 2011, Shah 2019). The World Bank has estimated that textile industry effluents comprise 17 to 20% of industrial wastewater generated worldwide (Kant 2012) and textiles are considered to be one of the most polluting industries due to effluent composition and the significant amount of wastewater that it produces. The operations of dyeing, desizing, mercerization, washing, and finishing processes are the main sources of textile wastewater (Lotito *et al* 2014, Bapat *et al* 2020). In particular, a range of dyes are used for the production of fabrics, such as direct, reactive, indigo, and dispersed dyes. Reactive azo dyes cover more than half of the dyes produced annually and are the most used dyes in the textile industry (Popli and Patel 2015). The major pollutants in colored effluents of textile industry are synthetic dyes, dispersants, surfactants, heavy metals, and auxiliary chemicals, generating a complex wastewater composition with high chemical oxygen demand (COD), suspended solids, total dissolved solids (TDS), toxicants, and recalcitrant organics (Lotito *et al* 2014, Suryawan *et al* 2018). Typical COD, biochemical oxygen demand (BOD), total suspended solids (TSS), and total Kjeldahl nitrogen (TKN) concentrations of textile industry wastewater are observed in the range of 150 to 12 000 mg l^{-1} COD, 80 to 6000 mg l^{-1} BOD_5, 2900 to 3100 mg l^{-1} TSS, and 70 to 80 mg l^{-1} N, respectively (Oller *et al* 2011). The low BOD/COD ratio shows the high fraction of nonbiodegradable organics in textile industry wastewater. Textile effluents adversely affect photosynthesis in aquatic life because colored textile wastewater reduces the light penetration and oxygen consumption (Holkar *et al* 2016). Moreover, the synthetic dyes that are used in the textile industry possess mostly toxic, mutagenic, and carcinogenic properties (Aquino *et al* 2014, Kaykhai *et al* 2018). Thus, the complex matrix and intense color of textile wastewater is an environmental challenge and necessitates sustainable wastewater treatment technologies.

Physical, chemical, biological, and integrated or hybrid treatment technologies have been developed and applied to textile wastewaters (Vandevivere *et al* 1998, Holkar *et al* 2016, Shah Maulin 2020). The physicochemical technologies used to treat textile effluents include flocculation–coagulation, membrane filtration, chemical precipitation, ozonation, adsorption, ion exchange, chemical or advanced oxidation, and flotation (Bapat *et al* 2020). Studies of color removal from textile wastewater are particularly important because colored effluent is a serious environmental and aesthetic problem. Synthetic textile dyes may have special functional groups or isomers with various chemical composition and are mostly persistent to degradation (Shah 2019, Singh *et al* 2019, Shah 2021a). Consequently, the differences in their chemical structure greatly affect the efficiency of color removal in various treatment technologies.

Biological treatment systems are generally preferable to physicochemical processes because they are cost-competitive, produce less sludge, and are referred to as an eco-friendly alternative (Hayat *et al* 2015, Singh *et al* 2019). Biological technologies applied to textile wastewaters include single or combined aerobic, anoxic, and anaerobic treatment processes and enzymatic degradation (Shah 2019). Various configurations have been operated effectively to treat textile industry wastewater, such as upstream anaerobic sludge blanket (UASB), expanded granular sludge bed (EGSB), anaerobic/aerobic sequential batch reactor (SBR), and conventional activated sludge (CAS) process (Popli and Patel 2015, Shah 2021a). Among them, CAS systems have been commonly used to treat textile effluents. However, they do not sustain efficient treatment because most of the dyes, surfactants, and auxiliary chemicals used in the process are resistant to biological treatment (Lapertot *et al* 2006, Grekova-Vasileva and Topalova 2009). Anaerobic-aerobic biological treatment is generally implemented in large-scale textile wastewater treatment plants (Wang *et al* 2011). Biogas generation through anaerobic treatment is only applicable for textile effluents with high COD composition, such as desizing process wastewater (Rongrong *et al* 2011). Numerous studies have reported the removal of textile dyes, specifically azo dyes, with combined anaerobic-aerobic processes using mixed microbial culture (Kusvuran *et al* 2004, Shah 2019). However, the formation of hazardous aromatic amines after anaerobic degradation of azo dyes has been reported (Popli and Patel 2015). The microbial degradation of dyes depends on the adaptability of mixed microbial consortia or single microbial strains and enzyme activity such as lignin peroxidase, laccase, tyrosinase, and NADH-DCIP reductase (Solís *et al* 2012). Bacteria, fungi, and algae species have been reported to degrade various types of textile dyes. Recently, the isolation and cultivation of single strains for the degradation of certain recalcitrant dyes has been gaining attention (Holkar *et al* 2016). However, the contamination problem in full-scale applications should be considered for pure cultures (Popli and Patel 2015). Mixed microbial cultures have the advantage of providing process stability for full-scale applications and reducing the toxicity effect compared to pure cultures (Popli and Patel 2015). Overall, the variable composition of textile wastewater; the operational conditions; the sulfur derivatives, salts, and toxicants present in textile effluents; and the composition of the dyes greatly affect the efficiency of the microbial decolorization (Singh *et al* 2019).

Coagulation and flocculation have been reported to remove disperse dyes efficiently but are inadequate to remove reactive dyes, while the high amount of sludge production limits their use (Liang *et al* 2014). Membrane filtration has been used to recycle reactive dyes and to reuse water with an effective removal of COD. However, high investment cost, membrane fouling, and the generation of insoluble metabolites limits its large-scale application (Saratale *et al* 2011, Koyuncu and Güney 2013). Adsorption has also gained attention due to its high decolorization efficiency (Singh *et al* 2019). Activated carbon has been used for decolorization but high adsorbent cost, regeneration problems, and difficulties in handling produced sludge are the major drawbacks (Galán *et al* 2013, Siddique *et al* 2017).

Oxidation processes are categorized as chemical oxidation, in which ozone or hydrogen peroxide agents are used, and advanced oxidation, which uses cavitation, Fenton's chemistry, and photocatalytic oxidation (Gogate and Pandit 2004). The hydroxyl radicals produced in advanced oxidation processes oxidize most of the dyes and complex organic and inorganic chemicals detected in textile wastewater (Holkar *et al* 2016). Fenton's chemistry has been reported to be an effective process to remove both soluble and insoluble dyes, together with complex organic pollutants (Siddique *et al* 2017). Fenton oxidation proceeds through the reaction between Fe^{3+} ions and hydrogen peroxide, and produces iron containing sludge that requires further treatment (Shah 2019, Mirza *et al* 2020). Cavitation is another promising technology for the treatment of textile wastewater (Gogate and Bhosale 2013). High removal rates of COD and color with lower process time are the major advantages of cavitation technology; however, high operational cost restricts its large-scale applications (Holkar *et al* 2016). Cavitation has successfully been applied as a pretreatment process integrated with various chemical oxidants for the decolorization of textile effluents (Gogate and Bhosale 2013). Photocatalytic oxidation has been also used as pretreatment for efficient removal of dyes from textile effluents (Shindhal *et al* 2020).

Ozone and hydrogen peroxide have high oxidation potential due to the formation of strong hydroxyl radicals. Decolorization is effectively achieved because they can break the conjugated double bonds and functional groups of dye chromophores in textile effluents (Holkar *et al* 2016). Ozone has high reactivity with a wide range of dyes and is the most widely used oxidant to remove color from textile industry wastewater (Arslan-Alaton *et al* 2002, Koch *et al* 2002). The ozonation process does not increase the wastewater amount or sludge production due to use of ozone in gaseous form (Holkar *et al* 2016, Miralles-Cuevas *et al* 2017). However, ozonation may produce toxic by-products from textile wastewater (Miralles-Cuevas *et al* 2017). Moreover, the refractory materials present in textile effluents may react with ozone and pretreatment before ozonation is suggested (Shah 2019).

Overall, biological treatment processes have been widely used to treat textile wastewaters but they do not sustain adequate removal of color and recalcitrant pollutants (Oller *et al* 2011, Mirza *et al* 2020). In addition, physicochemical technologies have various limitations (Singh *et al* 2019). The reported studies and results of full-scale applications clearly show that single treatment approaches do not provide environmentally sustainable treatment for textile wastewaters (Bapat *et al* 2020). Hence, integrated treatment technologies should be evaluated to provide adequate removal of color and recalcitrant organics together with conventional pollutants, as well as to minimize operational costs. A combination of chemical and advanced oxidation processes and biological treatment technologies has mostly been studied (Lotito *et al* 2014).

6.2 Effective COD and color removal with integrated ozonation and biological treatment approaches in textile wastewater

In this section, the results of the current literature of effective COD and color removal with integrated ozonation and biological treatment approaches in textile wastewater will be discussed. Ozonation is an important alternative to remove recalcitrant dyes and integration with biological treatment will reduce the operational cost. Partial oxidation to biodegradable intermediates through ozonation will reduce the ozone dose and biological intermediates will be further treated through biological processes (Fu *et al* 2011, Blanco *et al* 2014, Lotito *et al* 2014). One of the major drawbacks of ozonation is the possible formation of toxic by-products. Meanwhile, integration with biological treatment can decrease the toxicity of the effluent. Table 6.1 summarizes the COD and color removal results of integrated ozonation and biological treatment studies applied to textile wastewaters. Souza *et al* (2010) applied ozonation as a pretreatment to biological treatment on synthetic wastewater containing reactive dye of Remazol Black B and achieved 96% color removal. The results of acute toxicity tests with *Daphnia magna* revealed that ozonation generated toxic by-products; however, subsequent biological treatment eliminated the toxicity of the pretreated effluent (Souza *et al* 2010). The integration of chemical oxidation with various types of biological treatment technologies has also been investigated. Dias *et al* (2019) studied the combination of ozonation as pretreatment and moving-bed biofilm reactor (MBBR) as post-treatment for the removal of reactive azo dye 239 (RR 239), which has a complex structure and high molecular weight. In total, 95% of color removal was reported at 40 and 20 mg l^{-1} of ozone dose within 4 and 12 min, respectively. A rapid increase in toxicity to *Aliivibrio fischeri* was observed within the first minutes of ozonation and neutralize after 4 min. Aniline, catechol, phenol, phthalic acid, and a chlorinated amino triazine were detected in the synthetic wastewater after ozonation, whereas aniline and catechol were not detected after MBBR. However, phenol, phthalic acid, and a chlorinated amino triazine were not metabolized in the MBBR system. In addition, deterioration on the nitrification process was observed and the authors claimed that ozonation by-products might cause inhibition to nitrification (Dias *et al* 2019). The influence of anaerobic MBBR and ozonation followed by aerobic MBBR on the degradation of azo dye Reactive Orange 16 (RO16) was compared (Castro *et al* 2020). Both reactors were fed with an auxiliary carbon source, i.e., glucose. High color removal efficiencies could be observed at very low influent dye concentrations (5 mg l^{-1}) and a longer hydraulic retention time (HRT: 12 h) was required for the anaerobic degradation of RO16. In the case of ozone pretreatment, 97% of color removal was achieved within 20 min at the highest influent dye concentration (500 mg l^{-1}).

The effect of integrated ozonation-biological treatment on full-scale textile effluents has also been reported. The effect of ozonation as pretreatment prior sequencing batch biofilter granular sludge reactor (SBBGR) was studied with yarn dyeing effluents (Lotito *et al* 2014). The COD removal efficiencies of single biological treatment were in the range 63.6%–78.5% with effluent residual concentrations of 142–250 mg l^{-1}. Low removal efficiency was attributed to the recalcitrant pollutants of textile effluents. Integration of ozonation with biological treatment explicitly improved the removal efficiency. COD, total nitrogen (TN) and surfactants removal were increased to 89.8%, 88.2% and 90.7% with a 110 mg ozone dose/l. Color removal was reported as decolorization percentage through the measurement of the absorbance of the samples at three wavelengths of 426, 558, and 660 nm and 99% of color removal was achieved (Lotito *et al* 2014). Numerous studies have investigated the influence of

Table 6.1. COD and color removal with integrated ozonation and biological treatment approaches in textile wastewater.

No.	Wastewater	Combined system	Results	References
1	Real textile wastewater	Ozonised membrane bioreactor with photocatalytic process	• 94% color removal • 93% COD removal	Sathya *et al* (2019)
2	Real textile wastewater (washing with bleaching, dyeing, and first rinse after dyeing)	Biodegredation (SBR) followed by ozonation and post-biodegredation (SBR)	• >83% color removal • 71.8% COD removal	Paździor *et al* (2017)
3	Real textile wastewater (washing with bleaching, dyeing, and first rinse after dyeing)	Biodegredation (Horizontal Continuous Flow Bioreactor; HCFB) followed by ozonation and post-biodegredation	• Around 90% color removal • 74.6% COD removal	Paździor *et al* (2017)
4	Artificial textile wastewater (Remazol Red)	Anaerobic biofilm reactor followed by ozonation	• 99% color removal • 85%–90% COD removal	Punzi *et al* (2015)
5	Real textile wastewater	Anaerobic biofilm reactor followed by ozonation	• Around 65% color removal • 70% COD removal	Punzi *et al* (2015)
6	Yarn dyeing wastewater	Sequencing batch biofilter granular sludge reactor connected to an ozone column	• >85% color removal • 78.5%–84.3% COD removal	Lotito *et al* (2014)
7	Textile effluents from cotton mills	Integrated ozone biological aerated filters (BAFs)	• 87.5% color removal • 62.5% COD removal	He *et al* (2013)
8	Printing wastewater	SBBGR connected to an ozone column	• Almost complete color removal • >60% COD removal	Lotito *et al* (2012)

9	Printing and dyeing combined wastewater	Anoxic filter bed and biological wriggle bed-ozone biological aerated filter	• Around 80% color removal • 74.1%–84.1% COD removal	Fu *et al* (2011)
10	Real textile wastewater	Ozonation followed by BAF	• 87.5% color removal • >60% COD removal	Wang *et al* (2008)
11	Dyebath effluent	Anaerobic, aerobic, ozonation, aerobic	• >95% color removal • 75% DOC removal	Libra and Sosath (2003)

different continuous flow configurations of biological and chemical oxidation processes with full-scale textile effluents. Libra and Sosath (2003) set up two different configurations: (i) in the first configuration (two-stage), pretreatment with ozonation followed by aerobic degradation was applied to segregated dyebath wastewater obtained from textile finishing industry; (ii) in the second configuration (four-stage), dyebath effluent was mixed with yeast extract and acetic acid, and then fed to a sequence of anaerobic and aerobic reactors. The effluent was ozonated and finally conveyed to aerobic reactor. The dyebath effluent contained a reactive dye of CI Reactive Black 5. An average of 95% of color removal was achieved with a range of 0.1–0.6 g m^{-3} s^{-1} ozone feed rates in the two-stage configuration. In the four-stage configuration, color removal was observed to be 74% at the effluent of anaerobic-aerobic biological reactors and increased to 95% after ozonation with an ozone feed rates of 0.25 g m^{-3} s^{-1}. The final aerobic stage improved the color removal up to 99%. An average of 80% of dissolved organic carbon (DOC) removal was achieved at ozone feed rates of 0.4 g m^{-3} s^{-1} within two-stage configuration, though DOC removal was decreased to 75% in four-stage configuration with 0.25 g m^{-3} s^{-1} ozone feed rate (Libra and Sosath 2003). The integration of membrane bioreactors with ozonation and photocatalyst was evaluated for real textile effluent (Sathya *et al* 2019). Maximum color removal of 93% was achieved when ozone is combined with UV(16W) at a dosage of 5 g h^{-1}. The biodegradability index was increased from 0.25 to 0.4 with COD removal efficiency of 94%.

The aim of the integration of chemical oxidation processes as pretreatment is partial oxidation of the persistent compounds to biodegradable reaction intermediates, which will lower the operational costs (Oller *et al* 2011). However, formation of toxic by-products, specifically in the case of azo dyes, has initiated the integration of ozonation after biological treatment as post-treatment (Punzi *et al* 2015). The use of an anaerobic biofilm reactor and subsequent ozonation as post-treatment has been studied to treat synthetic and real textile effluents containing azo dye (Punzi *et al* 2015).

Acute toxicity to test organism of *Artemia salina* and mutagenicity were determined, as well as COD and color removal efficiency to evaluate the performance of post-ozonation. The results of the integrated anaerobic-ozonation process were reported as 99% of color removal and a range of 85%–90% of COD removal for the synthetic wastewater which contains 100–1000 mg l^{-1} azo dye Remazol Red. COD removal efficiency was lower (70%) for real textile effluent due to the radical scavenging organic molecules present in the textile wastewater. The integrated treatment efficiently reduced the toxicity; however, mutagenicity of the effluent was increased after biological treatment and ozonation. The authors suggested that azo dyes and their metabolites, specifically alkylating agents formed during ozonation, are mutagenic. It was concluded that toxicity and mutagenicity with different test organisms should be performed to optimize the ozonation process (Punzi *et al* 2015).

6.3 Integrated anaerobic-ozonation-aerobic process: a case study

The textile industry was selected as a case study for applying different treatment approaches to investigate COD and color removal. The factory that was chosen has a fabric printing capacity of 30,000 meters with 40 tons of fabric and 10 tons of yarn dyeing per day. Sodium chloride (NaCl) and urea are used as fixing agents. Dyeing is performed using reactive and indigo dyes in the factory. The wastewater produced in the facility is treated biologically in the wastewater treatment plant with an average daily flow capacity of 4000 m^3/day. The raw wastewater containing dyestuffs was subjected to two different treatment configurations, as shown in figure 6.1. The composite raw wastewater samples were taken from the fabric dyeing and printing industry located in Lüleburgaz, Turkey. The temperature of the dyeing effluent in the existing wastewater treatment plant of the facility is around 35 °C.

Laboratory experiments were conducted at the Environmental Biotechnology Group Laboratory of Istanbul Technical University. The treatability study was initiated with 1 l of semi-continuously fed anaerobic reactor with an effective volume as 600 ml setup, which was operated for 3 months. The pH in the reactor was adjusted to 7.2 adding 1 M HCl and 1 M NaOH solutions. A laboratory-scale reactor was operated with an organic load of 0.07 kg m^{-3}.day. The biomass seed used in the anaerobic reactor was taken from a facility with an anaerobic treatment plant treating textile wastewater. Suspended solids (SS) and volatile suspended solids

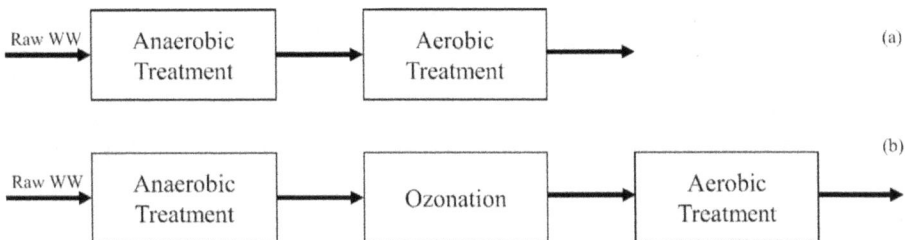

Figure 6.1. Treatment configurations applied to raw textile wastewaters.

(VSS) concentrations of seed were measured as 60 and 45 g l^{-1}. During the anaerobic phase, nitrogen gas was added to the reactor to ensure that there was no oxygen intrusion. The system was operated in mesophilic conditions at 35±1 °C.

In aerobic acclimation experiments, the activated sludge was obtained from the aeration basin of the wastewater treatment plant. The activated sludge plant was operated at 3000 mg SS/L with a ratio of 80% VSS/SS. The biomass cultivation was achieved in fill and draw reactors for a sludge age of 20 days. During the acclimation period, the reactors were operated at a HRT of 24 h. Reactors were fed with anaerobic effluent and ozonated anaerobic effluents. The working volumes of the reactors were adjusted to 1 l by discharging an appropriate volume of aliquots at the end of the aeration phase. The minimum dissolved oxygen concentration in the bulk was maintained above 2.0 mg O_2 l^{-1} by air compressor. The process temperature of the systems was maintained around 35±1 °C. The total and soluble COD, Color, pH, SS, and VSS measurements were carried out to observe steady-state conditions of the reactors.

Ozonation of the anaerobic effluent samples was performed in a 1000-mL capacity borosilicate glass column in semi-batch mode. Ozone was produced by a Sander S-1000 model corona discharge ozonator (Erwin Sander, Germany) (capacity: 1000 $mgO_3 \cdot h^{-1}$) fed with high-purity oxygen (99.5%). Ozonation experiments were conducted for 0–30 min range at 13.6 mg min^{-1} of ozone dose. Dissolved ozone was measured via indigo colorimetry according to standard methods (APHA/AWWA/WEF 2012). Details of the experimental setup are given elsewhere (Arslan-Alaton and Caglayan 2005). The COD, SS, color, TKN, ammonium nitrogen, and pH parameters were analyzed according to standard methods (2012). The effluent wastewaters were subjected to a toxicity test. The tests were performed in raw, anaerobic, aerobic, and ozonated anaerobic effluents with the photobacteria *Vibrio fischeri* in accordance with the ISO 11348–3 test protocol (2008). The bacterial assay was carried out with the commercially available BioTox test kit (*V. fischeri* code 1243–500; Aboatox Oy, Finland). The toxicity was measured as the percent of relative inhibition of the photobacteria with respect to the light emitted under test conditions in the absence of the sample. All bioassays were organized as triplicate experiments.

6.3.1 Results and discussion

6.3.1.1 Full scale treatment performance

The factory has a wastewater treatment plant configured as a CAS system with surface aerators followed by final clarifiers. The aerobic system was configured as four-reactor in series with 1250 m^3 of volume for each reactor. The CAS system has 20 days of solid retention time and 1.2 days of HRT. The average influent COD, TKN, and total phosphorus (TP) concentrations were 700 mg l^{-1}, 40 mgN l^{-1} and 6.2 mg P l^{-1}, respectively. The effluent wastewater had COD, TKN, and TP average concentrations of 220 mg l^{-1}, 28 mgN l^{-1}, and 2.8 mgP l^{-1}, respectively. The effluent TSS concentration was resported as 35 mg l^{-1}. The CAS plant achieves 71% of COD and 30% of TKN removal. The effluent NH_4-N concentration was around 13 mgN l^{-1}, where nearly half of the nitrogen is still in organic form. The color parameters measured at the influent and effluent of the plant were around 1000 in Pt/Co unit. The average plant effluent shows that the system was able to meet discharge limits, except for the color parameter, according to the local regulations.

6.3.1.2 Laboratory-scale treatment studies

6.3.1.2.1 Sequential anaerobic-aerobic system

The biodegradation and decolorization of dyestuffs in textile industry wastewaters can be carried out by applying anaerobic bioprocesses (Van der Zee and Villaverde 2005, Işık and Sponza 2008). The degradation of dyestuffs under anaerobic conditions varies depending on the nature of the dyes. In the case of reactive azo dyes, adsorption to biomass and azo-bond reduction occur during anaerobic treatment (O'Neill *et al* 2000). Based on the studies in the literature, the aerobic reactor is not sufficient for color removal from textile wastewater because most textile dyes are resistant to aerobic biodegradation (Bapat *et al* 2020). Anaerobic treatment is required to break the azo bonds as a pretreatment. However, toxic aromatic amines can be generated during anaerobic degradation of azo dyes (Popli and Patel 2015), which can be efficiently removed through aerobic processes (O'Neill *et al* 2000, Popli and Patel 2015). In this respect, the use of a combined sequential anaerobic and aerobic system is suggested for textile effluents, especially those containing azo dyes.

In this study, the improvement of the existing wastewater treatment plant in terms of color removal by including anaerobic phase prior to aerobic system was investigated. In this context, the raw textile wastewater was subjected to anaerobic pretreatment. After an acclimation period, the aerobic system was fed with anaerated effluents. The results are summarized in table 6.2.

In the anaerobic-aerobic sequential system, an overall of 60% of COD (47% in anaerobic phase) and 43% of color (32% in anaerobic phase) removals were achieved. These results are in accordance with the reported studies in the literature. O'Neill *et al* (2000) applied combined anaerobic-aerobic treatment to stimulate textile effluent containing azo dye and achieved an overall 57% of COD removal and 68% of color removal. The integration of an anaerobic packed column reactor and aerobic activated sludge reactor was studied for real textile wastewater, and 85% of color removal and 90% of COD removal were achieved (Kapdan and Alparslan 2004). It has been emphasized that the sequential system is more advantageous in achieving better performance in color removal (Kapdan and Alparslan 2004). However, biological processes may not sustain adequate removal of color and recalcitrant pollutants in textile effluents, and the integration of physicochemical process could be considered to improve decolorization and removal of recalcitrant organics.

6.3.1.2.2 Integration of the ozonation system

Additional studies were carried out on the anaerobic effluents at ozone (O_3) dose of 13.6 mg min^{-1} with a retention time up to 15 min (figure 6.2) to improve COD and color removal. During the ozonation batch tests, aliquots were taken from the reactor at certain time intervals; color and COD analyses were determined as a function of time. As shown in figure 6.2, the efficiency in color and COD removal nearly reached to the steady-state level within first 5 min. Nearly 60%–80% of color removal efficiencies were achieved depending upon different wavelenghts (figure 6.2). Therefore, the optimum time was determined to be 5 min in batch ozonation experiments carried out on anaerobic effluents. It should be noted that 620 nm yielded the maximum removal efficiency, while 436 nm remained at 60% removal after ozonation.

6.3.1.2.3 The anaerobic-ozonation-aerobic system

The anaerobic effluent, which was ozonated for 5 min at ozone dose of 13.6 mg min^{-1}, was then fed to the aerobic reactor. Together with anaerobic treatment following ozonation, COD and color removals were found to be 69%, 62% (Pt-Co), 56% (436 nm), 76% (525 nm), and 86% (620 nm), respectively. The performance of the integrated system is given in table 6.2. In this case study, the anaerobic-ozonation-aerobic system in series yielded 79% of COD and 81%–86% of color removal, respectively. In this way, it was possible to meet discharge regulation securely (WPCR 2004). The various configurations set for integration of ozonation with biological systems were reported in the literature and the results of these studies are summarized in table 6.1. Exactly the same configuration applied in this study cannot be found in the literature; however, integration of biological treatment with ozonation mostly yielded 60 to 95% of color removal and 60 to 90% of COD removal (table 6.2). The type of the textile dyes and variations in the configurations of the integrated systems and operational conditions significantly affect the treatment efficiencies.

6.3.1.3 Ecotoxicological tests

The acute inhibition tests were conducted on (a) raw wastewater, (b) anaerobic effluent, (c) ozonated anaerobic effluent, and (d) anaerobic-O_3-aerobic effluents using the standard *V. fischerie* test. The toxicity levels with respect to different stages of treatment including raw influent are illustrated in figure 6.3(a). The raw influent exhibited 17% toxicity level. Applying only anaerobic and anaerobic-O_3 treatment increased the toxicity level up to 26% and 76%, respectively. However, the effluent became non-toxic after a final aerobic stage as post-treatment. The toxicity on aquatic life is of great importance in addition to meeting the discharge parameters. Therefore, the arrangement of treatment schemes should be well established for industrial effluents considering the toxicity level to the environment.

Batch respirometric tests were applied on raw wastewater and anaerobic effluent by adjusting the initial F/M ratio with acclimated biomass. The results of respirometry are illustrated in figure 6.3(b). It should be noted that the influent wastewater sampled from the plant for respirometry had relatively diluted characteristics compared to average plant influent. The endogenous decay model was used to calibrate the experiments with the aid of the Aquasim program (Reichert 1998).

Table 6.2. Performance of anaerobic-aerobic and anaerobic-ozonation-aerobic systems.

Parameter	Unit	Raw	Anaerobic-aerobic		Anaerobic-ozonation-aerobic			Discharge limit (WPCR 2004)
			Anaerobic effluent	Aerobic effluent	Anaerobic effluent	Ozonated effluent	Aerobic effluent	
COD	mg l^{-1}	712±52	373±119	282±25	373±119	222±19.8	149±10	200
Sol. COD	mg l^{-1}	604±88	301±102	241±15	301±102	182±12.7	94±8	—
Color	Pt-Co	703±258	481±180	400±55	481±180	270±11.3	120±7	260
	436 nm	23.2 ±10.6	17.9±7.3	17.0±8.1	17.9±7.3	10.2±0.85	4.5±0.85	—
	525 nm	17.4±3.5	10.64±5	10.5±5.2	10.64±5	4.1±0.42	2.1±0.45	—
	620 nm	21.3±7.8	13.8±4.7	12.5±3.8	13.8±4.7	2.9±0.71	2.9±0.82	—

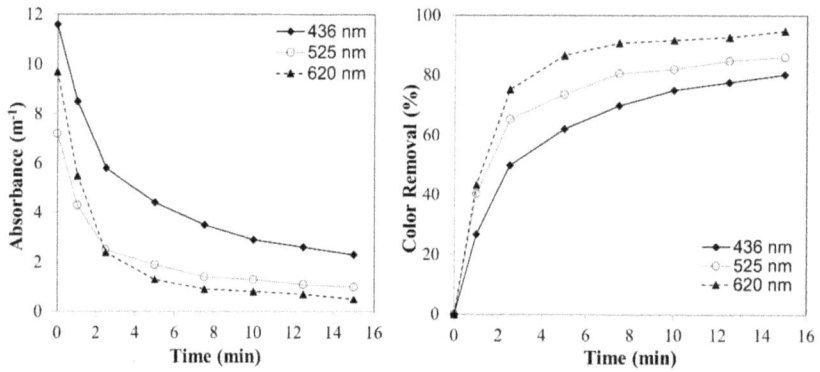

Figure 6.2. Absorbance values and color removal efficiency at a dose of 13.6 mg min^{-1} applied ozone.

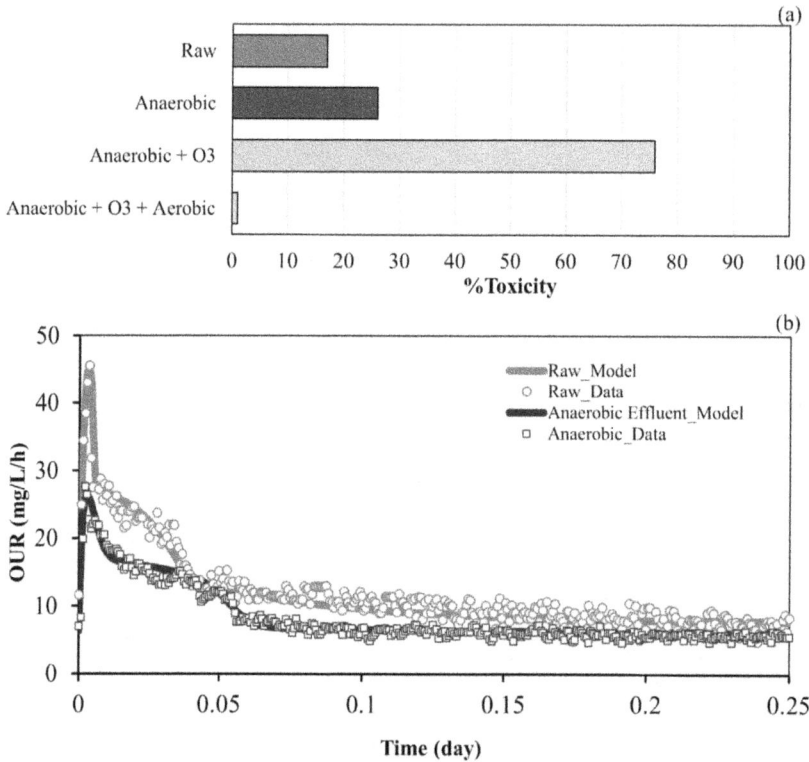

Figure 6.3. Toxicity levels (a) and respirometric results (b).

The influent COD fractionation, as well as growth and hydrolysis kinetics were identified using automatic parameter estimation (Insel *et al* 2003, Orhon and Okutman 2003, Katipoglu-Yazan *et al* 2012). Table 6.3 shows the influent COD fractions and kinetic parameters extracted from the respirograms. The modeling results showed that the influent total COD contains 6% readily biodegradable and

Table 6.3. The kinetic parameters estimated by respirometric modeling

Parameter	Unit	Raw WW	Anaerobic effluent	Orhon *et al* (2009)
Kinetic parameters				
Maximum heterotrophic growth rate, $\hat{\mu}_H$	day^{-1}	5	5	3.2–6.0
Half-saturation constant for growth, K_S	mg COD/L	5	5	1.0–25
Maximum hydrolysis rate, k_h	day^{-1}	2.5	2.5	0.8–3.8
Hydrolysis half-saturation constant, K_X	g COD/g COD	0.06	0.06	0.02–0.70
COD fractions				
Total COD, C_{T1}		664	500	
Biodegradable COD, C_{S1}		559	414	
Readily biodegradable COD, S_{S1}		41	24	
Slowly hydrolysable COD, S_{H1}		518	390	
Inert COD, C_{I1}		105	86	

Default parameters: Y_H = 0.62 gcellCOD/gCOD, f_E = 0.20; b_H = 0.24 d^{-1}.

78% slowly biodegradable COD of the total COD. Meanwhile, the degradation kinetics were close to the values reported for municipal wastewaters (Henze *et al* 2008, Gunes *et al* 2019). In addition, the estimated kinetic parameters for raw textile wastewater stayed in their suggested ranges cited according to the literature (Orhon *et al* 2009). Applying anaerobic treatment provided 40% and 23% of readily biodegradable COD (S_{S1}) and slowly biodegradable COD (S_{H1}) removal, respectively. However, the impact of anaerobic treatment on raw textile wastewater was found to be ineffective considering the aerobic degradation kinetics of raw textile wastewater (table 6.3).

6.4 Conclusion

The effluent of a large textile plant applying yarn dying, knitting, and printing operations was studied by means of anaerobic, aerobic, and intermediate ozonation treatment for color and organic carbon removal. Sequential anaerobic-aerobic process applied on raw wastewater was found to be insufficient to meet stringent discharge limits with respect to color and organic carbon removal. However, intermediate oxidation with ozone between anaerobic and aerobic processes provided considerable performance amendment for the achievement of color and COD removals.

COD fractionation and kinetic evaluation were experimentally obtained for (a) raw textile wastewater and (b) effluent of anaerobic treatment by means of batch aerobic respirometry. Respirometric modeling revealed that only 25% biodegradable COD removal could be obtained after anaerobic treatment. However, no considerable change in aerobic heterotrophic activity with respect to growth and hydrolysis kinetics was reported based on the parameter estimation applied on oxygen uptake rate profiles.

Environmental toxicity has been highlighted on effluents from different treatment stages by means of the standard *V. fischerie* test. The toxicity considerably increased after applying anaerobic and ozonation processes up to 26% and 76%, respectively. Finally, the treated effluent was found to have non-toxic characteristics after applying aerobic post-treatment. Thus, in addition to discharge limits, the selection of treatment alternatives for industrial effluents should be elaborated considering the toxicity impact on receiving bodies.

Acknowledgments

This study was conducted in the Environmental Biotechnology Group Laboratory of Istanbul Technical University, and supported by ITU BAP (Project No. 35045).

References

APHA/AWWA/WEF 2012 *Standard Methods for the Examination of Water and Wastewater* 22nd edn (Washington, DC: American Public Health Association)

Arslan-Alaton I, Balcioglu I A and Bahnemann D W 2002 Advanced oxidation of a reactive dyebath effluent: comparison of O3, H2O2/UV-C and TiO2/UV-A processes *Water Res.* **36** 1143–54

Arslan-Alaton I and Caglayan A E 2005 Ozonation of Procaine Penicillin G formulation effluent Part I: Process optimization and kinetics *Chemosphere* **59** 31–9

Aquino J M, Rocha-Filho R C, Ruotolo L A, Bocchi N and Biaggio S R 2014 Electrochemical degradation of a real textile wastewater using β-PbO$_2$ and DSA® anodes *Chem. Eng. J.* **251** 138–45

Babuna F G 2011 Wastewater reuse in textile industry *Security of Industrial Water Supply and Management* pp 131–9 (Dordrecht: Springer)

Bapat S, Jaspal D and Malviya A 2020 Integrated textile effluent treatment method *Water Environ. Res.* **93** 1060–76

Blanco J, Torrades F, Morón M, Brouta-Agnésa M and García-Montaño J 2014 Photo-Fenton and sequencing batch reactor coupled to photo-Fenton processes for textile wastewater reclamation: Feasibility of reuse in dyeing processes *Chem. Eng. J.* **240** 469–75

Castro F D, Bassin J P, Alves T L M, Sant'Anna G L and Dezotti M 2020 Reactive Orange 16 dye degradation in anaerobic and aerobic MBBR coupled with ozonation: addressing pathways and performance *Int. J. Environ. Sci. Technol.* **18** 1991–2010

Dias N C, Bassin J P, Sant'Anna Jr G L and Dezotti M 2019 Ozonation of the dye Reactive Red 239 and biodegradation of ozonation products in a moving-bed biofilm reactor: Revealing reaction products and degradation pathways *Int. Biodeterior. Biodegrad.* **144** 104742

Fu Z, Zhang Y and Wang X 2011 Textiles wastewater treatment using anoxic filter bed and biological wriggle bed-ozone biological aerated filter *Bioresour. Technol.* **102** 3748–53

Galán J, Rodríguez A, Gómez J M, Allen S J and Walker G M 2013 Reactive dye adsorption onto a novel mesoporous carbon *Chem. Eng. J.* **219** 62–8

Gogate P R and Bhosale G S 2013 Comparison of effectiveness of acoustic and hydrodynamic cavitation in combined treatment schemes for degradation of dye wastewaters *Chem. Eng. Process.* **71** 59–69

Gogate P R and Pandit A B 2004 A review of imperative technologies for wastewater treatment I: oxidation technologies at ambient conditions *Adv. Environ. Res.* **8** 501–51

Grekova-Vasileva M and Topalova Y 2009 Biological algorithms for textile wastewater management *Biotechnol. Biotechnol. Equip.* **23** 442–7

Gunes G, Hallac E, Ozgan M, Erturk A, Okutman Tas D, Cokgor E, Guven D, Takacs I, Erdincler A and Insel G 2019 Enhancement of nutrient removal performance of activated sludge with a novel hybrid biofilm process *Bioprocess. Biosyst. Eng.* **42** 379–90

Hayat H, Mahmood Q, Pervez A, Bhatti Z A and Baig S A 2015 Comparative decolorization of dyes in textile wastewater using biological and chemical treatment *Sep. Purif. Technol.* **154** 149–53

He Y, Wang X, Xu J, Yan J, Ge Q, Gu X and Jian L 2013 Application of integrated ozone biological aerated filters and membrane filtration in water reuse of textile effluents *Bioresour. Technol.* **133** 150–7

Henze M, van Loosdrecht M C, Ekama G A and Brdjanovic D (ed) 2008 *Biological Wastewater Treatment* (IWA Publishing)

Holkar C R, Jadhav A J, Pinjari D V, Mahamuni N M and Pandit A B 2016 A critical review on textile wastewater treatments: possible approaches *J. Environ. Manage.* **182** 351–66

Insel G, Orhon D and Vanrolleghem P A 2003 Identification and modelling of aerobic hydrolysis —application of optimal experimental design *J. Chem. Technol. Biotechnol.: Int. Res. Process, Environ. Clean Technol.* **78** 437–45

Işık M and Sponza D T 2008 Anaerobic/aerobic treatment of a simulated textile wastewater *Sep. Purif. Technol.* **60** 64–72

ISO 2008 *11348-3 Water Quality—Determination of the Inhibitory Effect of Water Samples on the Light Emission of Vibrio fischeri (LUMINESCENT BACTERIA TEST). Method Using Freeze-Dried Bacteria* (Geneva: ISO,)

Kant R 2012 Textile dyeing industry an environmental hazard *J. Natl. Sci.* **4** 22–6

Kapdan I K and Alparslan S 2004 Application of anaerobic–aerobic sequential treatment system to real textile wastewater for color and COD removal *Enzyme Microbial Technol.* **36** 273–9

Katipoglu-Yazan T, Cokgor E U, Insel G and Orhon D 2012 Is ammonification the rate limiting step for nitrification kinetics? *Bioresour. Technol.* **114** 117–25

Kaykhai M, Sasani M and Marghzari S 2018 Removal of dyes from environment by adsorption process *Chem. Mater. Eng.* **6** 31–5

Koch M, Yediler A, Lienert D, Insel G and Kettrup A 2002 Ozonation of hydrolyzed azo dye reactive yellow 84 (CI) *Chemosphere* **46** 109–13

Koyuncu I and Güney K 2013 Membrane-based treatment of textile industry wastewaters *Encycl. Membr. Sci. Technol.* 1–12

Kusvuran E, Gulnaz O, Irmak S, Atanur O M, Yavuz H I and Erbatur O 2004 Comparison of several advanced oxidation processes for the decolorization of Reactive Red 120 azo dye in aqueous solution *J. Hazard. Mater.* **109** 85–93

Lapertot M, Pulgarín C, Fernández-Ibáñez P, Maldonado M I, Pérez-Estrada L, Oller I, Gernjak W and Malato S 2006 Enhancing biodegradability of priority substances (pesticides) by solar photo-Fenton *Water Res.* **40** 1086–94

Liang C Z, Sun S P, Li F Y, Ong Y K and Chung T S 2014 Treatment of highly concentrated wastewater containing multiple synthetic dyes by a combined process of coagulation/flocculation and nanofiltration *J. Membr. Sci.* **469** 306–15

Libra J A and Sosath F 2003 Combination of biological and chemical processes for the treatment of textile wastewater containing reactive dyes *J. Chem. Technol. Biotechnol.: Int. Res. Process, Environ. Clean Technol.* **78** 1149–56

Lotito A M, De Sanctis M, Rossetti S, Lopez A and Di Iaconi C 2014 On-site treatment of textile yarn dyeing effluents using an integrated biological–chemical oxidation process *Int. J. Environ. Sci. Technol.* **11** 623–32

Lotito A M, Fratino U, Bergna G and Di Iaconi C 2012 Integrated biological and ozone treatment of printing textile wastewater *Chem. Eng. J.* **195** 261–9

Miralles-Cuevas S, Oller I, Agüera A, Llorca M, Pérez J S and Malato S 2017 Combination of nanofiltration and ozonation for the remediation of real municipal wastewater effluents: acute and chronic toxicity assessment *J. Hazard. Mater.* **323** 442–51

Mirza N R, Huanga R, Dub E, Pengb M, Panc Z, Dingc H, Shan G, Ling L and Xie Z 2020 A review of the textile wastewater treatment technologies with special focus on advanced oxidation processes (AOPs), membrane separation and integrated AOP-membrane processes *Desalin. Water Treat.* **206** 83–107

Orhon D, Babuna F G and Karahan O 2009 *Industrial Wastewater Treatment by Activated Sludge* (IWA Publishing)

Orhon D and Okutman D 2003 Respirometric assessment of residual organic matter for domestic sewage *Enzyme Microb. Technol.* **32** 560–6

Oller I, Malato S and Sánchez-Pérez J 2011 Combination of advanced oxidation processes and biological treatments for wastewater decontamination—a review *Sci. Total Environ.* **409** 4141–66

O'Neill C, Lopez A, Esteves S, Hawkes F R, Hawkes D L and Wilcox S 2000 Azo-dye degradation in an anaerobic-aerobic treatment system operating on simulated textile effluent *Appl. Microbiol. Biotechnol.* **53** 249–54

Paździor K, Wrębiak J, Klepacz-Smółka A, Gmurek M, Bilińska L, Kos L, Sójka-Ledakowicz J and Ledakowicz S 2017 Influence of ozonation and biodegradation on toxicity of industrial textile wastewater *J. Environ. Manage.* **195** 166–73

Popli S and Patel U D 2015 Destruction of azo dyes by anaerobic–aerobic sequential biological treatment: a review *International J. Environ. Sci. Technol.* **12** 405–20

Punzi M, Nilsson F, Anbalagan A, Svensson B M, Jönsson K, Mattiasson B and Jonstrup M 2015 Combined anaerobic–ozonation process for treatment of textile wastewater: removal of acute toxicity and mutagenicity *J. Hazard. Mater.* **292** 52–60

Reichert P 1998 AQUASIM 2.0-user manual Swiss Federal Institute for Environmental Science and Technology. Dubendorf, Switzerland.

Rongrong L, Xujie L, Qing T, Bo Y and Jihua C 2011 The performance evaluation of hybrid anaerobic baffled reactor for treatment of PVA-containing desizing wastewater *Desalination* **271** 287–94

Saratale R G, Saratale G D, Chang J S and Govindwar S P 2011 Bacterial decolorization and degradation of azo dyes: a review *J. Taiwan Inst. Chem. Eng.* **42** 138–57

Sathya U, Nithya M and Balasubramanian N 2019 Evaluation of advanced oxidation processes (AOPs) integrated membrane bioreactor (MBR) for the real textile wastewater treatment *J. Environ. Manage.* **246** 768–75

Shah M P 2019 Bioremediation of azo dye *Microbial Wastewater Treatment* pp 103–26 (Amsterdam: Elsevier)

Shah M P 2020 *Microbial Bioremediation and Biodegradation* (Berlin: Springer)

Shah M P 2021a *Removal of Refractory Pollutants from Wastewater Treatment Plants* (Boca Raton, FL: CRC Press)

Shindhal T, Rakholiya P, Varjani S, Pandey A, Ngo H H, Guo W, Ng H Y and Taherzadeh M J 2020 A critical review on advances in the practices and perspectives for the treatment of dye industry wastewater *Bioengineered* **12** 70–87

Siddique K, Rizwan M, Shahid M J, Ali S, Ahmad R and Rizvi H 2017 Textile wastewater treatment options: a critical review *Enhancing Cleanup of Environmental Pollutants Volume 2: Non-biological Approaches* (Springer) pp 183–207

Singh R P, Singh P K, Gupta R and Singh R L 2019 Treatment and recycling of wastewater from textile industry *Advances in Biological Treatment of Industrial Waste Water and their Recycling for a Sustainable Future* (Singapore: Springer) pp 225–66

Solís M, Solís A, Pérez H I, Manjarrez N and Flores M 2012 Microbial decolouration of azo dyes: a review *Process Biochem.* **47** 1723–48

Suryawan I W K, Helmy Q and Notodarmojo S 2018 Textile wastewater treatment: colour and COD removal of reactive black-5 by ozonation *IOP Conf. Ser.: Earth Environ. Sci.* **106** 012102

Souza S M D G U, Bonilla K A S and de Souza A A U 2010 Removal of COD and color from hydrolyzed textile azo dye by combined ozonation and biological treatment *J. Hazard. Mater.* **179** 35–42

Van der Zee F P and Villaverde S 2005 Combined anaerobic–aerobic treatment of azo dyes—a short review of bioreactor studies *Water Res.* **39** 1425–40

Vandevivere P C, Bianchi R and Verstraete W 1998 Treatment and reuse of wastewater from the textile wet-processing industry: Review of emerging technologies *J. Chem. Technol. Biotechnol.: Int. Res. Process, Environ. Clean Technol.* **72** 289–302

Wang X J, Chen S L, Gu X Y, Wang K Y and Qian Y Z 2008 Biological aerated filter treated textile washing wastewater for reuse after ozonation pre-treatment *Water Sci. Technol.* **58** 919–23

Wang Z, Xue M, Huang K and Liu Z 2011 Textile dyeing wastewater treatment *Adv. Treat. Text. Effl.* **5** 91–116

WPCR 2004 Water Pollution Control Regulation, Official Gazette, Issue date: 31.12.2004, Issue number: 25687.

Chapter 7

Mycoremediation of wastewater: sustainable approaches

Maitri Nandasana and Sougata Ghosh

In recent years, the rapid rate of industrialization and anthropogenic activities have caused a serious threat to the environment. Hazardous dyes, pesticides, oil spills, and heavy metals are released into the environment through industrial effluents, which affects human health and the ecosystem. The conventional physical and chemical methods for the removal of toxic pollutants are not always efficient and have several limitations, such as the formation of toxic by-products, high energy consumption, and involvement of an expensive set up. As an alternative, microbe-based bioremediation techniques are widely accepted due to several advantages, such as eco-friendly nature, easy operation, and cost-effectiveness. In view of this background, this chapter gives an elaborate account of fungi-mediated bioremediation and its associated mechanisms. Fungi such as *Agaricus bisporus*, *Aspergillus flavus*, *Aspergillus fumigatus*, *Aspergillus niger*, *Galactomyces geotrichum*, *Myrothecium roridum*, *Penicillium simplicissimum*, and others are able to degrade toxic dyes such as acid blue 43, acid orange II, acid red 97, acid yellow 7, bromophenol blue, Congo red, cotton blue, crystal violet, direct black, direct blue 71, direct violet, malachite green, methyl violet, methylene blue, reactive black 5, reactive blue 19, reactive red 24, and reactive turquoise blue. Fungi such as *A. fumigatus*, *Aspergillus oryzae*, *Aspergillus terreus*, *Cladosporium cladosporioides*, *Cladosporium tenuissimum*, *Fusarium begoniae*, *Fusarium oxysporum*, *Penicillium citrinum*, *Penicillium glomerata*, *Penicillium melanoconidium*, *Trichoderma harzianum*, and *Trichoderma longibrachiatum* can effectively degrade pesticides, including atrazine, chlorfenvinphos, diuron, imidacloprid, isoproturon, and malathion. Furthermore, *Aspergillus ustus*, *Candida pseudointermedia*, *Fusarium solani*, *Meyerozyma guilliermondii*, *Penicillium funiculosum*, *Phanerochaete chrysosporium*, *Purpureocillium lilacinum*, *Rhodotorula taiwanensis*, *Trematophoma* sp., and *Yamadazyma mexicana* can remove petroleum, such as crude oil, and associated hydrocarbons, such as pyrene, octane, naphthalene, and phenanthrene. The key enzymes responsible for the bioremediation by fungi include lignin peroxidase, manganese peroxidase, tyrosinase, triphenylmethane reductase, glucose oxidase, laccase, and carboxylesterase. Careful optimization of various process parameters such as density of the fungal inoculum, initial pollutant concentration, time, temperature, and pH can help in achieving maximum efficiency of the bioremediation potential. Hence, fungus mediated environmental clean-up seems to be a promising sustainable approach.

7.1 Introduction

Water bodies throughout the globe have been speedily depleted in the last few decades due to the extensive growth of industrialization, agriculture, and urbanization (Kumar et al 2016). A group of chemical compounds existing at trace-level (μg l^{-1}, ng l^{-1}) in the current environment are generally named micropollutants (MPs). Some of the common examples of these pollutants are metallic trace elements, dyes, polycyclic aromatic hydrocarbon (PAH), pesticides, cosmetics, industrial chemicals, pharmaceuticals, and personal and household care products, etc (Sathe et al 2020). Among these pollutants, industrially important synthetic dyes are potent carcinogens and mutagens. These hazardous dyes can penetrate through the skin at low concentration, causing irritation, digestion irregularities, kidney impairment, breathing problems, and even blindness (Sarma et al 2016). The recalcitrant nature of these dyes makes them nondegradable once they enter the body, and thus can cause severe metabolic disorders in plants and animals (Bankole et al 2018, Kumar et al 2019).

Pesticides are another group of highly hazardous pollutants that cannot be degraded easily. Pesticides are mainly classified as organochlorines, organophosphorus, carbamates, pyrethrins, and pyrethroids (Sadeghi et al 2019). These toxic chemicals can enter the food chain due to long term use and pass to humans through various crops and vegetables. Upon accumulation in the human body, pesticides can cause severe health hazards, including cancer, immune suppression, neurodegeneration, and hormonal defects (Cao et al 2013). Additionally, petroleum hydrocarbons (PHs) that are widely used in industries and for energy generation are potential environmental pollutants. The PHs can be broadly classified as aliphatic, aromatic, asphaltene-based, and resin-based hydrocarbons. The PHs, particularly PAHs, are immensely toxic to humans because they are carcinogens, mutagens, and teratogens (Liu et al 2017).

Various chemical and physical methods are conventionally employed to remove refractory pollutants, such as ozonation, adsorption by activated carbon and nanomaterials, advanced oxidation process (AOP), flocculation, electrocoagulation, fenton treatment, photocatalysis, and others. However, they are often expensive, labor intensive, time consuming, and inefficient. Hence, there is an urgent requirement to develop a new strategy or method to address water pollution due to the release of dyes, pesticides, and oils from industries. It is important to note that biological methods are environmentally friendly, inexpensive, biocompatible, and more effective when compared to chemical and physical methods. Biological methods generally use aerobic, anaerobic, membrane, and activated sludge bioreactors, in addition to constructed wetlands, biosorption, and enzyme treatment (Bind et al 2018, Goswami et al 2019).

This chapter describes the degradation of various recalcitrant pollutants, such as dyes, pesticides, and PAHs using fungi, which is referred as mycoremediation, which are listed in table 7.1. It also discusses the major mechanism behind the removal of pollutants.

Table 7.1. Removal of various refractory pollutants by fungi.

Type of pollutant	Fungal strain	Removal efficiency	References
Acid red 97 (AR97), crystal violet (CV)	*A. bisporus*	372.69 mg g^{-1} (AR97), 228.74 mg g^{-1} (CV)	Drumm et al (2021)
Congo red (CR)	*A. niger*	96%	Asses et al (2018)
Crystal violet (CV), methyl violet (MV), malachite green (MG), cotton blue (CB)	*P. simplicissimum*	CV (98.7%), CB (97.5%), MG (97.1%), MV (97.7%)	Chen et al (2019)
Crystal violet (CV)	*A. niger*	96.1%	Ali et al (2016)
Direct blue 71 (DB71), direct blue 86 (DB86), reactive blue 19 (RB19)	*A. flavus* (A5p1)	100% (DB71), 100% (DB86), 71.8% (RB19)	Ning et al (2018)
Malachite green (MG)	*M. roridum*	90%	Jasinska et al (2015)
Methylene blue (MB)	*G. geotrichum* KL20A	71.5%	Contreras et al (2019)
Methylene blue	*A. niger, A. flavus, A. fumigatus*	92%	Karaghool *et al* (2021)
Atrazine, diuron, isoproturon, chlorfenvinphos	*P. citrinum, A. fumigatus, A. terreus, T. harzianum*	—	Oliveira et al (2015)
Atrazine	*A. niger* AN 400	72 ± 2	Marinho et al (2017)
Diazinon	*A. niger* MK640786	82	Hamad (2020)
Imidacloprid	*T. longibrachiatum, A. oryzae*	85% (*A. oryzae*), 81% (*T. longibrachiatum*), 92% (consortium)	Gangola et al (2015)
Malathion	*F. oxysporum* JASA1	100%	Peter et al (2015)
Octane, pyrene	*R. taiwanensis* KKUY-0162, *M. guilliermondii* KKUY-0214, *C. pseudointermedia* KKUY-0192, *R. ingeniosa* KKUY-0170, *Y. mexicana* KKUY-0160, *M. guilliermondii* KKUY-0214	—	Hashem et al (2018)

(*Continued*)

7-3

Table 7.1. (*Continued*)

Type of pollutant	Fungal strain	Removal efficiency	References
Crude oil	*Penicillium* sp. RMA1 and RMA2	57% (RMA1), 55% (RMA2)	Al-Hawash et al (2018)
Crude oil	*P. chrysosporium*	58.1 ± 1.3%	Behnood et al (2014)
Crude oil, diesel, used engine oil	*A. ustus* and *P. lilacinum*	30.43% of crude oil, 21.27% of diesel, 16.00% of used engine oil (*A. ustus*) 44.55% of crude oil, 27.66% of diesel, 14.39% of used engine oil (*P. lilacinum*)	Benuguenab and Chibani (2021)
Crude oil	*A. niger, A. fumigatus, F. solani, P. funiculosum*	95% (*A. niger*), 75% (*A. fumigatus*), 55% (*F. solani*), 65% (*P. funiculosum*)	Al-Jawhari (2014)
Crude oil, naphthalene, phenanthrene, pyrene	*Aspergillus* sp. RFC-1	51.8% (crude oil), 84.6% (naphthalene), 51.3% (phenanthrene), 55.1% (pyrene)	Al-Hawash et al (2019)
Pyrene, anthracene, phenanthrene	*Trematophoma* sp. UTMC 5003	56% (pyrene), 87% (anthracene), 90% (phenantherene)	Moghimi et al (2017)

7.2 Dye

Various fungi are employed to remove highly toxic industrial dyes from the environment. Drumm *et al* (2021) used the *Agaricus bisporus* (ABR) for the removal of acid red 97 (AR97) and crystal violet (CV) dye. The fungal biomass was collected and washed to remove impurities, followed by 48 h of drying at 50 °C and pulverization into a gray powder with a 150 μm particle diameter. The biodegradation experiments were carried out using the different dosages of ABR (0.25, 0.5, 0.75, 1.0, 1.25 g l^{-1}) with 50 ml dye (100 mg l^{-1}) that was incubated for 2 h under shaking conditions (150 rpm) at 25 °C. Scanning electron microscope (SEM) images showed rough surface morphology of the ABR. X-ray diffraction (XRD) analysis determined the presence of a non-crystalline phase and unorganized structures, which aid in the adsorption of dyes on the fungal surface. Fourier transform infrared spectroscopy (FTIR) analysis revealed peaks at 3420 cm^{-1} associated with -OH, 2932 cm^{-1} representing the CH bond, 1640 and 1465 cm^{-1} for the amide as well as ether group, 1015 cm^{-1} for the C–OH and C–O– vibrations, and 710 and 498 cm^{-1} for the C–H and C–N–C bonds. The point of zero charge (pH$_{pzc}$) was 4.6 for the surface of the adsorbent. The optimum parameters for the adsorption of dye were 0.5 g l^{-1} ABR dosage at initial pH 2 for AR97 and pH 8 for CV. The kinetic study revealed the adsorption efficiency of 165.91 mg g^{-1} AR7 within 210 min, and the maximum CV adsorption of 259.57 mg g^{-1} was acquired within 125 min of treatment. The Elovich kinetic model was most suitable for the adsorbate/adsorbent systems. The statistical indicators evaluated in the case of AR97 were $R^2_{adj} \geqslant 0.975$ and MSE $\leqslant 24.6$ (mg g^{-1})^2, while for CV it was was $R^2_{adj} \geqslant 0.981$ and MSE $\leqslant 37.0$ (mg g^{-1})^2. An isotherm study showed the Langmuir model as the better-fitted model for the adsorption process and the highest efficiencies of adsorption for AR97 ($R^2_{adj} \geqslant 0.941$ and an MSE $\leqslant 523.45$ (mg g^{-1})^2) and CV ($R^2_{adj} \geqslant 0.984$ and an MSE $\leqslant 48.83$ (mg g^{-1})^2) were 372.69 and 228.74 mg g^{-1}, respectively. The thermodynamic studies confirmed the spontaneous nature of the adsorption process from the negative value of $\Delta G°$ and the endothermic mode of reaction was identified from $\Delta H°$ value of 9.53 kJ mol^{-1} (AB97) and 10.69 kJ mol^{-1} (CV). The magnitude of $\Delta H°$ suggested the role of dipole–dipole, ion-dipole, or hydrogen bonds in the adsorption of CV as well as AR97 on ABR.

Asses et al (2018) developed a new approach to study the decolorization and toxicity of Congo red (CR) dye using *Aspergillus niger*. The fungal isolates grown on the potato dextrose agar (PDA) at 28 ± 2 °C. The biomass was generated by inoculating three mycelial plugs (0.7 cm diameter) in synthetic nutrient broth medium (100 ml) comprising of 10 g l^{-1} glucose, 1 g l^{-1} yeast extracts, and 2 g l^{-1} peptone. The pH 6.0 ± 0.2 was maintained and was followed by 8 days of incubation at 30 °C. After incubation, the fungal biomass was subjected to homogenization, filtration, and washing using sterile distilled water. A 20 g l^{-1} of fresh fungal biomass was inoculated in 100 ml synthetic broth medium containing CR dye (200 mg l^{-1}) followed by 10 days of incubation at 150 rpm and 30 °C. After centrifugation for 15 min at 5000 rpm to separate the fungal mycelium, the pellet was washed two times. The CR biosorption was performed by adding 2 g autoclaved fresh fungal biomass (inactive biomass) to the mixture of 100 ml broth and 200 mg l^{-1} dye followed by incubation at 30 °C for 48 h at constant stirring (120 rpm). UV-visible spectroscopy showed a reduction in the intensity of the absorption peak at 495 nm, which suggested the structural transformation in chromophore region of the dye after treatment. The enhancement in the absorbance peak at 340 nm suggested the formation of a new aromatic compound or the partial destruction of the existing interactions among aromatic or polycyclic aromatic groups and chromophores, as evident from figure 7.1. FTIR analysis revealed the distinctive absorbance peaks for bending vibrations of O–H, C–H of –CH$_2$ group, N–H group of chitin, amide associated –CH$_3$, –SO$_3$, C–OH, and C–O at 3275, 2937, 1641, 1559, 1420, 1139, 1090, and 579 cm^{-1}, respectively, The different parameters played a crucial role in CR dye elimination. The enhancement in decolorization efficiency from 45% to 98% was noted on increasing the shaking speed from 0 to 150 rpm. The maximum lignin peroxidase (LiP) and manganese peroxidase (MnP) activities of about 54 and 452 U l^{-1} were obtained at a speed of 100 rpm. However, the increase in speed from 150 rpm to 200 rpm resulted in the reduction of MnP activity, while no change was noted in the LiP activity. The higher CR elimination (85%) and enhanced enzyme activities were observed at slight acidic conditions (pH 5–6). The reduction in the MnP activity was observed at a pH value higher than 7, which suggested its stability at a lower pH (acidic). The optimum temperature for dye decolorization was between 25 and 30 °C. About 96% CR dye removal with LiP (56 U l^{-1}) and MnP

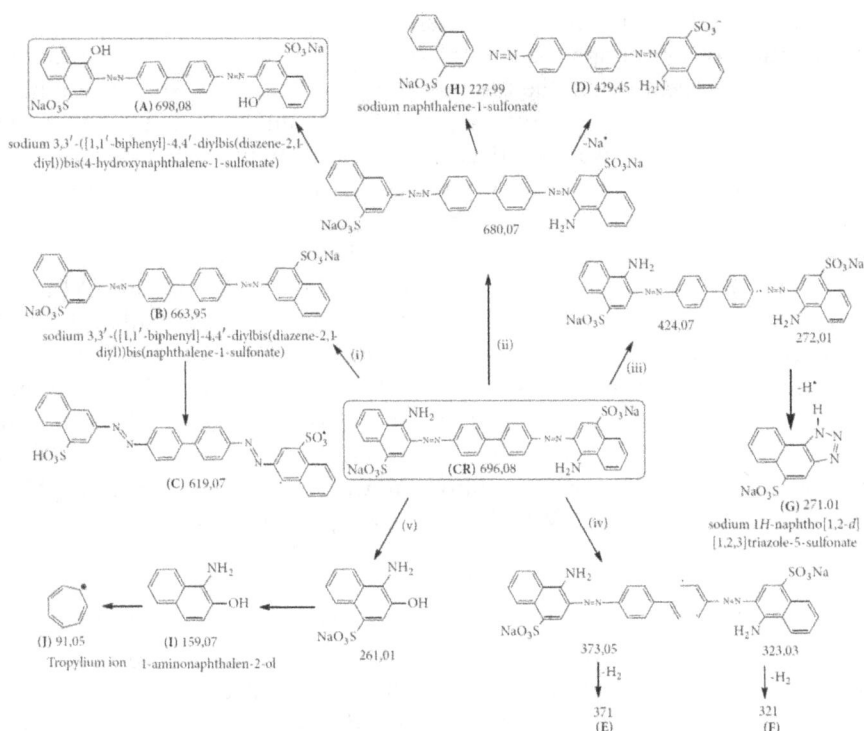

Figure 7.1. Proposed biodegradation pathway of CR using *A. niger*, with the identification of different degradation intermediates by LCMS/MS. Adapted from Asses et al (2018). CC BY 4.0.

$(470 \ U \ l^{-1})$ activity was noticed with 0.25 g. l^{-1} dye concentration. The maximum mycelium dry weight was around 6.4 g l^{-1}, which resulted in 97% decolorization within 6 days. However, the dead cells showed 27% elimination of dye after 6 days of treatment. The maximum LiP and MnP activities were $450.13 \pm 11 \ UI \ l^{-1}$ and $53.2 \pm 1.41 \ UI \ l^{-1}$ after the fifth and sixth day of treatment, respectively.

Chen et al (2019) investigated the reduction of toxic triphenylmethane (TPM) dyes such as CV, cotton blue (CB), malachite green (MG), and methyl violet (MV) employing *Penicillium simplicissimum*. The 7-day-old mycelial plugs were added into the 100 ml potato dextrose broth (PDB), followed by 5 days of incubation at $25 \pm 2 \ ^\circ C$ to develop the fungal biomass. The fungal biomass was washed, and homogenized, which was followed by filtration. The degradation experiment was carried out by mixing 100 ml of each dye (50 mg l^{-1}) solution with 4.0 ± 0.1 g biomass (for MG, MV, and CV solutions) and 8.0 ± 0.1 g biomass (in case of CB), which was followed by incubation under agitated conditions (150 rpm) at $25 \pm 2 \ ^\circ C$. The aliquots were collected at an interval of 2 h in the first 8 h, and afterwards at 24 h. The maximum decolorization efficiencies after 2 h of treatment for CB, MG, and CV, were 97.5%, 97.1%, and 98.7%, respectively. Treatment for 8 h resulted in 97.7% decolorization of MV. UV-visible spectroscopy showed major absorption peaks at 580 nm for untreated CV, 600 and 310–320 for CB, 580–590 nm for untreated MV, and 610–620 nm for MG, which disappeared completely after fungal treatment. The successful biodegradation of chromophoric groups of dyes was suggested by the fading of major absorbance peak or by the formation of new peak of absorbance. Enzymes such as MnP (23.31 U ml^{-1}), tyrosinase (16.18 U ml^{-1}), and triphenylmethane reductase (1.15 U ml^{-1}) were produced at higher amounts during the MG dye biodegradation and the activity of enzyme tyrosinase increased up to 20.35 U ml^{-1} and 18.74 U ml^{-1} for MV and CB removal, respectively. However, no role of enzyme activity was noticed in the case of CV biodegradtion.

Ali et al (2016) evaluated CV dye elimination using *A. niger* strains (A1 and P1) that were grown at 30 °C on Czapek-Dox (CzD) agar medium with pH 5.5. The spore suspension was prepared in sterile 1% glucose solution. The final concentration was equivalent to 10^8 spores/ml. A 100 ml of liquid culture media supplemented with dye solution and pH 5.5 was inoculated with 1 ml spore suspension, followed by static incubation for different time periods at 30 °C. The mixture was centrifuged followed by filtration, washing, and adjusting the pH to 7. The evaluation of the dye by UV-visible spectroscopy at 585 nm confirmed that both the fungal strains decolorized up to 80.9% and 75.9% of dye within 10 days of treatment, which further remained statistically constant. The dry weight of A1 (0.53 g/100 ml) and P1 (1.13 g/100 ml) after 10 days suggested the decolorization was the priority for the isolate A1 whereas, microbial growth was the priority in the case of isolate P1. Slight morphological alterations, such as increment in mycelial thickness, were noted at 40 ppm concentration of dye, whereas complete inhibition of the microbial growth was observed at 50 ppm of dye. An increment in the reduction of color (84.6%) was noted at 40 ppm of dye along with the inhibition in growth of microbe (0.47 g/100 ml), suggesting the focus of the fungal isolate towards CV degradation rather than the growth. Sucrose showed no effect on the decolorization of dye, while a lower rate of decolorization was observed in presence of ammonium sulfate when used instead of sodium nitrate as the nitrogen source. However, enhanced decolorization up to 96.1% was noted in the presence of corn steep liquor as nitrogen source. The decolorization efficiency against CV was 96.1% at 40 ppm dye concentration and pH 5.5 under static conditions when sucrose and corn steep liquor were used as the carbon and nitrogen sources, respectively.

Ning et al (2018) reported the mycoremediation of 15 different dyes by *Aspergillus flavus* (A5p1). Initially, PDA was inoculated with fungal strain and incubated for 4 days at 39 °C. The obtained fungal biomass was washed and inactivated employing 20 min of pyrolysis at 121 °C. The crude enzyme extraction procedure was initiated with inoculation of 4.3×10^4 spores/ml into 60 ml of autoclaved culture media with the desired dye concentration, followed by 4 days of incubation at 39 °C and 150 rpm. The pellets were separated using centrifugation, followed by homogenization in a phosphate buffer (0.05 mol l^{-1}, pH 7.0). The extracellular enzyme was collected by ultrafiltration and the intracellular enzyme was collected via centrifugation of the cell culture filtrate. The mixture of intracellular and extracellular enzymes resulted in the formation of the crude enzyme solution. The decolorization experiment was initiated by inoculating 4.3×10^4 live spores/ml in 60 ml of autoclaved culture media with the desired dye concentration, followed by agitation (150 rpm) at 39 °C. After 8 days, about 1 ml sample was subjected to 10 min of centrifugation at 8000 rpm. Additionally, 60 ml of culture medium with desired dye concentration was treated with 0.3 g inactivated fungal biomass at 39 °C under shaking condition (150 rpm). About 1 ml sample was withdrawn after 1 day and centrifuged for 10 min at 8000 rpm. The decolorization using crude enzyme solution at different dye concentrations (100 mg l^{-1} C_{RB19}, 200 mg l^{-1} C_{DB71}, and 300 mg l^{-1} C_{DB86}) was carried out using 60 ml culture medium, and 1 ml crude enzyme under the aforementioned reaction conditions. The decolorization efficiencies for all 15 dyes at an 100 mg l^{-1} initial concentration were: 100% in case of direct blue 71 (DB71), reactive red 24, reactive black 5, direct blue 86 (DB86), and direct violet, and 99% for direct black 19, 93.1% for acid red 18, 79.2% for acid yellow 7, 61.7% for acid orange II, 87% for reactive blue 19 (RB19), 64% for acid blue 43, 83.6% for bromophenol blue, 74.2% for malachite green, 72.9% for CV, and 98.9% for reactive turquoise blue. The decolorization efficiencies of all 15 dyes at 500 mg l^{-1} dye concentration were: 99.9% for DB71, 95.1% for reactive red 24, 93.4% for reactive black 5, 90.7% for direct blue 86, 83.5% for direct violet, 80.2% for direct black 19, 78.8% for acid red 18, 40.6% for acid yellow 7, 31.1% for acid orange II, 71.1% for reactive blue 19, 50.3% for acid blue 43, 20.7% for bromophenol blue, 18.5% for malachite green, 15.2% for CV, and 88.5% for reactive turquoise blue. Around 90% decolorization was achieved under acidic conditions with inactive biomass, and under basic conditions with live biomass for DB71 and DB86. Similarly, 90% decolorization was noted for DB19 under neutral conditions. Decolorization up to 75.8% and 89.8% was noticed for DB71 and DB86, respectively, at 1000 mg l^{-1} concentration, and 72.5% for RB19 at 500 ml^{-1} concentration. The decolorizations of DB71, DB86, and RB19 with the inactivated cells were 83.4%, 100% and 49.4%, respectively while the decolorization of RB19 with the live biomass was twice that of the inactivated biomass. However, no significant difference was noted for DB71 and DB86. The reduction of DB71, DB86, and RB19 by the crude enzyme was 100%, 100%, and 71.8%, respectively. The glucose oxidase (GOD) and manganese peroxidase (MnP) enzymes were involved in the dye degradation. It is interesting to note that both the enzymes showed higher activities in the presence of dye. Around 3.2-fold increase in the activity of MnP and 138-fold in GOD activity were noted during the presence of DB86 and DB7, thus indicating the crucial role of the metabolic enzyme in the dye bioremediation process.

Jasinska et al (2015) also isolated a filamentous fungus from the dye-contaminated area, which was identified as *Myrothecium roridum*. The fungus was grown in CzD medium supplemented with 0.75% glucose at different temperatures (28, 38, or 48 °C), pH values (4,5,6, or 7), and agitation speeds (0, 75, or 150 rpm). The mycelium was separated from the culture at appropriate time intervals by 15 min of centrifugation at 15,000 ×g, followed by freeze-drying. No decolorization of MG occurred in presence of the cytochrome P-450. The MnP and LiP enzymes were not involved in the MG decolorization process. It was noted that laccase production (44–46 U l^{-1}) increased the MG decolorization activity. The leucomalachite green was reduced to N-demethylated metabolites, which was attributed to the laccase activity. Furthermore, the role of the laccase in decolorization was confirmed by inactivating the same using 0.5 mM sodium azide or by boiling followed by ultrafiltration. Even after chemical or heat inactivation, the enzyme showed MG decolorization. It is interesting to note that even the nonenzymatic fraction separated after ultrafiltration with compounds with mass lower than 3 kDa exhibited MG decolorization. However, the decolorization was 8%–11% higher compared to inactivated laccase, which indicates the involvement of the low-molecular-weight (<3 kDa) nonenzymatic factors (resistance to higher temperature) in MG decolorization. The toxicity study revealed no significant change in the bacterial growth by fungal culture without MG, whereas the MG-containing sample reduced the bacterial growth. The inhibition was highest against *Staphylococcus aureus* ATCC 61538 whereas, better tolerance towards MG (more than 40%) was exhibited by *Escherichia coli* ATCC 25922 and *Pseudomonas aeruginosa* ATCC 15442. These results indicate that the degradation products of MG were non-toxic in nature. The MG removal efficiency was blocked in the presence of 0.75% glucose, and the MG removal efficiencies of 58% and 57% were observed in the culture containing 1.5% glucose and 0.75% sucrose, respectively. It was speculated that, due to rapid exhaustion of the carbon sources at their low concentrations, the microorganisms started using dye as a source of energy in the form of carbon, which aided in the enhanced dye degradation. The best degradation of MG was obtained under acidic conditions. The highest removal up to 93% was obtained at 28 °C, whereas the 40% removal was observed at 38 °C or 48 °C. The data revealed that static or agitated (75 and 150 rpm) conditions resulted in similar rates of MG removal. The average color removal was more than 35%, 56%, 94%, and 90% when incubated for 4, 6, 18, and 24 h, respectively.

Contreras et al (2019) analyzed the elimination of methylene blue (MB) dye from aqueous solution using the yeast, *Galactomyces geotrichum* KL20A. The fungus was cultivated on yeast extract peptone dextrose (YPD) medium and 3.2×10^7 cell/ml yeast cell suspension was used for the MB removal, as shown in figure 7.2. Different concentrations of dye (50, 100, and 200 ppm) were added into 1 ml YPD-broth medium (pH-7) and incubated with cells (3.2×10^7 cells/ml) for 48 h at various temperatures (25, 30, and 35 °C). The decrease in the rate of dye degradation was noted with the increase in concentration of the dye. The highest MB removal up to $71.5 \pm 0.8\%$ was achieved at 35 °C with 50 ppm dye concentration and at pH 7. The dye degradation was mainly dependent on initial MB concentration, pH, and temperature. The kinetic studies revealed that removal rate constant was 2.2×10^{-2} h^{-1} and the biotransformation half-time was 31.2 h. The cytotoxicity test indicated that the by-product obtained after the biodegradation resulted in 22% hemolysis when compared to the negative control (100%). The MB removal process followed a pseudo-first-order kinetics.

Karaghool *et al* (2021) discovered an advanced method for the biodegradation of MB dye using *Aspergillus* spp. (*A. niger*, *A. flavus*, and *Aspergillus fumigatus*). The fungi were inoculated on the PDA medium (2% w/v of mineral salt medium), supplemented with dye, and were incubated at 28 °C. The decolorization experiment

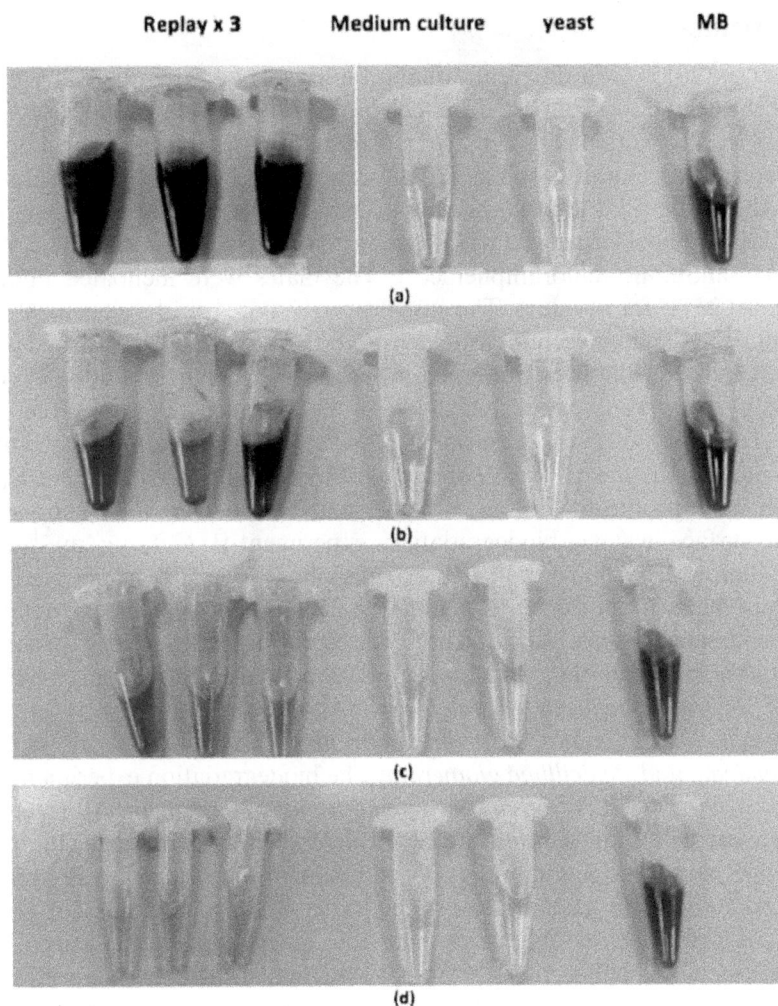

Figure 7.2. Tests of biodegradation of MB dye by the action of the yeast *Galactomyces geotrichum* KL20A. (a) Time zero hours, (b) time 12 h, (c) time 24 h, and (d) time 48 h (MB concentration was 50 ppm and temperature was 30 °C). The images on the left-hand side correspond to replays (×3) of the degradation experiments. The images on the right-hand side correspond to the controls used in order of yeast and culture medium (YPD-broth). Adapted from Contreras et al (2019). CC BY 4.0.

was carried out in batch mode, where the fungal biomass was initially suspended in an SDA medium that was eventually transferred to the BMSM. The reaction mixture was subjected to 4 h of incubation under shaking conditions (150 rpm) at 30 °C. The reaction mixture was comprised of MB solution (100 ml) and BMSM (25 ml). Around 92% dye removal was achieved at 36 h of incubation. The optimum temperature, pH, inoculums size, and initial dye concentration were 30 °C, 9, 2 g l^{-1}, and 150 g l^{-1}, respectively.

7.3 Pesticides

Oliveira et al (2015) used different aquatic fungi isolated from the water of Tagus River for the degradation of pesticides such as diuron, chlorfenvinphos, isoproturon, and atrazine. The untreated river water was initially filtered and the filtrate was transferred into dichloran rose-Bengal chloramphenicol (DRBC) agar pre-supplemented with 100 mg l^{-1} chloramphenicol (Oxoid). The existing fungal population was checked by inoculating 100 µl sample in malt extract agar (MEA) containing 100 mg l^{-1} antibiotic (chloramphenicol). The plates were incubated under dark conditions at 25 °C for 5–7 days. The suspension of spore was developed by scraping and filtering the fungal mycelium using glass wool. The filtrate was washed (0.5 g l^{-1} Tween 80) and the pellet was preserved in the cryoprotectant solution (3 g l^{-1} beef extract, 5 g l^{-1} tryptone pancreatic digest of casein, and 150 g l^{-1} glycerol). The initial concentration of the spore suspension was 9×10^5 spores/ml. Three different biodegradation experiments were conducted: biodegradation experiment-I (250 µg ml^{-1} each of isoproturon, diuron, atrazine, and chlorfenvinphos + non-sterilized untreated surface water), biodegradation experiment-II (250 µg ml^{-1} each of atrazine, diuron, isoproturon, and chlorfenvinphos + sterilized untreated surface water), and biodegradation experiment-III (250 µg ml^{-1} of chlorfenvinphos + sterilized untreated surface water). The biodegradation experiment-I contained the natural microbial population. The biodegradation experiment-II contained 9×10^5 spores/ml of *A. fumigatus*, *Cladosporium tenuissimum*, *Aspergillus terreus*, *Cladosporium cladosporioides*, *Penicillium citrinum*, *Fusarium begoniae*, *Penicillium melanoconidium*, and *Penicillium glomerata*. The biodegradation experiment-III had *Trichoderma harzianum*, *A. terreus*, *A. fumigatus*, and *P. citrinum*. All three experiments were carried out in darkness at 27 °C under stirring conditions (100 rpm). In experiment-I, slower reduction in the concentration of the pesticide with approx. 20, 50, 70, and 100% of degradation was noted after 82 days of treatment. The only pesticide degraded below its direct injection detection limit (10 µg l^{-1}) was chlorfenvinphos. On comparing the initial and final fungal population after the experiment, the existence of only two fungal species *P. citrinum* and *T. harzianum* was noted in the water samples after 82 days of treatment. In the case of experiment-II, only the reduction of chlorfenvinphos was noted after around 50 days, the concentration of which showed a decrease until the level below the detection limit was obtained after a further 3 months, while no change in the concentrations of atrazine, diuron, and isoproturon was observed. In the case of experiment-III which was conducted for 165 days, the results showed that only *A. terreus*, *T. harzianum*, *P. citrinum*, and *A. fumigatus* were resistant and able to degrade chlorfenvinphos in the aquatic environment.

In another study, Marinho et al (2017) reported atrazine remediation from wastewater employing *A. niger* AN 400. The fungus was grown on Saboraund Dextrose Chloramphenicol (ASDC) agar for 7 days at 28 °C followed by recovery of the spores in Tween 80 solution. Around 2×10^6 spores/ml were inoculated for atrazine (30 mg l^{-1}) biodegradation under aerobic conditions. Air was supplied by minicompressors with a flow of 20 l h^{-1} into the reactors that operated under different conditions, such as without glucose (RG0) and with glucose in range from 0.5 to 5 g l^{-1}, namely RG0.5, RG1, RG2, RG3, RG4, and RG5. The incubation was carried out for a period of 8 days. However, there was no effect on the diameters of the spores even after 6 days of treatment, which suggested the superior tolerance of *A. niger* towards the atrazine, even at the high levels. The highest atrazine removal was observed in the R3 reactor, having a removal efficiency of 72% ± 2% and removal rate of 0.48 ± 0.06 d^{-1}. In RG0, only 40% atrazine degradation was noted with a rate constant equivalent to 0.29 ± 0.06 d^{-1}.

Hamad (2020) studied the biodegradation of diazinon using the fungal strain *A. niger* MK640786. The fungal biomass was prepared in 100 ml Sabouraud broth for 48 h at 30 °C. The biomass was recovered by centrifugation for 15 min at 4000 rpm, followed by the collection of cell pellets that were washed with phosphate buffered saline (pH 7.3). The pesticide removal experiments were performed with 3 g of fungal biomass, pH 3 to 11, temperature of 25 to 45 °C, diazinon concentration ranging from 15 to 55 mg l^{-1}, and retention times varying between 1 to 21 days. The experiments were designed on the basis of response surface methodology (RSM). The highest diazinon degradation was 82%, which was obtained for *A. niger* at 30 ° C, pH 5, initial pesticide concentration of 25 mg l^{-1}, and an incubation time of 7 days.

In another interesting study, Gangola et al (2015) investigated the mycoremediation of imidacloprid using the *Trichoderma longibrachiatum* (FIII) and *Aspergillus*

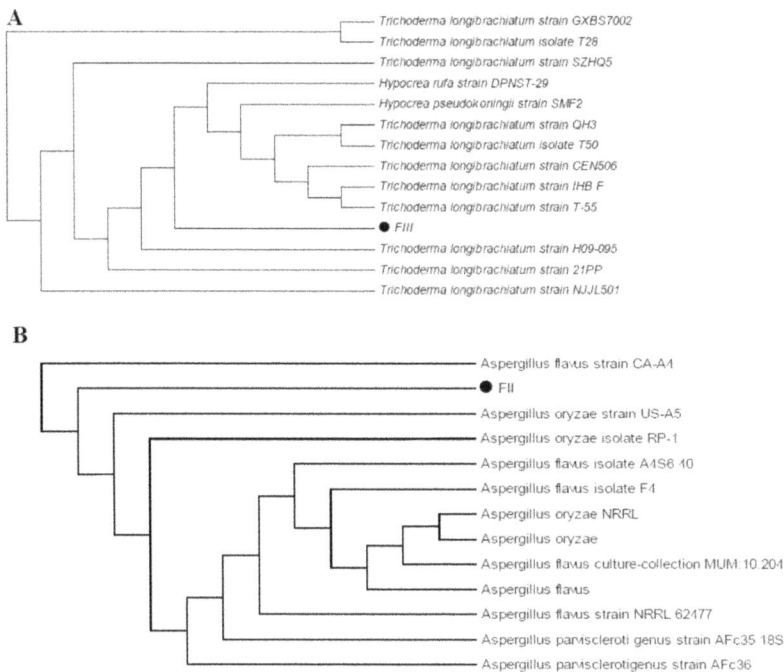

Figure 7.3. (A) Phylogenetic tree of FIII and (B) FII. Adapted from Gangola et al (2015). CC BY 4.0.

oryzae (FII) (figure 7.3). The microbial inocula were developed by mixing 20 g soil (from pesticide-contaminated agricultural fields) in 100 ml PDB, which was followed by overnight incubation under shaking conditions (150 rpm) at 28 °C. The imidacloprid (10 ppm) supplemented PDA plates were inoculated with the filtered soil suspension and incubated for 5–10 days at 28 °C. The developed fungal colonies were further transferred to a similar medium containing higher imidacloprid concentration for enrichment. The purification of the recovered fungal colonies was carried out by cultivating them on a sterile CzD medium (pH 6). The fungal colonies showing higher imidacloprid tolerance were further cultivated in the minimal medium supplemented with imidacloprid (30 ppm to 350 ppm). Only the selected fungal isolates were used for the evaluation of imidacloprid biodegradation in the minimal medium. Around 20 ppm imidacloprid was added to the 50 ml of CzD broth followed by addition of 1 ml fungal culture (4 days old) to the reaction mixture. The analysis of degradation process was carried out for individual fungal isolates and also for the fungal consortium. The inoculated medium was further subjected to incubation under shaking conditions (150 rpm) at 30 °C. Another procedure for studying the biodegradation of imidacloprid used immobilized fungal culture in sodium alginate beads and agar discs. The sodium alginate beads were formed by mixing 4% sodium alginate and homogenized fungal culture, followed by dropwise addition of the mixture in the autoclaved pre-chilled 0.4 M $CaCl_2$ by using a syringe to prepare fine beads in the laminar airflow. The agar discs were prepared by autoclaving the 4% agar solution and cooling it to 44 °C. The mixture of homogenized fungal cultures and molten agar was poured into a plate in an aseptic condition, followed by the formation of an equal-sized agar disc using a cork borer. The biodegradation experiment was carried out by adding sodium alginate beads and the agar discs in separate flasks containing the 50 ml of CzD medium and imidacloprid (20 ppm). The highest degradation of imidacloprid in the minimal medium was 92% on the 15th day where the consortium was used for the treatment, followed by *A. oryzae* (85%) and *T. longibrachiatum* (81%). Enhanced imidacloprid biodegradation by the consortium immobilized in the sodium alginate and agar discs was noted, which was equivalent to 95% and 97%, respectively.

Peter et al (2015) studied the mineralization of malathion using *Fusarium oxysporum* JASA1 obtained from fields used for growing sugarcane. The screening of the isolated fungal strain was performed in the CzD broth composed of yeast extract (3 g l^{-1}), dextrose (2 g l^{-1}), peptone (10 g l^{-1}), malathion (100 mg l^{-1}), and 20 g of soil pre-exposed to malathion. The enrichment was carried out in 100 ml media supplemented with 100 mg l^{-1} malathion with an incubation time of 5 days under shaking conditions (100 rpm) at room temperature. The biodegradation of malathion was initiated by inoculation of 1 ml spore suspension into malathion (400 mg l^{-1}) supplemented M1 medium (100 ml) followed by incubation at 30 ± 2 °C under shaking condition (120 rpm). The ability of the JASA1 to degrade malathion was determined by conducting two trials: (i) 400 mg l^{-1} pesticide + fungal spore + nutrients (100 carbon: 10 nitrogen: 1 phosphorous) and (ii) 400 mg l^{-1} pesticide + fungal spore without nutrients. The MIC of malathion against the JASA1 strain was 500 mg l^{-1}, whereas the confluent growth of fungi was observed at 400 mg l^{-1}

malathion. The mean malathion recoveries from the M1 medium at levels of 100, 200, 300, and 400 mg l^{-1} were around 96.2% ± 4.3%, 97.6% ± 4.8%, 98.4% ± 3.2%, and 97.6% ± 1.2%, respectively. The degradation of malathion by the fungal strain JASA1 was analyzed using FTIR. The malathion degraded sample showed absorbance bands attributing N-H stretching (3446 cm^{-1}), C=C stretching (1641 cm^{-1}), acid dimer band (991 cm^{-1}), PH bending (1085 cm^{-1}), and C-H deformation (678 cm^{-1}). The acid dimer band supported the degradation of malathion-to-malathion mono-acid or malathion diacid, which occurred due to the action of carboxylesterase.

7.4 Oils

Hashem et al (2018) developed a method for the elimination of aliphatic as well as aromatic hydrocarbons using novel yeast species. Around 67 yeasts were isolated, purified, and preserved on the Yeast-Malt agar medium (YMA) comprising of 0.3% yeast extract, 1% glucose, and 1.5% agar. The primary screening of hydrocarbon degradation was carried out on yeast nitrogen base agar (YNBA) having a pH 5.4 ± 0.2 at 25 °C. Each species of yeast (10^8/ml) was inoculated in the 15 ml of YNBA medium and poured into the Petri plates. Yeast inoculum was prepared by cultivating each isolate for 2 days on YMA plates at room temperature. The hole in the agar was made after solidification using a cork borer, which was further filled with 50 µl of either octane or pyrene (1% concentration) that acted as the sole carbon source. The plates were further incubated for 1 week at 25 °C. The evaluation of the hydrocarbon degradation was checked 15 days using 50 ml of minimal salt medium (MSM) comprising of $MgSO_4.7H_2O$ (0.2), KH_2PO_4 (1 g l^{-1}), $CaCl_2.2H_2O$ (0.02 g l^{-1}), NH_4NO_3 (1 g l^{-1}), K_2HPO_4 (1 g l^{-1}), $FeCl_2$ (0.02 g l^{-1}), and Tween 80 (0.1%) in distilled water (1 l). The medium was autoclaved at 121 °C for 20 min and octane or pyrene was added so that the final concentration of the same was 1%, which was eventually inoculated with 1 ml of yeast suspension. The flasks were incubated at 30 °C under shaking conditions (150 rpm). The strain KKUY-0214 showed superior degradation of both the hydrocarbons, while KKUY-0160, KKUY-0162, KKUY-0163, and KKUY-0192 exhibited higher growth on octane compared to pyrene. However, the growth of KKUY-0170 was more on pyrene compared to that of octane. The yeast strain *Rhodotorula taiwanensis* KKUY-0162 was considered to be the best organism for the degradation of octane after 15 days of incubation, followed by *Meyerozyma guilliermondii* KKUY-0214 and *Candida pseudointermedia* KKUY-0192. The most efficient strain *Rhodotorula ingeniosa* KKUY-0170 grew gradually until it reached an optical density (OD) equivalent to 2.23 after 15 days of incubation. Both *Yamadazyma mexicana* KKUY-0160 and *M. guilliermondii* KKUY-0214 grew well on pyrene supplemented medium showing their superior ability to metabolize the same as the sole carbon source.

Al-Hawash et al (2018) isolated crude oil-degrading fungi from an oil field using the MSM and PDB medium. About 5 g of soil sample was added in the freshly prepared MSM of pH 7 composed of 0.5 g l^{-1} of NaCl, 0.1 g l^{-1} of $(NH_4)_2SO_4$, 0.2 g l^{-1} of $NaNO_3$, 0.025 g l^{-1} of $MgSO_4.7H_2O$, 1 g l^{-1} of $K_2HPO_4.3H_2O$, 0.4 g l^{-1} of

KH_2PO_4 and 1% crude oil followed by shaking (150 rpm) at 30 °C. The aliquots from the medium were transferred on the MSM plates containing 1% crude oil and 1.5% agar followed by incubation for 3–7 days at 30 °C. The pure colonies were collected and stored in PDA slants. The biodegradation experiments were carried out by using 100 ml MSM containing 1% crude oil as a sole carbon source. Five fungal plugs derived from the outside edge of actively growing culture on PDA were further transferred onto the MSM (supplemented with crude oil), followed by 14 days of incubation at 30 °C. The crude oil degradation efficiency by *Penicillium* sp. RMA1 and RMA2 were estimated by the modified 2,6-dichlorophenol indophenol (DCPIP) method. In this method, the inoculation of five mycelial plugs of RMA1 and RMA2 on the degradation media containing 0.4 μg ml^{-1} DCPIP was carried out, which was followed by 14 days of incubation at 30 °C. The change in color of DCPIP from blue to colorless confirmed the crude oil elimination by both RMA1 and RMA2. The emulsifying ability of *Penicillium* sp. RMA1 was slightly higher than that of RMA2. During 7 days of incubation, about 52% and 49% elimination of crude oil was noted for RMA1 and RMA2, respectively which further increased to 57% and 55% during 14 days of incubation. The rate of degradation by RMA1 and RMA2 were around 0.040 and 0.039 per day.

Behnood et al (2014) studied the elimination of crude oil-polluted saline wastewater employing white rot fungi *Phanerochaete chrysosporium*. The fungal biomass obtained after 48 h of incubation under shaking conditions at 37 °C was homogenized and used as the inoculum. The concentration of the pollutant along with its elimination efficiencies were determined by two different approaches. In the first approach, vigorous shaking of 30 ml sample and 60 ml dichloromethane mixture was carried out for 10 min to extract residual hydrocarbons. The extracts formed were further filtered by anhydrous sodium sulfate for removal of water, followed by filtration, drying, and weighing the samples. In the second approach, the residual oil concentration was evaluated via infrared spectroscopy because all of the oils and greases exhibited the absorption band at 2930 cm^{-1}. The percentage of crude oil removal increased with the increase in salinity from 0 to 10 g l^{-1}, while the removal efficiencies were reduced at higher salinities. The highest removal of crude oil was 58.1 ± 1.3%, which was noted at 10 g l^{-1} salinity. After 12 days, the highest elimination efficiency of 600 ppm concentration of the contaminants in the presence of Tween 80 at 0, 20, and 40 g l^{-1} salinities was around 63.3%, 64.8%, and 38.7%, whereas in the presence of NAR-111-2 it was around 82.8%, 75.5%, and 40.3%, respectively.

Benuguenab and Chibani (2021) reported the degradation of the PHs by the filamentous fungi *Aspergillus ustus* and *Purpureocillium lilacinum*. Uncontaminated soil (4 kg) was collected and artificially converted to contaminated soil by the addition of 100 ml used engine (UE) oil, and was then exposed to outer environmental conditions for a period of 3 months. The dilution plate method was employed to enumerate the heterogenic fungi present in the oil-polluted soil. The mixture of soil (1 g) in 9 ml sterile distilled water was vortexed followed by preparation of serial dilutions (10^{-5}). About 0.5 ml of each dilution was inoculated on PDA having 250 mg l^{-1} chloramphenicol, followed by 5 days of incubation at 25 °C. Visible fungal colonies were further transferred to new PDA plates to get pure cultures. The preliminary evaluation for the utilization of crude oil, diesel, and UE oil by the fungi was based on the redox indicator DCPIP, where Bushnell-Haas (BH) broth medium ($0.2 \, g \, l^{-1}$ $MgSO_4$, $1 \, g \, l^{-1}$ KH_2PO_4, $0.02 \, g \, l^{-1}$ $CaCl_2$, $1 \, g \, l^{-1}$ K_2HPO_4, $0.05 \, g \, l^{-1}$ $FeCl_2$, and $1 \, g \, l^{-1}$ NH_4NO_3 in 1000 ml distilled water) with pH 7 was used. Two culture plugs were transferred from PDA plate to 50 ml BH medium in which crude oil (1% v/v), diesel, or UE oil, and the redox indicator ($0.016 \, mg \, ml^{-1}$) were added and incubated for 2 weeks under shaking conditions (110 rpm) at 30 °C. The conversion of DCPIP color from blue (oxidized) to colorless (reduced) confirmed that the fungi were able to degrade hydrocarbons. The tolerance of the fungi against PHs was tested using different concentrations of crude oil, diesel, and UE oil on BH agar. The DCPIP assay confirmed the complete conversion of the color of the solution, along with the notable fungal proliferation in the media containing crude oil and UE oil. The isolates denoted as HM3 and HM4 were identified as *A. ustus* HM3.aaa and *P. lilacinum* HM4.aaa, respectively. The highest tolerance was observed for *A. ustus* against 2% crude oil along with a growth rate of 0.94 cm/day. *P. lilacinum* and *A. ustus* eliminated 44.55% and 30.43% of crude oil, 27.66% and 21.27% of diesel, and 14.39 and 16.00% of UE oil. The rate constant (K) and half-lives ($t_{1/2}$) for crude oil removal by *P. lilacinum* were 0.02/day and 34.66 days, and by *A. ustus* were 0.015/day and 46.2 days, respectively.

Al-Jawhari (2014) investigated the removal of petroleum hydrocarbon using the four fungal strains isolated from the indigenously polluted soil. The fungi were isolated and their tolerance against petroleum was checked on PDA, as discussed in the previous section. The biodegradation studies were performed by taking 2 ml of crude oil and 98 ml of mineral salts medium, followed by inoculation of a 5 mm disk from the mycelia (7-day old fungi colony). The reaction mixture was incubated at 25 °C for 7 days, after which the microbial activity was stopped by the addition of 1N HCl. The fungi were identified as *A. niger*, *A. fumigatus*, *Fusarium solani*, and *Penicillium funiculosum*. Maximum resistance against 2% crude oil was confirmed from the growth of *A. niger*, *F. solani*, *A. fumigates*, and *P. funiculosum*, with colony diameters equivalent to 8.5 cm, 5.9 cm, 4.5 cm, and 3.6 cm, respectively, while the mycelia dry weights were 1.20 g, 0.81 g, 0.61 g, 0.56 g, respectively, after 7 days of incubation. After 28 days of biodegradation, the highest removal of crude oil was about 95% and 75% by *A. niger* and *A. fumigates*, respectively, followed by *F. solani* (55%) and *P. funiculosum* (65%). It is interesting to note that the highest crude oil elimination efficiency was 90% when a mixed culture of *A. niger* and *A. fumigatus* was used, while the lowest was 70% when a mixed culture containing all four fungal strains was used.

Al-Hawash et al (2019) removed various PHs using *Aspergillus* sp. RFC-1. A 100 ml MSM containing 1% crude oil was used as the enrichment medium for the microorganisms isolated from 5 g of soil sample. After 7 days of incubation under shaking conditions (130 rpm) at 30 °C, about 10 ml of inoculums was transferred to the fresh MSM (containing 1% crude oil), followed by 7 days of incubation. The optimum conditions required in the cultivation of the fungi were 30 °C of temperature, pH 7, $50 \, mg \, l^{-1}$ naphthalene, $250 \, mg \, l^{-1}$ crude oil, $20 \, mg \, l^{-1}$ phenanthrene, and 20 mg l^{-1} pyrene. The biodegradation studies of naphthalene, crude oil, pyrene, and phenanthrene were carried out by inoculating 10% v/v RFC-1 mycelial suspension in the degradation medium containing 50 mg l^{-1} naphthalene, 250 mg l^{-1} crude oil, 20 mg l^{-1} phenanthrene, or 20 mg l^{-1} pyrene. The reaction mixture was incubated at 30 °C under shaking conditions (120 rpm). The highest surface adsorption for crude oil by the live mycelial pellets was 43.8% at 180 min, while 50.5%, 59.0%, and 67.2% were recorded for naphthalene, phenanthrene, and pyrene, respectively, at 90 min. The cellular absorption of crude oil, naphthalene, phenanthrene, and pyrene was 35.7%, 67.7%, 83.9%, and 88.9%, respectively. The adsorption efficiency of the heat-killed mycelial pellets remained constant after 10 and 40 min for PAHs and crude oil, respectively. The biodegradation efficiencies within 7 days of incubation were 51.8%, 84.6%, 51.3%, and 55.1% against crude oil, naphthalene, phenanthrene, and pyrene, respectively.

Moghimi et al (2017) reported PAHs degradation by *Trematophoma* sp. UTMC 5003. Crude oil-contaminated soil was collected, dried, grounded, and sieved with a mesh to homogenize and remove the large particles. The modified MSM containing 50 μg ml^{-1} chloramphenicol and 1% light crude oil was utilized as the enrichment medium with pH 6.8 ± 0.2. The fungus was selected for biodegrading PAHs such as anthracene, pyrene, and phenanthrene, along with their mixtures. The residual PAHs were extracted in hexane and were detected by high-performance liquid chromatography (HPLC). The crude oil was degraded up to 70% during the 15 days of treatment and biomass production of G-05 was noted. The molecular identification of the G-05 isolate revealed that the isolate was from the genus *Trematophoma*. About 56%, 87%, and 90% removal by the fungal isolate *Trematophoma* sp. UTMC 5003 was recorded for the pyrene, anthracene, and phenanthrene, respectively. In the case of a mixture of PAHs, the fungus showed significant degradation of anthracene, pyrene and phenanthrene, respectively, which was up to 87%, 56%, and 90%, respectively. The surface tension of the medium was reduced from 44.5 ± 0.03 m N m^{-1} to 25.95 ± 0.1 m N m^{-1} which was attributed to the fungal growth and biosurfactant production (table 7.1).

7.5 Conclusion and future perspectives

Fungi mediated remediation of hazardous pollutants seems to be an efficient and eco-friendly bioremediation strategy. Fungi from various ecological niches have been employed for the effective removal of dyes, pesticides, heavy metals, PAHs, and other toxic pollutants that are released into the environment, mostly through industrial effluents. Hence, more high throughput screening should be done to detect and isolate fungi that are highly resistant to such pollutants. Integrated multidisciplinary approaches using molecular genetics, bioinformatics, and metabolomics will help to understand the mechanism behind the fungal ability to degrade these pollutants. It will also enable us to find the genes and enzymes involved in the bioremediation process. Additionally, enzymes such as lignin peroxidase (LiP), manganese peroxidase (MnP), tyrosinase, triphenylmethane reductase, glucose oxidase, laccase, carboxylesterase, and others that play a significant role in the detoxification of the pollutants can be immobilized on nanomaterials. Functionalization of such enzymes on to the surface of nanoparticles will help to stabilize the enzymes, and ensure their reusability and recyclability. Nanozymes can be fabricated that mimic these enzymes for enhanced degradation of hazardous dyes, pesticides and PAHs. Entrapment of enzymes in polymeric beads will help the easy recovery of enzymes from bioreactors after completion of the treatment process so that they can be used in successive batches.

Genetically engineered fungi can be developed with superior pollutant removal capability using advanced recombinant DNA technology. Furthermore, fungal consortium or mixed cultures can be employed for enhanced bioremediation due to their synergistic bioconversion. Fungal biomass generated biochar can also act as a superior biosorbent for the removal of refractory contaminants. Used fungal biomass from the wineries, bakeries, food industries and others can be collected and converted to biochar, which will serve the dual purpose of waste biomass disposal and bioremediation.

In view of this background, there is immense scope for development in this field. In particular, fungi-based wastewater treatment strategies can be developed with optimized process parameters, such as duration, temperature, pH, inoculum density, initial pollutant concentration, and aeration.

References

AI-Jawhari I F H 2014 Ability of some soil fungi in biodegradation of petroleum hydrocarbon *J. Appl. Environ. Microbiol.* **2** 46–52

Al-Hawash A B, Alkooranee J T, Abbood H A, Zhang J, Sun J, Zhang X and Ma F 2018 Isolation and characterization of two crude oil-degrading fungi strains from Rumaila oil field, Iraq *Biotechnol. Rep.* **17** 104–9

Al-Hawash A B, Zhang X and Ma F 2019 Removal and biodegradation of different petroleum hydrocarbons using the filamentous fungus *Aspergillus* sp. RFC-1 *Microbiol. Open* **8** e00619

Ali H M, Shehata S F and Ramadan K M A 2016 Microbial decolorization and degradation of crystal violet dye by *Aspergillus niger Int. J. Environ. Sci. Technol.* **13** 2917–26

Asses N, Ayed L, Hkiri N and Hamdi M 2018 Congo red decolorization and detoxification by *Aspergillus niger*: removal mechanisms and dye degradation pathway *BioMed Res. Int.* **2018** 3049686

Bankole P O, Adekunle A A and Govindwar S P 2018 Enhanced decolorization and biodegradation of acid red 88 dye by newly isolated fungus, *Achaetomium strumarium J. Environ. Chem. Eng.* **6** 1589–600

Behnood M, Nasernejad B and Nikazar M 2014 Biodegradation of crude oil from saline waste water using white rot fungus *Phanerochaete chrysosporium J. Ind. Eng. Chem.* **20** 1879–85

Benguenab A and Chibani A 2021 Biodegradation of petroleum hydrocarbons by filamentous fungi (*Aspergillus ustus* and *Purpureocillium lilacinum*) isolated from used engine oil contaminated soil *Acta Ecol. Sin.* **41** 416–23

Bind A, Goswami L and Prakash V 2018 Comparative analysis of floating and submerged macrophytes for heavy metal (copper, chromium, arsenic and lead) removal: sorbent preparation, characterization, regeneration and cost estimation *Geol. Ecol. Landsc.* **2** 61–72

Cao X, Yang C, Liu R, Li Q, Zhang W, Liu J, Song C, Qiao C and Mulchandani A 2013 Simultaneous degradation of organophosphate and organochlorine pesticides by *Sphingobium japonicum* UT26 with surface-displayed organophosphorus hydrolase *Biodegradation* **24** 295–303

Chen S H, Cheow Y L, Ng S L and Ting A S Y 2019 Biodegradation of triphenylmethane dyes by non-white rot fungus *Penicillium simplicissimum*: enzymatic and toxicity studies *Int. J. Environ. Res.* **13** 273–82

Contreras M, Grande-Tovar C D, Vallejo W and Chaves-Lopez C 2019 Bio-removal of methylene blue from aqueous solution by *Galactomyces geotrichum* KL20A *Water* **11** 282

Drumm F C, Franco D S P, Georgin J, Grassi P, Jahn S L and Dotto G L 2021 Macro-fungal (*Agaricus bisporus*) wastes as an adsorbent in the removal of the acid red 97 and crystal violet dyes from ideal colored effluents *Environ. Sci. Pollut. Res.* **28** 405–15

Gangola S, Khati P and Sharma A 2015 Mycoremediation of imidaclopridin the presence of different soil amendments using *Trichoderma longibrachiatum* and *Aspergillus oryzae* isolated from pesticide contaminated agricultural fields of Uttarakhand *J. Bioremed. Biodeg.* **6** 310

Goswami L, Kumar R V, Manikandan N A, Pakshirajan K and Pugazhenthi G 2019 Anthracene biodegradation by Oleaginous *Rhodococcus opacus* for biodiesel production and its characterization *Polycycl. Aromat. Compd.* **39** 207–19

Hamad M T M H 2020 Biodegradation of diazinon by fungal strain *Apergillus niger* MK640786 using response surface methodology *Environ. Technol. Innov.* **18** 100691

Hashem M, Alamri S A, Al-Zomyh S S and Alrumman S A 2018 Biodegradation and detoxification of aliphatic and aromatic hydrocarbons by new yeast strains *Ecotoxicol. Environ. Saf.* **151** 28–34

Jasinska A, Paraszkiewicz K, Sip A and Długonski J 2015 Malachite green decolorization by the filamentous fungus *Myrothecium roridum*—mechanistic study and process optimization *Bioresour. Technol.* **194** 43–8

Karaghool H A K 2021 Biodecolorization of methylene blue using *Aspergillus* consortium *IOP Conf. Ser: Earth Environ. Sci.* **779** 012111

Kumar M, Goswami L, Singh A K and Sikandar M 2019 Valorization of coal fired-fly ash for potential heavy metal removal from the single and multi-contaminated system *Heliyon* **5** e02562

Kumar RV, Goswami L, Pakshirajan K and Pugazhenthi G 2016 Dairy wastewater treatment using a novel low cost tubular ceramic membrane and membrane fouling mechanism using pore blocking models *J. Water Process Eng.* **13** 168–75

Liu S H, Zeng G M, Niu Q Y, Liu Y, Zhou L, Jiang L H, Tan X F, Xu P, Zhang C and Cheng M 2017 Bioremediation mechanisms of combined pollution of PAHs and heavy metals by bacteria and fungi: a mini review *Bioresour. Technol.* **224** 25–33

Marinho G, Barbosa B C A, Rodrigues K, Aquino M and Pereira L 2017 Potential of the filamentous fungus *Aspergillus niger* AN 400 to degrade Atrazine in wastewaters *Biocatal. Agric. Biotechnol.* **9** 162–7

Moghimi H, Heidary Tabar R and Hamedi J 2017 Assessing the biodegradation of polycyclic aromatic hydrocarbons and laccase production by new fungus *Trematophoma* sp. UTMC 5003 *World J. Microbiol. Biotechnol.* **33** 136

Ning C, Qingyun L, Aixing T, Wei S and Youyan L 2018 Decolorization of a variety of dyes by *Aspergillus flavus* A5p1 *Bioprocess. Biosyst. Eng.* **41** 511–8

Oliveira B R, Penetra A, Cardoso V V, Benoliel M J, Barreto Crespo M T, Samson R A and Pereira V J 2015 Biodegradation of pesticides using fungi species found in the aquatic environment *Environ. Sci. Pollut. Res.* **22** 11781–91

Peter L, Gajendiran A, Mani D, Nagaraj S and Abraham J 2015 Mineralization of malathion by *Fusarium oxysporum* strain JASA1 isolated from sugarcane fields *Environ. Prog. Sustain. Energy* **34** 112–6

Sadeghi A, Ehrampoush M H, Ghaneian M T, Najafpoor A A, Fallahzadeh H and Bonyadi Z 2019 The effect of diazinon on the removal of carmoisine by *Saccharomyces cerevisiae Desalin. Water Treat.* **137** 273–8

Sarma G K, Gupta S S and Bhattacharyya K G 2016 Retracted: adsorption of crystal violet on raw and acid-treated montmorillonite, K10, in aqueous suspension *J. Environ. Manage.* **171** 1–10

Sathe S S, Goswami L, Mahanta C and Devi L M 2020 Integrated factors controlling arsenic mobilization in an alluvial floodplain *Environ. Technol. Innov.* **17** 100525

IOP Publishing

Trends in Biological Processes in Industrial Wastewater Treatment

Maulin P Shah

Chapter 8

Advances in the remediation of xenobiotics using microbes

Anne Bhambri, Santosh Kumar Karn* and Navneet Joshi

The environment is heavily polluted due to the increased release of untreated contaminants, hazardous chemicals, and xenobiotics. This environmental pollution dramatically impacts the health of all living things and their habitats, attributed to limited degradability, and high toxicity, which is still not fully comprehended. The complex nature of emerging compounds and their various physicochemical properties, such as ionizability, polarity, size, solubility, lipophilicity, and volatility, make identifying, quantifying, and degradation difficult. Several conventional and modern remediation technologies are being engaged to remove xenobiotics from the environment. Some of which are not effective in the removal of xenobiotics, while others are not available for use on a large-scale and a few are expensive. The recent advances in the removal of xenobiotic molecules rely on nano-based technologies coupled with biochar, phytoremediation, phycoremediation, nanocomposites, biomaterials etc. This chapter discusses the variety of xenobiotic molecules used primarily with their bioavailability and spotlights the use of nano-based techniques in removing such complex xenobiotic molecules. It also discusses direct sources of xenobiotics, environmental pollutants, food additives, oil mixtures as well as indirect sources, such as drugs. Finally, it justifies the remediation process of xenobiotics using different methods.

8.1 Xenobiotics: an overview

The environment consists of everything that surrounds us naturally and influences our everyday life. A healthy and safe environment is necessary for life to survive. In the era of rapid industrial development and alarming population expansion, human activities are primarily responsible for the introduction of harmful and dangerous

---

* Corresponding author: santoshkarn@gmail.com, santoshkarn.sbsu@gmail.com.

doi:10.1088/978-0-7503-5678-7ch8

chemicals, such as environmental xenobiotics [1]. Xenobiotics are chemical substances that represent arbitrarily synthesized chemicals that are not produced naturally [1]. Since xenobiotics are comparatively new substances, it is very difficult to eliminate them from the environment. Growing public concern has been expressed on the large variety of xenobiotic substances being intentionally or unintentionally introduced into the environment, which pose a severe danger to both people, animals, and all forms of life. For example, high concentrations of xenobiotic compounds such as drug molecules, dyes, and pesticides are released without prior treatment into the municipal sewage system, which remains a significant hurdle in remediation [1, 2].

These substances are made up of hundreds of known elements. Toxic substances such as heavy metals are well acknowledged in the environment. Elemental forms of essential moelcules can be toxic and result in environmental damage [1]. Compounds can either be organic or inorganic, which are further segregated depending on the existence of a variety of functional groups. Man-made inorganic or organic pollutants spread in every part of the hydrosphere, lithosphere, atmosphere, and tend to transform into other compounds that can be poisonous to flora and fauna [1, 2]. However, water and soil are the primary recipients of xenobiotic deposits that are released into the environment from industries and other resources. Polycyclic aromatic hydrocarbons (PAHs), heavy metals, pesticides, flame retardants, solvents, polychlorinated aromatics etc are the classes of xenobiotics [3, 4]. Xenobiotics and heavy metals have higher stability in the environment due to their bio-refractory or recalcitrant nature. These compounds accumulate and magnify in the food web, which causes severe disorders in humans and other biotas [5–8]. A number of xenobiotics are known to have carcinogenic, teratogenic, and mutagenic effects on aquatic life and human beings [8].

Various methods are used to reduce the xenobiotics from the environment, including bioremediation, physical removal, incineration, recycling, and chemical detoxification. Nevertheless, for the removal of some xenobiotics, these methods are not effective while others are inconvenient to utilize because they are costly [9]. Living organisms have mutagenic and deleterious effects on degradation of xenobiotic products and these modifications lead to abrupt microbial community profiles [10, 11]. As higher concentrations of pesticides are added in the agricultural fields, it hampers the health of the soil. There are a large number of organic compounds that persevere for long time in aquifers, soils, aquatic sediments, subsoils, and surface waters. To control agricultural pests, the most stable compounds that have been used worldwide are the Organochlorine pesticides (OCP), Chlordane, Mirex, Metolachlor, Dichlorodiphenyltrichloroethane (DDT), Endosulfan, Metoxychlor, Hexachlorocyclohexane (HCH), Aldrin, Toxaphene, Heptachlor, Hexachlorobenzene (HCB), and Dieldrin [10–15].

Emerging contaminants may form co-products, mixtures, and byproducts. This makes the assessment of chemicals even more complex because they might become visible in extremely low concentrations. These compounds can be quantified and identified by small equipment. For emerging contaminant analysis coupled with adequate equipment, appropriate methodologies and sample preparation are essential for the understanding of mixture concentrations, as well as byproducts

[16, 17]. In the environment, so-called endocrine disrupters for the occurrence of xenobiotics and pharmaceuticals, urban development, industrialization, and agricultural growth have been contributed. Frequently, these activities are responsible for the partial contamination of water through the introduction of numerous synthetic compounds [18]. During application, agricultural xenobiotics are deposited in a given compartment of the environment. Consequently, they quickly tend to disperse into other adjacent compartments, such as estuaries, groundwater, lakes, plant atmosphere, rivers, plant soil, soil groundwater, etc. Hollert et al [19] assert that, for effective concentration, the molecules must reach equilibrium, facilitating the formation of firmly fixed chemical linkages during development. This is an essential tool to predict the distribution of these xenobiotics in the environment. This chapter focusses on xenobiotic compounds, different sources, bioavailability, and their remediation.

8.2 Extensively used xenobiotics

Since the twentieth century, xenobiotics comprise many types of compounds predominantly used in industries, such as pharmaceuticals, plastics, pesticides, polymers, and dyes. Other sources include electronic waste, energy generation with conventional fossil fuel burning, transportation, (textile, agrochemical, paints), natural emissions, hospital effluents, mining of precious minerals, agricultural practices, radioactive materials, transformer oils leakages from the installations of electrical, cigarette smoking, oil and gas production, etc [20]. Emerging compounds have complex molecular structures and physiochemical properties (i.e., ionizability, low molecular size, lipophilic, volatility, polarity, and solubility of water) that create difficulties in their degradation and quantification.

8.3 Xenobiotic sources

Xenobiotic sources are separated into five categories [2]: (a) pulp and paper bleaching (in the environment they are the main artificial sources and natural organic compounds of chlorine), (b) rigorous agriculture (which uses a large amount of herbicides, pesticides, and fertilizers), (c) mining (which allows heavy metals to enter the biogeochemical cycles), (d) pharmaceutical and chemical industries (which produce a wide array of synthetic polymers and xenobiotics), and (e) fossil fuels (which can be accidentally released into the ecosystem in large amounts).

8.3.1 Direct sources

Xenobiotics have primary direct sources, which are solid residual and industrial released wastewater such as pharma and chemical, paper mills, agriculture (an increase products such as pesticides, herbicides etc), textile mills and plastics. Wastewater and other effluents include some remaining compounds, such as different dyes, hydrocarbons, effluents of paints, pesticides, and insecticides and phenol etc. The collection of dye is the major cause for the persistence of xenobiotics, and due to reduced penetration of light it will affect the activity of photosynthesis due to their presence in aquatic life, even at low concentrations.

8.3.1.1 Dyes

Several industrial processes, such as paper printing, photography, and textile industries, use synthetic dyes, which generally have complex aromatic molecular structures. Dyes of the class phthalocyanine, azo and anthraquinone are commonly used and contribute to environmental pollution. Aromatic amines are produced by the degradation of these dyes, which can be mutagenic and carcinogenic.

8.3.1.2 Phenol and phenol compounds

Various industries, such as pharmaceuticals, petrochemical, coal refineries, dying, phenol manufacturing, pulp mill, etc, use a vast diversity of organic phenols, together with several alternative phenols.

8.3.1.3 Hydrocarbons

Hydrocarbons are derived from hydrogen and carbon. These hydrocarbons have toxic as well as carcinogenic effects on humans, as well as on microbial diversity. Due to the combustion of byproducts, these chemicals are synthesized. Petroleum effluents mainly contain nitrogen-sulfur-oxygen compounds, saturated hydrocarbons, and PAHs.

8.3.1.4 Pesticides and insecticides

Chemicals such as pesticides are used as fungicides, insecticides, and herbicides together with associated pollutants of organics. These pesticides can be obtained from two large categories: they are (a) anthropogenic or rather industrial activities without any intention or byproducts, or (b) accidentally or unnaturally produced several compounds. Furthermore, from natural processes of pesticides can also be produced [21].

Intentionally produced persistent organic pollutants (POP) can be divided into various subgroups and with many families of brominated and chlorinated aromatics, these chemicals correlate that comprises polychlorinated biphenyls (PCBs), polychlorinated naphthalenes (PCNs), organochlorine pesticides (OCPs), and polychlorinated dibenzo-p-dioxins/furans (PCDD/FS) with DDT and its metabolites such as chlordane and toxaphene. Chemicals such as hexachlorobenzene (HCB) is used in industries and is also a by-product that is produced without any intention [22]. Another example is PCB that is produced from both sources [23, 24]. The comparative significance is ambivalent when it comes to the formation of by-products [25]. Insecticides and pesticides that are widely used, such as morpholine, methyl parathion, benzimidazoles, and organophosphorus compounds, slowly degrade, which contributes to the load of pollution.

8.3.1.5 Paints

Additives and volatile organic mixtures such as texturizers as well as emulsifiers in paint are regarded as dangerous and may be broken down by various means like hygroscopic stresses, microbial sources, and chemicals (water as solvent).

8.3.1.6 Plastics

Plastic polymers are also called synthetic polymers. Because of their interactions and molecular bonds, plastics are stable and are degraded slowly. They are made up of polyethylene and their derivatives, as well as polyvinyl chloride along with polystyrene. Recently, because they degrade into liquid hydrocarbons, plastics are utilized as industrial fuels. Bioplastic types are polylactic acid plastics, cellulose-based plastics, starch-based plastics, and bio-derived polyethylene. Synthetic polymer such as nylon is very common and in cooked food transportation it is frequently used in packaging, garments, as well as wrapping.

8.3.1.7 Paper and pulp effluents

Effluents discharged from paper mill industries contribute to pollution in the environment, which cannot be ignored. During pulp bleaching, lots of chlorinated organic compounds are synthesized randomly, which is the cause for increased pollution level. The situation can be made even worse when the effluents from the pulp mill industries are set free to oxygen-limited or rather oxygen-depleted waters. To control pollution, the more restrictive laws and increased public awareness have forced the paper industries to limit the release of absorbable organic halides, as well as to search the cleaner production technologies.

8.3.1.8 Environmental pollutants

The functioning of a normal ecosystem can be changed by any compound and produced intentionally or naturally. Environmental xenobiotics mostly exist as synthetic or unnatural compounds.

8.3.1.9 Food additives

Food additives can have hazardous consequences on animals and people.

8.3.1.10 Oil mixtures

Natural productsuch as oil which can be subject to the degradation of microbial and have many degradation rates for several mixtures in it. Little seepages can be handled by the capacity of biodegradation and pollution problem becomes extremely acute for the large oil spills. Mostly because few of its constituents are toxic and insolubal, oil is recalcitrant in nature. Oil spills from ships, tankers, or leaks in marine oil pipes are something that is needed to deal with to make the earth a better place.

8.3.2 Indirect sources

Xenobiotics have some indirect sources, such as pharmaceutical compounds, pesticide residues, nonsteroidal anti-inflammatory drugs etc. Pharmaceutically active compounds are released directly in the effluents from hospitals or pharmaceutical manufacturers that have performed the biologically planned effect and are either fragmented or complete when passed into the environment. These mainly include antibiotics, hormones, and anesthetics, which are passed on the common food chain and bioaccumulate in organisms [26]. Biocompatible biomaterials have been developed from synthetic polymers but they degrade in the body into poisonous substances, which are the main cause for worry. The ecological cycle is adversely affected by these indirect sources. The bioaccumulation of pesticides and biomagnification processes has a toxic effect on both people and animals.

8.3.2.1 Drugs

Pharmaceutical medicines, known as drugs, are regulated for treating disorders or enhancing body functionality; they are considered a subclass of xenobiotics. They are natural mixtures of exogenous origin. The metabolism of drugs converts the lipophilic chemical mixtures into the more readily excreted hydrophilic products. The metabolism rate regulates the intensity and duration of the pharmacological actions of drugs.

8.4 Bioavailability of xenobiotics

When synthetic xenobiotic compounds such as industrial chemicals and agro-chemicals are effectively used, they arrive in the soil where they dealt with by sorption, volatilization, assimilated by organisms, degradation, and leaching. The simple presumption is that the chemicals that are present in the soil are totally available to plant roots, soil fauna, and microorganisms through contact and direct exposure. Due to bioaccumulation, exposure to these xenobiotics increases for organisms higher up the food chain.

The chemical residues that are present in the habitat are not entirely bioavailable, and therefore their absorption by biota is below the total quantity of soil that is present [27, 28]. Nevertheless, the efficacy, toxicity, and biodegradability of xeno-biotics depend on the bioavailability of soil, which renders this concept essential to pesticide risk evaluation.

Bioavailability is the quantity of a chemical that is present in the soil that has the power to interact with the microorganisms inhibiting the habitat of the soil. Many factors affect bioavailability, such as lifetime in the climate, soil, and the micro-organisms that are involved, and the properties of the soils and chemicals. Bioavailability displays differences and changes over time when thought about in distinct contexts, such as toxicity vs. degradation or efficacy and when indirect exposure happens via the food chain.

Chemical bioavailability is the integration of several dynamic processes, such as food web uptake, advection, degradation, volatilization, and sorption and desorption etc. The chemical that is available to an organism or organism in a determined time or environment is known as the bioavailability of a chemical. Chemicals that are usually vaporized and dissolved are entirely bioavailable. However, if the chemicals are in touch with sediments or soil, then bioavailability decreases by sorption. The area of sequestration as well as sorption differs with both soild and chemical type. The idea of bioavailability can be put in any habitat compartment, i.e., soil, water, sediments, and atmosphere. Nevertheless, the idea is essential when put in the sediments and soil impurities. The bioavailability of chemicals is strongly affected by these solid environmental matrices as a result of sequestration, sorption etc. Sediments have been shown to have a notable effect on the degrading rates of chemicals, even in aqueous environments [29].

8.5 Remediation of xenobiotics

8.5.1 Phytoremediation

To eliminate or detoxify habitats of pollutants, plants, and their related organisms can be used to facilitate water or rather soil reclamation, particularly xenobiotic compounds that are available at the polluted sites. Plants take up organic pollutants such as tetrachloroethylene and trinitrotoluene, PAHs, pesticides, or herbicides and gasoline via the roots present in the inner side of tissues, where they may be degenerated with intracellular enzymes, which is called phyto-degradation. For extracellular enzymes, the rhizospheric degradation of contaminants secreted by the microbes in rhizosphere or plant roots is known as rhizodegradation. Inorganics such as metalloids or heavy metals, plant fertilizers, and radionuclides do not undergo degeneration, unlike the organic contaminants, and survive in the soil for a long period and may only change position. Plants can be used to prevent or decrease the bioavailability as well as migration of contaminants in the habitat through stabilization or rather immobilization of the soil by the roots, which is known as phyto-stabilization.

In phytoextraction (phytoaccumulation) from contaminated sites, plants absorb inorganic pollutants and then accumulate them in the harvestable parts of the plants, such as leaves, shoots,etc. These plants are known as hyperaccumulators. In hydroponic systems, a few plants are involved in the adsorption or rather inorganic absorption by the roots of the plant, which is called rhizofiltration and is used to treat industrial emissions, agricultural runoff, and radioactive and metal pollution. In the process of photo-volatilization, plants take up metalloids such as selenium (Se) and mercury (Hg), and volatile organic compounds from the soil and transform them into evaporative form, they are also released via the leaf stomata into the atmosphere [30, 31].

For phytoremediation, the genetic engineering approaches towards xenobiotics mostly consist of plant-associated microorganisms and genetic engineering of plants to improve their metabolism and increase the removal of xenobiotics.

8.5.1.1 Transgenic plants

In molecular biology, the recent progress from several technologies of sequencing and omics platforms offers a molecular tool for the characterization, identification, and isolation of structural genes. This involves not only the removal of xenobiotics but also their intergeneric transfer.

Recent developments in cell and omics technologies, molecular biology, cell imaging, genetics, and next-generation sequencing technology bioinformatics assist us in the quick identification and screening of novel proteins or genes, and also help to determine the role of the tolerance of xenobiotics and degradation. Many genes belonging to the phytochelatins, metal chelators, membrane transporters, translocations, sequestration of inorganic contaminants, and enzymes responsible for uptake have been isolated from mammals, plants, and bacteria to find the most suitable plants for phytoremediation. Plants that lack catabolic genes and an autotrophic system are responsible for the degeneration of organic compounds. Once these catabolic genes are isolated from fungi and bacteria, they can be moved to plants to convey the catabolic pathways or genes that are used in the process of improved phytoremediation for organic xenobiotics [32].

Many transgenic plants have been produced using the well-established biolistic transformation and agrobacterium-mediated methods. These techniques are unrefined because these integrated transgenes are placed at random positions in the genome [33]. Recently, there have been some advances in the editing tools for genomes, such as mega-nucleases, zinc-finger nucleases (ZFN) and clustered regularly interspaced short palindromic repeats-associated (CRISPER/Cas) genes of transcription activator-like effectors nucleases (TALEN), which provide emerging ways to targeted the moderation of genome in a wider range of organisms [34].

The technology used for the remediation of chemicals depends on the utilization of chemical substances in diminishing or rather detoxifying the deleterious effects of xenobiotics in the soil and materials such as oxidizing chemicals and solvents. Abdel-Salam et al [35] discussed chemical remediation methods for chemo-remediating heavy-metal pollution in soils, including chemicals such as chelators, immobilizers, and acids. The acids and acidic materials that are utilized for the removal of polluted soils are organic acids having low molecular weight, synthetic amino poly-carboxylic acids, acidic humic substances, and natural acids, [36]. Examples of these materials are diethylenetriaminepentaacetic acid (DTPA), N-(2-hydroxyethyl)-iminodiacetic acid (HEIDA), ethylene glycol tetra-acetic acid (EGTA), ethylenediaminetetraacetic acid (EDTA), ethylenediamine-N,N0-bis (o-hydroxyphenyl) acetic acid (EDDHA), hydroxyethylenediaminetetraacetic acid (HEDTA), trans-1,2-cyclohexylene dini-trilo-tetra-acetic acid (CDTA) and N, N0-bis(2-hydroxybenzyl) ethylenediamine-N, N0-di-acetic acid (HBED), and also solvents that dissolve insoluble water substances such as polychlorinated biphenyl (PCBs) that are associated with these materials.

Hechmi et al [37] used an organo-mineral complex (OMC) consisting of zeolite orclinoptilite and humic acid clay minerals to consume pentachlorophenols (PCP) contaminants from polluted soils. The consequences show that polluted soils have high efficiency, particularly with the use of bioremediation when combined with special microbial isolates. Dolomitic pulverized limestones were utilized to remove the polluted soils with heavy metals, with successful results. This proves that the efficacy of this technique in the stabilization of these contaminants along with biologically friendly alkali reservoirs for the long- term stabilization of metalloids and heavy metals. Consequently, Mg-carbonates can be a practical means in the remediation of polluted soils, in the form of dolomitic limestones.

8.5.2 Removal of xenobiotics by adsorption method

Activated carbon features microporous pores that help the adsorption process remove xenobiotics more effectively. The activated carbon and the process's diffusion mechanism reduce the pollutant sorption rate in the soil. The surface load density changes based on the substance, and the xenobiotic's electrostatic response with the pollutant allows the contaminant to be removed. This method effectively reduces waste by regulating pH as an important element in xenobiotic adsorption. The demand for this technology is increasing as a result of its high removal efficiency, but the primary limiting factor is the expense of the adsorption process. As a result, alternative sources with superior adsorption as well as for component regeneration are required. This has been accomplished by combining surface phenomena with bioremediation approaches to remove xenobiotics from saline wastewater [38]. Because of their high solubility, pharmaceutical-based xenobiotic chemicals are frequently non-degradable. Due to its process features, the adsorption method remains a viable alternative in this situation. Because of their capacity to totally remove pollutants without generating any byproducts, biosorbents are now popular for the treatment of pharmaceutically active substances. The sorption process is also cost-effective due to its high regeneration capacity and ability to aid in waste bioaccumulation and biomagnification [39]. Due to their microporous structure, which offers very high specific areas of up to 2000 m^2 g^{-1}, activated carbons (ACs) provide effective xenobiotic retention [40]. For a majority of xeno-biotics, such as carbamazepine and diclofenac, granular AC (GAC) filtration has achieved removal efficiencies of up to 90% [41]. Powder-activated carbon (PAC) has also been used in membrane bioreactors to achieve high removal efficiencies, with removal efficiencies of over 90% for atrazine, naproxen, and estrone [42]. Nonetheless, using ACs in the tertiary treatment stage raises a number of difficulties, including their cost and the environmental impact of the manufacturing and regeneration process [43].

8.5.3 Removal of xenobiotics by biochar

Biochar has been shown to be a potential sorbent to remove heavy metals from industrial and municipal wastewater [44, 45]. Biochar produced from agricultural residues such as rice husks and stover of maize can help to remove toxicants from the environment. The special properties of biochar are negative charge, higher surface area, resistance to degradation, functional groups, and high cation exchange capacity [46–48], which make it ideal for the remediation of habitats. In addition, it does not readily decompose because of its resistance to bacterial attack [49, 50]. Consequently, it may be reused several times. For the production of biochar at various temperatures, rice husks (RH) and maize stover (MS) as discrete feedstocks have been pyrolyzed, and have been shown to be active in the elimination of toxicants. RH and MS are mainly found in developing countries.

MS biochar yield of 28.21% was obtained by [51], whereas MS has made biochar with 35% yield [52–56]. Between the pesticides and soil, the most essential mode of interaction is probably adsorption. This process can be entirely physical (Van der Waals forces) or rather chemical (electrostatic interactions), which is the conse-quence of electrical pull between the charged particles.

The hydrological system is an essential pathway for pesticide distribution. A few of the studies based on simulation and modeling strongly suggest that the preserving OCP can also have an effect on the neighboring regions of the application site. Therefore, there are certain effects of pesticides on wildlife, which are excessively complex. For several types of pollutants, lakes and rivers are especially exposed and are known as an end-point because they are highly fragile and at low altitudes [57–59]. Bioengineering has an important role in bioremediating contaminated sites by taking into consideration OCP dynamics in water as well as soil. There is evidence that with less toxicity, plants may store, absorb, and break down particles into other products. In a built wetland system, organic matters with higher concentration can give the perfect habitat to bind the molecules of xenobiotics. The compounds that are obtainable to interact with biota decrease due to the nature of binding forces, which also decreases the compound toxicity and immobilizes it due to a decrease in their transport as well as leaching properties.

Even at very low concentration, an analytical instrument is an indispensable element to enable the detection of pharmaceutical active substance compounds. Gas chromatography (GC) and reverse phase high performance liquid chromatography (RP-HPLC) can be used alone or coupled with mass spectrometry, which are commonly used systems having less common flame detection in GC or rather less commonly used, in RP-HPLC cases, detection with diodes and fluorescence system [60, 61]. Recently, the separation techniques that are most widely used are GC and high-performance liquid chromatography (HPLC) with a C18 column [62, 63]. Another technique that is commonly used is an enzyme-linked immunosorbent assay (ELISA), which is an extremely sensitive immunoassay that involves an enzyme that is linked to an antibody or rather an antigen for specific detection of proteins as a marker.

To achieve considerable removal efficiency, aerobic and anaerobic biological processes and activated sludge are usually used. The natural attenuation processes are sorption, microbial degradation, volatilization, and dilution [64]. Bioremediation is one of the most sustainable as well as effective techniques, in which the bacterial processes decrease the exponential quantity of xenobiotics by converting them into less harmful mixtures. Consequently, the development of tools for the evaluation of *in situ* bioremediation may assist site managers to monitor the progress of remediation.

8.5.4 Nanotechnology-based approaches for removal of xenobiotics

Xenobiotics have been rapidly emerging in aquatic habitats in recent years, posing significant health risks. Because traditional treatment procedures fail to successfully remove contaminants at low dose levels, nanomaterials with substantial physico-chemical features may provide a solution to overcome these obstacles [65]. Therefore, researchers have studied the use of engineered nanoparticles and their specificity in the treatment of these pollutants. The development of innovative nanomaterials from a variety of sources for the degradation of contaminants has increased. Since nanomaterials have inherent physiochemical and magnetic proper-ties, choosing the right nanomaterial to trap a certain contaminant is critical.

The use of nanomaterials in conjunction with either plants or microorganisms is also efficient in the modification of contaminants such that they are completely removed from the ecosystem.

8.5.4.1 Nanoadsorbents

Activated carbon, porous carbon, graphitized carbon, and carbon nanotubes are all examples of carbon-based nanoadsorbents. Carbon-based nanoadsorbents interact with pollutants through hydrophobic, electrostatic, hydrogen bonding, and covalent bonding interactions. Each form contains many adsorption sites that can absorb organic contaminants due to its flexibility and physical chemistry [66]. In recent years, single-walled and multi-walled carbon nanotubes have been modified to form high-energy sites by changing their surface chemistry [67]. Toxic substances that operate efficiently as porosity increases are effectively removed from usable sorption sites using magnetic nanoparticles. Magnetic carbon nanotubes, which can be made in a variety of ways, can be utilized to change the surface via carbon dots. According to the findings of previous research, there is a high adsorption rate with a great potential for reuse [68]. The sol–gel methods are typically utilized to remove cationic xenobiotic from multiple-walled carbon nanotubes having a negatively charged surface [69]. Polymeric nanoadsorbents are porous and linked to magnetic nano-particles that form core shells [70]. This method of synthesis resulted in the successful deployment of hybrid structures for the removal of xenobiotics with strong absorption over a wide pH range.

8.5.4.2 Nanofilters

The ionic strength of filtration has increased since the electrospinning technique was implemented, lowering the organic pollutant hardness [69]. Polyvinyl fluoride cellulose acetate, polypropylene, polyacrylonitrile, and other natural and synthetic polymers are utilized to make nanoporous membranes [69]. The focus of the nanofiltration research was on a pressure-driven approach for eliminating components with low molecular weight and sizes between 10 and 1 nm. This approach filters water using membrane nanopores and hydrodynamics on the membrane surface. Filtration efficiency is largely dependent on filter membrane charge, porosity, and membrane surface concentration polarization [71]. High-quality interconnected 3D membranes are manufactured using electro-spinner technology for the development of nanofibrous membranes [72]. Separation and elimination are normally achieved through physicochemical interactions between membranes and contaminants; however, in the case of trace components such as pesticides, only nanofiltration is possible. Because of the range of molecular weights and the hydrophobicity of pesticide molecules, nanofilters had limitations for dissolved organic chemicals and uncharged insecticides. Pesticides of 11 distinct types, including aromatic, hydrophobic, and phenol chemicals, were rejected by the nanofilter membrane, and the polarity of the pesticides affected membrane filtration capacity. Because it is close to membranes, the separation of charged pesticides is successful. To avoid organic fouling, nanofilters with high hydrophilicity are made with non-polar membranes [71].

8.5.4.3 Nanofibers

Traditional water treatment procedures such as flocculation, sedimentation, coagulation, and activated carbon cannot remove organic contaminants to fulfill the required conditions, hence membrane filtration plays a significant role in water purification [73]. Nanopore membranes are made from a variety of natural and synthetic polymers, including polyvinyl fluoride, polypropylene, and polyacrylonitrile, and these nanofibers are good at removing wastewater micropollutants. Nanofibers have stable adsorption structures due to their loose bundles compared to tubes and particles. Pesticide pollutants adsorption via molecular propagation pathways is particularly important. Atrazine herbicide-adsorbing nanofibrous membranes polymerize to pyrrole. For the treatment of industrial harmful substances, nanofibers made of semiconducting materials with a photocatalytic ability can be utilized. On the breakdown of various dye compounds, composite nanofibers comprised of titanium dioxide and polymers with graphene exhibit a very significant photocatalytic impact. The elimination of arsenic had a significant impact on the presence of sodium ions and calcium ions because the ionic potential of the membrane changed when the ions came into contact with it. These nanofibers are successful in the purging of micropollutants that are endocrinologically active in the environment in wastewater treatment [74]. The adsorption of pharmaceutical antibiotics utilizing electrical spinning techniques (Ciproflaxin and Bisphenol) results in carbon nanofibers constructed of polyacrylonitrile, and the total absorbance ability is improved by a decrease in molecular dimensions [75]. The PA6 nanofibers are polymerized with nanofibers for the adsorption of atrazine herbicide [74]. The bio-polysaccharide cellulose produced from bacterial cells is used to prepare ultra-thin nanofilms. Excellent adsorbents for removing diverse POPs such as phenol, bisphenol A (BPA), and glyphosate are discovered in this film (2,4-DCP). This substance had outstanding reusable properties and showed perfect adsorption over a wide pH range [74].

8.5.4.4 Nanocomposites

In contrast to micro-composite and monolithic agents, nanocomposites are organic matrices that are substantially influenced by the adsorption of pollutants [76]. Through the use of adjustable features such as electrical, mechanical, and magnetic capabilities, new hybrid matrices have been devised that efficiently store pollutants and release harmful payload. Nanocomposites are promising pollutant removal materials due to their recycling potential in comparison to other materials [77]. Nanocomposites can be made using a variety of methods, including co-precipitation [78], hydrothermal thermal deposition synthesis [79], sol–gel synthesis [80], microwave synthesis [81, 85], and chemical vapor deposition [82]. Although nanocomposite xenobiotic removal utilizing adsorbents, ion exchangers, or photocatalysts offers numerous advantages, it also has certain drawbacks in terms of application stability. Because there is a risk of metal ions leaching fromnanocomposites, the commercial manufacturing of nanocomposites is hampered by environmental maintenance and production costs.

8.5.4.5 *Advantages and drawbacks of nanotreatment methods*

Nanomaterials in the form of sorbents for pollutant removal are unique. The treatment procedures are influenced by temperature, pH, nanoadsorbent dosages, and contact duration. However, several limitations on the use of nanotechnology have arisen in recent years. Each government has its own set of structures, which are unique to each country. Many of the problems associated with pollution treatment are being addressed for the first time in nanotechnology. In the transport and processing of pollutants in the environment, nanoparticles provide bioavailability. In the preparation of nanomaterials, biodegradable materials are often used to avoid bioaccumulation. The new technology studies were carried out to focus on pollutant capture at the location, avoiding the challenges associated with non-targeted methods. Engineered nanoparticles are made with particular qualities, such as targeted pollutant capture and the use of biodegradable raw materials that can regenerate and recycle nanomaterials, mostly for the treatment of pollutants. Furthermore, the synthesis procedure should avoid the production of agglomerates and stable nanomaterials. The possibility for toxicity, the production of byproducts during synthesis, and the recovery costs should all be carefully considered. Knowing these processes for producing potential nanoparticles for pollutant removal in the environment can help with process design and optimization. Several studies have attempted to eliminate the halogenated xenobiotic chemicals that represent a serious environmental hazard. To do so, conventional bioremediation methods are combined with nanotechnology methodologies to remove such substances from the ecosystem's surface layers. The concentration levels of the xenobiotic may increase or decrease in response to severe natural climatic conditions, and organisms may become sensitive to the components' elimination. Advances in next-generation biotechnological technologies, as well as the functionalization of nanomaterials, provide a viable approach for pollution remediation in such circumstances [83, 84]. Even though the nanoparticles were removed, some were discharged unintentionally, which is undesirable for maintaining a sustainable ecosystem. However, there is no proof that employing the components at the nanoscale is safe. Meanwhile, nanobioremediation approaches are based on the concept of biological removal procedures, which have a degradation advantage over chemical treatment methods.

8.6 Challenges and perspectives of xenobiotics removal

Every year, the release of xenobiotic compounds increases in dose levels as the number of consumers grows. Conventional procedures have been used to treat contaminants at the ground and surface water levels, but they fall short of achieving zero waste removal. Despite the fact that chemical treatment technologies have a larger potential for removing xenobiotic compounds, most areas of practicality and safety have concerns from an economic standpoint. Following the introduction of nanotechnology principles, a slew of new treatment solutions for selective waste removal have emerged. Nanomaterial use poses environmental problems in terms of manufacturing techniques and uses. To avoid the release and transport of nanoparticles in the environment, many ways have been taken to improve the properties of the nanomaterial utilizing engineering concepts. Until now, most of the challenges to xenobiotic treatment have been overcome utilizing nanotechnology in laboratory tests. Based on the findings, it was determined that the research may be enhanced in the following areas to increase xenobiotic chemical elimination. The main findings are that the surface properties of nanoadsorbents can be altered and effectively used for contaminant removal from surface water, that photocatalysts with interface properties can be used for efficient degradation of dye-based components, and that nanomaterials can be improved in terms of reusability, material cost, and lifetime using computational methods.

Consent for publication

Not applicable.

Conflict of interest

The author confirms that this chapter contents have no conflict of interest.

Acknowledgments

Authors are thankful to Dr Gaurav Deep Singh, Chancellor,SardarBhagwan Singh University, Balawala, Dehradun, Uttarakhand, India, for providing facility, space, and resource to conduct this work.

References

[1] Thakur I S 2006 *Environmental Biotechnology* (New Delhi: IK International)
[2] Thakur I S 2008 Xenobiotics: pollutants and their degradation-methane, benzene, pesticides, bioabsorption of metals. https://nsdl.niscpr.res.in/bitstream/123456789/664/1/Xenobiotics.pdf
[3] Jagtap U B and Bapat V A 2017 Transgenic approaches for building plant armor and weaponry to combat xenobiotic pollutants: current trends and future prospects *Xenobiotics in the Soil Environment* (Berlin: Springer) pp 97–215
[4] Qadir A, Hashmi M Z and Mahmood A 2017 Xenobiotics, types, and mode of action *Xenobiotics in the Soil Environment* (Berlin: Springer) pp 1–7
[5] Hayyat A, Javed M, Rasheed I, Ali S, Shahid M J, Rizwan M, Javed M T and Ali Q 2016 Role of biochar in remediating heavy metals in soil *Phytoremediation* (Berlin: Springer) pp 421–37
[6] Khan Aand Rao T S 2019 Molecular evolution of xenobiotic degrading genes and mobile DNA elements in soil bacteria *Microbial Diversity in the Genomic Era* (Amsterdam: Elsevier) pp 657–78
[7] Yuan P, Wang J, Pan Y, Shen B and Wu C 2019 Review of biochar for the management of contaminated soil: preparation, application and prospect *Sci. Total Environ.* **659** 473–90
[8] Chakraborty I, Sathe S, Khuman C and Ghangrekar M 2020 Bioelectrochemically powered remediation of xenobiotic compounds and heavy metal toxicity using microbial fuel cell and microbial electrolysis cell *Mater. Sci. Energy Technol.* **3** 104–15
[9] Amin M, Alazba A and Manzoor U 2014 A review of removal of pollutants from water/wastewater using different types of nanomaterials *Adv. Mater. Sci. Eng.* **2014** 1–25

[10] Hoai P M, Ngoc N T, Minh N H, Viet P H, Berg M, Alder A C and Giger W 2010 Recent levels of organochlorine pesticides and polychlorinated biphenyls in sediments of the sewer system in Hanoi, Vietnam *Environ. Pollut.* **158** 913–20

[11] Al-Wabel M, El-Saeid M, Al-Turki A and Abdel-Nasser G 2011 Monitoring of pesticide residues in Saudi Arabia agricultural soils *Res. J. Environ. Sci.* **5** 269

[12] Abhilash P and Singh N 2009 Pesticide use and application: an Indian scenario *J. Hazard. Mater.* **165** 1–12

[13] Liu J, Cui Z, Xu H and Tan F 2009 Dioxin-like polychlorinated biphenyls contamination and distribution in soils from the Modern Yellow river delta, China *Soil Sediment Contam.* **18** 144–54

[14] Kumarasamy P, Govindaraj S, Vignesh S, Rajendran R B and James R A 2012 Anthropogenic nexus on organochlorine pesticide pollution: a case study with Tamiraparani river basin, South India *Environ. Monit. Assess.* **184** 3861–73

[15] Xu Y, Tian C, Ma J, Zhang G, Li Y F, Ming L, Li J, Chen Y and Tang J 2012 Assessing environmental fate of β-HCH in Asian soil and association with environmental factors *Environ. Sci. Technol.* **46** 9525–32

[16] Pessoa G d P, Santos A B d, Souza N Cd, Alves J A C and Nascimento R F d 2012 Development of methodology to determine estrogens in wastewater treatment plants *Quím. Nova* **35** 968–73

[17] Brack W, Altenburger R, Schüürmann G, Krauss M, Herráez D L, van Gils J, Slobodnik J, Munthe J, Gawlik B M and van Wezel A 2015 The SOLUTIONS project: challenges and responses for present and future emerging pollutants in land and water resources management *Sci. Total Environ.* **503** 22–31

[18] Klaschka U, von der Ohe P C, Bschorer A, Krezmer S, Sengl M and Letzel M 2013 Occurrences and potential risks of 16 fragrances in five German sewage treatment plants and their receiving waters *Environ. Sci. Pollut. Res.* **20** 2456–71

[19] Hollert H, CrawfordS E, Brack W, Brinkmann M, Fischer E, Hartmann K, Keiter S, Ottermanns R, OuelletJ D and Rinke K 2018 Looking back-looking forward: a novel multi-time slice weight-of-evidence approach for defining reference conditions to assess the impact of human activities on lake systems *Sci. Total Environ.* **626** 1036–46

[20] Kaur P and Parihar L 2014 Bioremediation: step towards improving human welfare *Annu. Res. Rev. Biol.* **4** 3150–64

[21] Breivik K, Alcock R, Li Y F, Bailey R E, Fiedler H and Pacyna J M 2004 Primary sources of selected POPs: regional and global scale emission inventories *Environ. Pollut.* **128** 3–16

[22] Bailey R E 2001 Global hexachlorobenzene emissions *Chemosphere* **43** 167–82

[23] Brown J, Frame G, Olson D and Webb J 1995 The sources of the coplanar PCBs *Organohalogen Compd.* **26** 427–30

[24] Lohmann R, Northcott G L and Jones K C 2000 Assessing the contribution of diffuse domestic burning as a source of PCDD/Fs, PCBs, and PAHs to the UK atmosphere *Environ. Sci. Technol.* **34** 2892–9

[25] Breivik K, Sweetman A, PacynaJ M and Jones K C 2002 Towards a global historical emission inventory for selected PCB congeners—a mass balance approach: 2. Emissions *Sci. Total Environ.* **290** 199–224

[26] Varsha Y, Naga Deepthi C and Chenna S 2011 An emphasis on xenobiotic degradation in environmental clean up *J. Bioremed. Biodegrad.* **2011** 1–10

[27] Alexander M 1995 How toxic are toxic chemicals in soil? *Environ. Sci. Technol.* **29** 2713–7

[28] Paine M D, Chapman P M, Allard P J, Murdoch M H and Minifie D 1996 Limited bioavailability of sediment PAH near an aluminum smelter: contamination does not equal effects *Environ. Toxicol. Chem: Int. J.* **15** 2003–18

[29] Rice P J, Anderson T A and Coats J R 2004 Effect of sediment on the fate of metolachlor and atrazine in surface water *Environ. Toxicol. Chem.* **23** 1145–55

[30] Doty S L 2008 Enhancing phytoremediation through the use of transgenics and endophytes *New Phytol.* **179** 318–33

[31] Pilon-Smits E 2005 Phytoremediation *Annu. Rev. Plant Biol.* **56** 15–39

[32] Van Aken B 2008 Transgenic plants for phytoremediation: helping nature to clean up environmental pollution *Trends Biotechnol.* **26** 225–7

[33] Shabbir R, Javed T, Afzal I, Sabagh A E, Ali A, Vicente O and Chen P 2021 Modern biotechnologies: innovative and sustainable approaches for the improvement of sugarcane tolerance to environmental stresses *Agronomy* **11** 1042

[34] Liu W, Yuan J S and Stewart Jr C N 2013 Advanced genetic tools for plant biotechnology *Nat. Rev. Genet.* **14** 781–93

[35] Abdel-Salam A A, Salem H M, Abdel-Salam M A and Seleiman M F 2015 Phytochemical removal of heavy metal-contaminated soils *Heavy Metal Contamination of soils* (Berlin: Springer) pp 299–309

[36] Evangelou M W, Ebel M and Schaeffer A 2006 Evaluation of the effect of small organic acids on phytoextraction of Cu and Pb from soil with tobacco *Nicotiana tabacum Chemosphere* **63** 996–1004

[37] Dercova K, Sejakova Z, Skokanova M, Barancikova G and Makovnikova J 2006 Potential use of organomineral complex (OMC) for bioremediation of pentachlorophenol (PCP) in soil *Int. Biodeterior. Biodegrad.* **58** 248–53

[38] Zhang Y, Sun X, Bian W, Peng J, Wan H and Zhao J 2020 The key role of persistent free radicals on the surface of hydrochar and pyrocarbon in the removal of heavy metal-organic combined pollutants *Bioresour. Technol.* **318** 124046

[39] Adewuyi A 2020 Chemically modified biosorbents and their role in the removal of emerging pharmaceutical waste in the water system *Water* **12** 1551–60

[40] Tahar A, Choubert J M and Coquery M 2013 Xenobiotics removal by adsorption in the context of tertiary treatment: a mini review *Environ. Sci. Pollut. Res.* **20** 5085–95

[41] Ruel S M, Choubert J, Esperanza M, Miege C, Navalón Madrigal P, Budzinski H, Le Ménach K, Lazarova V and Coquery M 2011 On-site evaluation of the removal of 100 micro-pollutants through advanced wastewater treatment processes for reuse applications *Water Sci. Technol.* **63** 2486–97

[42] Snyder S A, Adham S, Redding A M, Cannon F S, DeCarolis J, Oppenheimer J, Wert E C and Yoon Y 2007 Role of membranes and activated carbon in the removal of endocrine disruptors and pharmaceuticals *Desalination* **202** 156–81

[43] Verlicchi P, Galletti A, Petrovic M and Barceló D 2010 Hospital effluents as a source of emerging pollutants: an overview of micropollutants and sustainable treatment options *J. Hydrol.* **389** 416–28

[44] Demirbas A 2008 Heavy metal adsorption onto agro-based waste materials: a review *J. Hazard. Mater.* **157** 220–9

[45] Inyang M, Gao B, Ding W, Pullammanappallil P, Zimmerman A R and Cao X 2011 Enhanced lead sorption by biochar derived from anaerobically digested sugarcane bagasse *Sep. Sci. Technol.* **46** 1950–6

[46] Munera-Echeverri J L, Martinsen V, Strand L T, Zivanovic V, Cornelissen G and Mulder J 2018 Cation exchange capacity of biochar: an urgent method modification *Sci. Total Environ.* **642** 190–7

[47] Sizmur T, Quilliam R, Puga A P, Moreno-Jiménez E, Beesley L and Gomez-Eyles J L 2016 Application of biochar for soil remediation *Agric. Environ. Appl. Biochar: Adv. Barr.* **63** 295–324

[48] Zhang K, Sun P, Faye M C A and Zhang Y 2018 Characterization of biochar derived from rice husks and its potential in chlorobenzene degradation *Carbon* **130** 730–40

[49] Kamara A, Kamara A, Mansaray M M and Sawyerr P A 2014 Effects of biochar derived from maize stover and rice straw on the germination of their seeds *Am. J. Agric. Forest.* **2** 246–9

[50] Karn S K, Satya J E, Rajput V D, Kumar S and Kumar A 2017 Modeling of simultaneous application of *Vibrio* sp. (SK1) and biochar amendment for removal of pentachlorophenol (PCP) in soil *Environ. Eng. Sci.* **34** 551–61

[51] Mohan D, Kumar S and Srivastava A 2014 Fluoride removal from ground water using magnetic and nonmagnetic corn stover biochars *Ecol. Eng.* **73** 798–808

[52] Moyo G G, Hu Z and Getahun M D 2020 Decontamination of xenobiotics in water and soil environment through potential application of composite maize stover/rice husk (MS/RH) biochar—a review *Environ. Sci. Pollut. Res.* **27** 28679–94

[53] Yang T, Meng J, Jeyakumar P, Cao T, Liu Z, He T, Cao X, Chen W and Wang H 2021 Effect of pyrolysis temperature on the bioavailability of heavy metals in rice straw-derived biochar *Environ. Sci. Pollut. Res.* **28** 2198–208

[54] Yang X, Meng J, Lan Y, Chen W, Yang T, Yuan J, Liu S and Han J 2017 Effects of maize stover and its biochar on soil CO_2 emissions and labile organic carbon fractions in Northeast China Agric *Ecosyst. Environ.* **240** 24–31

[55] Cheng S, Chen T, Xu W, Huang J, Jiang S and Yan B 2020 Application research of biochar for the remediation of soil heavy metals contamination: a review *Molecules* **25** 3167

[56] Yang F, Sui L, Tang C, Li J, Cheng K and Xue Q 2021 Sustainable advances on phosphorus utilization in soil via addition of biochar and humic substances *Sci. Total Environ.* **768** 145106

[57] Gong P, Wang X, Sheng J and Yao T 2010 Variations of organochlorine pesticides and polychlorinated biphenyls in atmosphere of the Tibetan Plateau: role of the monsoon system *Atmos. Environ.* **44** 2518–23

[58] Wang X, Gong P, Yao T and Jones K C 2010 Passive air sampling of organochlorine pesticides, polychlorinated biphenyls, and polybrominated diphenyl ethers across the Tibetan Plateau *Environ. Sci. Technol.* **44** 2988–93

[59] Zhang G, Chakraborty P, Li J, Sampathkumar P, Balasubramanian T, Kathiresan K, Takahashi S, Subramanian A, Tanabe S and Jones K C 2008 Passive atmospheric sampling of organochlorine pesticides, polychlorinated biphenyls, and polybrominated diphenyl ethers in urban, rural, and wetland sites along the coastal length of India *Environ. Sci. Technol.* **42** 8218–23

[60] Ratola N, Cincinelli A, Alves A and Katsoyiannis A 2012 Occurrence of organic micro-contaminants in the wastewater treatment process. A mini review *J. Hazard. Mater.* **239** 1–18

[61] Dębska J, Kot-Wasik A and Namieśnik J 2004 Fate and analysis of pharmaceutical residues in the aquatic environment *Crit. Rev. Anal. Chem.* **34** 51–67

[62] Kolpin D W, Furlong E T, Meyer M T, Thurman E M, Zaugg S D, Barber L B and Buxton H T 2002 Pharmaceuticals, hormones, and other organic wastewater contaminants in US streams 1999–2000: a national reconnaissance *Environ. Sci. Technol.* **36** 1202–11

[63] Buxton H T and Kolpin D W 2005 Pharmaceuticals, hormones, and other organic wastewater contaminants in US streams *Water Encycl.* **5** 605–8

[64] Bombach P, Richnow H H, Kästner M and Fischer A 2010 Current approaches for the assessment of *in situ* biodegradation *Appl. Microbiol. Biotechnol.* **86** 839–52

[65] Cai Z, Dwivedi A D, Lee W N, Zhao X, Liu W, Sillanpää M, Zhao D, Huang C H and Fu J 2018 Application of nanotechnologies for removing pharmaceutically active compounds from water: development and future trends *Environ. Sci.: Nano* **5** 27–47

[66] Kurwadkar S, Hoang T V, Malwade K, Kanel S R, Harper W F and Struckhoff G 2019 Application of carbon nanotubes for removal of emerging contaminants of concern in engineered water and wastewater treatment systems *Nanotechnol. Environ. Eng.* **4** 1–16

[67] Rushi A D, Datta K P, Ghosh P S, Mulchandani A and Shirsat M D 2019 Functionalized carbon nanotubes for detection of volatile organic pollutant *Perspective of Carbon Nanotubes* (London: IntechOpen)

[68] Deng Y, Ok Y S, Mohan D, Pittman C U and Dou X 2019 Carbamazepine removal from water by carbon dot-modified magnetic carbon nanotubes *Environ. Res.* **169** 434–44

[69] Konicki W and Pełech I 2019 Removing cationic dye from aqueous solutions using as-grown and modified multi-walled carbon nanotubes *Pol. J. Environ. Stud.* **28** 717–27

[70] Moharrami P and Motamedi E 2020 Application of cellulose nanocrystals prepared from agricultural wastes for synthesis of starch-based hydrogel nanocomposites: efficient and selective nanoadsorbent for removal of cationic dyes from water *Bioresour. Technol.* **313** 123661

[71] Abdel-Fatah M A 2018 Nanofiltration systems and applications in wastewater treatment *Ain Shams Eng. J.* **9** 3077–92

[72] Vaz B, Costa J A V and Morais M G d 2018 Innovative nanofiber technology to improve carbon dioxide biofixation in microalgae cultivation *Bioresour. Technol.* **273** 592–8

[73] Obotey Ezugbe E and Rathilal S 2020 Membrane technologies in wastewater treatment: a review *Membranes* **10** 89

[74] Khalil A M and Schäfer A I 2021 Cross-linked β-cyclodextrin nanofiber composite membrane for steroid hormone micropollutant removal from water *J. Membr. Sci.* **618** 118228–38

[75] Li X, Chen S, Fan X, Quan X, Tan F, Zhang Y and Gao J 2015 Adsorption of ciprofloxacin, bisphenol and 2-chlorophenol on electrospun carbon nanofibers: in comparison with powder activated carbon *J. Colloid Interface Sci.* **44** 7120–127

[76] Srivastava V, Zare E N, Makvandi P, Zheng X, Iftekhar S, Wu A, Padil V V, Mokhtari B, Varma R S and Tay F R 2020 Cytotoxic aquatic pollutants and their removal by nanocomposite-based sorbents *Chemosphere* **258** 127324–33

[77] Guerra-Rodríguez S, Rodríguez E, Singh D N and Rodríguez-Chueca J 2018 Assessment of sulfate radical-based advanced oxidation processes for water and wastewater treatment: a review *Water* **10** 1828–37

[78] Sahu S, Sahu U K and Patel R K 2018 Synthesis of thorium–ethanolamine nanocomposite by the co-precipitation method and its application for Cr (vi) removal *New J. Chem.* **42** 5556–69

[79] Nasrollahi Z, Ebrahimian Pirbazari A, Hasan-Zadeh A and Salehi A 2019 One-pot hydrothermal synthesis and characterization of magnetic nanocomposite of titania-deposited copper ferrite/ferrite oxide for photocatalytic decomposition of methylene blue dye *Int. Nano Lett.* **9** 327–38

[80] Jaramillo-Fierro X, González S, Jaramillo H A and Medina F 2020 Synthesis of the ZnTiO$_3$/TiO$_2$ nanocomposite supported in ecuadorian clays for the adsorption and photocatalytic removal of methylene blue dye *Nanomaterials* **10** 1891–900

[81] Shah M P 2021 *Removal of Refractory Pollutants from Wastewater Treatment Plants* (Boca Raton, FL: CRC Press)

[82] Shah Maulin P 2021 *Removal of Emerging Contaminants through Microbial Processes* (Berlin: Springer)

[83] Omo-Okoro P N, Maepa C E, Daso A P and Okonkwo J O 2020 Microwave-assisted synthesis and characterization of an agriculturally derived silver nanocomposite and its derivatives *Waste Biomass Valorization* **11** 2247–59

[84] Wei X, Wang X, Gao B, Zou W and Dong L 2020 Facile ball-milling synthesis of CuO/biochar nanocomposites for efficient removal of Reactive Red 120 *ACS Omega* **5** 5748–55

[85] Vázquez-Núñez E, Molina-Guerrero C E, Peña-Castro J M, Fernández-Luqueño F and de la Rosa-Álvarez M 2020 Use of nanotechnology for the bioremediation of contaminants: a review *Processes* **8** 826–35

IOP Publishing

Trends in Biological Processes in Industrial Wastewater Treatment

Maulin P Shah

Chapter 9

Fungi-based biosensing platforms for detection of heavy metals: focus on the eukaryotic system

Ankur Singh, Vipin Kumar and Sarika

The past few decades have witnessed a rapid proliferation of biosensing devices in a wide variety of areas, including environmental applications. They provide an alternative platform that has the advantage of simplicity, portability, low cost, and low energy requirements over conventional techniques. The detection and quantification of heavy metals in environmental samples is one such example. The sensing platforms may vary from optical to electrochemical based on the type of biocomponent used in the biosensing device. Thus, in this chapter fungi-based sensors have been divided into two broad categories: the optical platform and the electrochemical platform. This chapter focuses on fungi-derived biosensors for the detection of heavy metals. Various aspects of fungi as a biomaterial or as a source of biomolecules for use in heavy metal sensing have been explored. The focus is on the fungi cells for adsorption-based electrochemical and optical sensors, fungi-derived enzyme-based sensors, fungi-derived protein, and organic polymer-based sensors. A brief review of the challenges and the lacunae in the research field in this direction have also been highlighted.

9.1 Introduction to biosensors and their environmental applications

Quantitative analysis of various contaminants in natural resources is essential, not only to ensure human health but also the health of the ecosystem. Thus, monitoring of the contaminant concentrations in the effluents discharged into the environment from various industrial and commercial sources has become an essential requirement under various legal constraints all over the world. The detection of contaminants such as agrochemicals, polycyclic aromatic compounds, chemical solvents, complex compounds, and heavy metals in environmental samples has become a vital part of environmental management and monitoring. The technologies used for the detection and quantification of the contaminants vary widely, from conventional analytical tools such as titration and spectrometry to modern technologies such as electrochemistry and atomic spectroscopy. However, a new branch of analytical devices, biosensors, is on the rise, and is considered to be the future of medicine as well as environmental health monitoring devices.

The past few decades have witnessed a rapid proliferation in biosensing technology. These analytical tools have been primarily used for medical and healthcare purposes. However, their versatile nature has made them useful for environmental applications as well. Interestingly, in the last 50 years (1970–2020) out of the total biosensors reported, 12.38% find environmental applicability (Siontorou and Georgopoulos 2021). In addition, of the total patented biosensors (Google patient) for the same duration, >56% are for environmental applications. This can be seen as the foreshadowing of a revolution in the field of environmental sensing, and monitoring technology. This can also be corroborated by the ever-rising contamination of natural resources, including soil, air, and water. Thus, the need for quality monitoring of these resources in day-to-day life has become essential, and biosensors are expected to play a key role.

9.1.1 Biosensors and their components

Biosensors can be defined as hybrid analytical devices comprising the characteristics of a physical and chemical reaction sensing technology (Buenger *et al* 2012). However, the International Union of Pure and Applied Chemistry (IUPAC) defines them as a device that uses specific biochemical reactions mediated by isolated enzymes, immunosystems, tissues, organelles, or whole cells to detect chemical compounds, usually by electrical, thermal, or optical signals (IUPAC 1997). Thus, a biosensor essentially has a biological component, which can be a single cell, a group of cells, tissue, or a molecule of biological origin, such as DNA, RNA, protein, enzyme, polymer, etc. A typical biosensor is characterized by its three-component system: (1) a receptor, which is essentially biological and may even be a single cell or molecule of biological origin; (2) a transducer, which transforms the signal from one form to another for better understanding; and (3) microelectronics, which function as a reporter and display the signal to the user in readable form (Velusamy *et al* 2022).

The function of the receptor is to generate a biochemical signal in response to an external stimulus, which can be the presence of or change in the concentration of ions or molecules. This generated signal is transformed by the transducer into another form that can be better understood by us. For instance, enzymatic activity is translated to an electrical signal, or an optical signal is transformed into an electrical form. This transformed signal is then amplified, processed, and compared to standard data, and displayed to the user as result by the reporter (Tripathi *et al* 2023). In the case of an optical signal, which involves a change in color, and bioluminescence, the need for a separate reporter is eliminated by the nature of the system—the microbial cells themselves act as receptors and processing also functions as a reporter, which can directly be visualized for drawing inference. The flow of the signal in a typical electrochemical sensor through its component can be seen in figure 9.1(A).

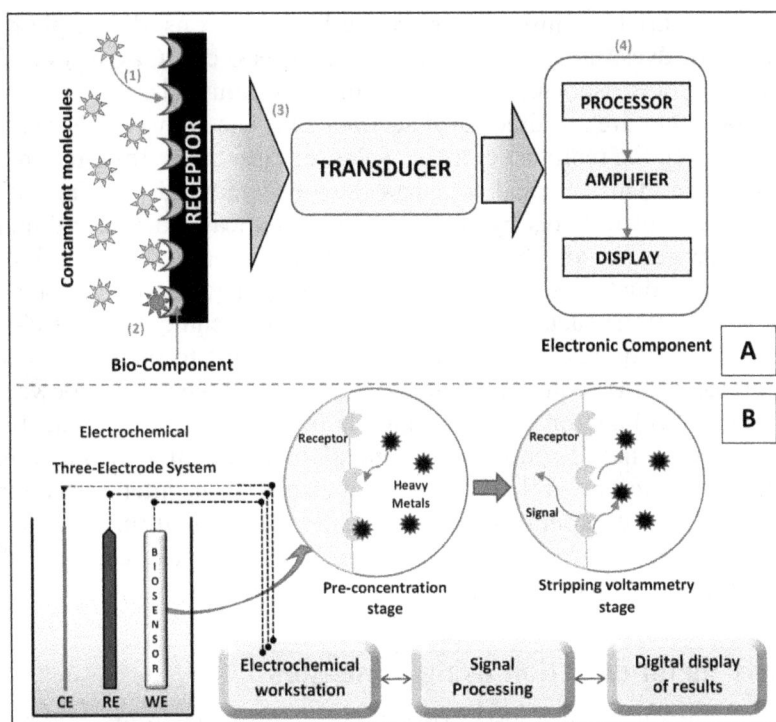

Figure 9.1. (A) Signal flow through the components of a typical biosensor: (1) The attachment of molecule of interest with the active receptor component. (2) Interaction of the molecules to produce signal. (3) Transduction of signal to electrical form. (4) Processing and the display of the signal into a readable form using the electronic component. (B) The components of an electrochemical sensor for heavy metals and its functioning.

The three components described above are also used as a basis to classify the sensors. For example, based on the recognition component, enzyme-based sensor, whole-cell-based sensor, aptamer sensor, etc, based on transducer there can be an optical sensor, electrochemical sensor, piezoelectric sensor, etc. The suitability and performance of a biosensor are often defined in terms of its, calibration curve (high correlation between analyte and response), the linear range of detection (of magnitude greater than two), sensitivity (higher slope of calibration curve), limit of detection (minimum concentration that can be detected), and the selectivity (inhibition of not more than 10%) (Siontorou and Georgopoulos 2021). However, the biosensor is expected to have good reproducibility and stability at conditions that are relatively easier to maintain, such as room temperature (Velusamy *et al* 2022). Additionally, reusability can significantly reduce the cost of analysis.

9.1.2 Advantages of biosensors over conventional technology

The era of biosensors has revolutionized the field of environmental analytical chemistry. Sensors can be designed for a wide range of target analytes, such as heavy metals, a variety of insecticides and herbicides, toxins, drugs, pharmaceutical compounds, dyes, nutrients, pathogenic microbes, etc (Gavrilaş *et al* 2022). The evolution of biosensors has led to the miniaturization and cost reduction of the analytical tools that are required for detection. This has allowed the migration of technology to even the remotest of areas, fields, or even house-to-house in the case of medical usage. This magical tool has thus also reduced the time taken for lengthy analytical procedures by conventional analytical tools. The miniaturization of these devices can also be considered to be an eco-friendly evolution thanks to the dramatic reduction in the per-sample energy required for the analysis. A conventional technique such as atomic spectroscopic technique or the chromatographic technique requires a huge amount of energy, due to the heavy machinery that is operated for the required results. In contrast, biosensors are powered by a small amount of voltage or current, which is often provided by small batteries, USB plugs, or even mobile phones. The energy requirement is also eliminated in the case of optical biosensors that rely on color changes. The interconnection of the biosensor with the computational system allows the real-time value of the analytical component or toxicity of the sample to be measured, thus allowing online monitoring (Kumar *et al* 2022).

9.2 Heavy metal detection using biosensors

Heavy metals are among the most persistent pollutants in environmental resources, including water, air, and soil. Thus, they remain in samples without degrading for very long durations. However, they might not be found in free ionic form, some fraction tends to bind with organic matter. Since most of the biosensors rely on the ionic form of the metals to be detected, a pretreatment of samples is required if the organic content is high in water. A wide number of sensors are used for heavy metal detection, from the catalysis group such as enzyme-based sensors (Ashrafi *et al* 2019) as well as affinity-based sensors such as aptamer-based sensors (McConnell *et al* 2020), protein-based sensors (Amin *et al* 2020), biomass-based sensors (Singh *et al* 2023), etc. Each category has its own advantages and disadvantages, depending on the specific elements used. Based on the transducing platform used for the signal relay, these sensors can be broadly categorized into two categories: electrochemical platforms and optical platforms. The optical platforms often include live whole-cell-based sensors, while the electrochemical sensors mostly rely on organic molecules for selective signaling. Even though the electrochemical sensors are more robust, most of the sensors fall into the optical biosensors, which mostly rely on biotechnological tools for development (Singh and Kumar 2021).

Electrochemical sensors can include whole cells, enzymes, proteins, and other biomolecules as sensing components. A generalized graphical design of an affinity-based electrochemical sensor is given in figure 9.1(B). In addition to these two platforms, other platforms such as microbial electrochemical systems (Do *et al* 2022) and mass-based sensors (Fu *et al* 2014) have also been reported for heavy metal detection.

9.2.1 The optical platform of heavy metal biosensing

The optical platform of heavy metal biosensors mostly relies on the whole-cell-based sensor, which uses metal-induced transcription of a reporter gene or activation of a protein for biosensing (Liu *et al* 2022). The activation reporter proteins that are synthesized or activated by the transcription of the reporter genes are often bioluminescent, fluorescent, or chromogenic, which leads to the development of specific colors (Adeniran *et al* 2015, Shah Maulin 2020). The specific combination of metal-specific receptors such as *MerR* and *ArsR*, with promoter genes such as *Pmer* and *Pars*, and reporters such as *mCherry*, *gfp*, and *RFP* require biotechnological amalgamation in genetic material from different organisms (Singh and Kumar 2021). This genetic material is then loaded into the host microbial cell for expression, resulting in a biosensor. These genetically engineered cells are often immobilized or entrapped in polymers such as agar-agar and gelatin, or encapsulated in microfluidic systems or a microstructure surface for final application with the test samples (Jarque *et al* 2016).

Some optical biosensors rely on the color change process induced by the metals in polymer composites instead of living cells. Macromolecules such as chitin, cellulose, nanocrystalline cellulose, and DNA may be combined with artificial conducting polymers such as polythiophene, polypyrrole, and polyaniline to achieve an optical sensor (Ramdzan *et al* 2020, Shah 2021). The working principle of electrochemiluminescence, fluorescence, and surface plasmon resonance.

9.2.2 The electrochemical platform for biosensing

Electrochemical sensors use biological material that can be immobilized or coated on the surface of an electrode, which serves as a working electrode in the three-electrode system. These receptor components can include a wide range of biological materials, such as dead microbial biomass (Singh *et al* 2023), live microbial cells, microbial polymers (Zhao *et al* 2022), DNA aptamers (Abu-Ali *et al* 2019), enzymes (Ashrafi *et al* 2019), etc. The working electrodes can also vary in shape and size, or be precoated to make them more compatible. The electrodes may be paper-based, silicon-chip based, or classical rod-shaped made with materials such as graphite, glassy carbon, metals, conducting composite, or conducting ink based, that can be printed onto surfaces to achieve a circuit. These electrodes are typically used in a three-electrode system for the detection of metals.

Several electrochemical techniques are considered suitable techniques depending on the biological material and the metal of interest. These techniques commonly include cyclic voltammetry (CV), differential pulse voltammetry (DPV), square wave voltammetry (SWV), and impedance spectroscopy (EIS), while techniques such as photoelectric electrochemistry (PEC) and electrochemical luminescence (ECL) are less commonly applied for electrochemical sensors (Chen *et al* 2022). The electrochemical sensor has three basic steps: the interaction of the biocomponent on the electrode with metal ions in the sample, signal generation by the technique used (voltammetry, impedance spectroscopy, etc), and signal transfer and processing by the electrochemical workstation (figure 9.1(B)).

9.3 Fungi-based biosensors for heavy metals

9.3.1 Fungi-based biosensors: eukaryotic systems vs the prokaryotic system

All fungal cells, including unicellular yeast, belong to the category of the eukaryotic system that possess double membrane bounded cell organelles. Thus, the response of sensors comprising these cells to heavy metals is often different compared to the prokaryotic cell-based systems. The eukaryotic whole-cell sensor is considered to more reliable than prokaryotic cells when the results are extrapolated for the higher organism. In other words, the relevance of prokaryotic-based sensors is low for the eukaryotic system, including higher organisms and humans (Jarque *et al* 2016). However, the permeability of the outermost layer is a challenge and the cell wall can be seen as a limiting factor, which tends to restrict the direct contact of the protoplast with the metal ions (Gutiérrez *et al* 2015). For this reason, cells with no cell wall, e.g., ciliates and animal cells, might be considered to give more reliable cells.

Fungal cells, both live and dead, can be excellent sorbents for heavy metals. In fact, dead fungal biomass is observed to possess better adsorption and desorption parameters compared to live cells (Ayele *et al* 2021). Thus, these dead cells have been used as a biological component in several studies. The abundance of functional groups coupled with the capability to undergo chemical modifications make them excellent sorbents, and thus help in the preparation of affinity-based electrochemical sensors.

9.3.2 Fungal cells based optical sensors

Optical sensors developed for heavy metals have mostly been developed using unicellular microbes (yeast) and some multicellular-based fungi myceliums. A single-cell-based sensor may include different species of yeast designed for optical transduction in the presence of specific heavy metals. The most commonly used eukaryotic fungi are *Saccharomyces cerevisiae*, which has been reported to be efficient due to its unicellular nature. *S. cerevisiae SEO1* has been used with a dual luciferase reporter gene for the detection of Cu(II) (Roda *et al* 2011). In a similar study, Shetty *et al* (2004) used fluorescent protein reporter gene *gfp*-based *S. cerevisiae* for Cu(II) and Ag(I) detection. Copper is the most common metal to be detected using yeast-based sensors (Khanam *et al* 2020, Fan *et al* 2021, 2022), using the prevent Cu-induced CUP1 gene. In another case, Park *et al* (2007) used methylotrophic yeast *Hansenula polymorpha* to detect Cd(II) in water samples.

The use of multicellular fungi can be equally effective and has been demonstrated using several species. Naturally, luminescent fungi *Armillaria mellea* and *Mycena citricolor* have been used to test the toxicity of Cu(II) and Zn(II) (Weitz *et al* 2002). Multicellular fungi can also be engineered for optical sensing, such as the *Botrytis cinerea*, which has been engineered with reactive oxygen species sensitive green fluorescent protein (*roGFP*) for biosensing (Heller *et al* 2012). It is important to note that genetically engineered multicellular fungi-based sensors are rare. This may be probably due to challenges in the biotechnological genetic engineering of these cells.

However, with the evolution of more robust genetic circuit designs and activator and transcription factors, their application may expand significantly to the heavy metal sensing domain. Today, *S. cerevisiae* remains one of the most explored microbes in genetic manipulation for industrial applications. The recombinant activator genes from microbes such as *Escherichia coli, Vibrio cholerae, Sinorhizobium meliloti, Herbaspirillum seropedicae, Acinetobacter* sp., *Klebsiella pneumonia, etc have shown significant*, etc have been efficiently demonstrated to work with different reporter genes (Wan *et al* 2019), showing a potential increase in yeast-based biosensor development.

9.3.3 Fungi biomass-based electrochemical sensors

A fungi biomass-based electrochemical sensor relies on the high adsorption capacity of the biomass to detect heavy metals. Fungi biomass is rich in several functional groups, such as the N^{-3}, NO^{-2}, SO^{-2}, CN^-, $C \, \hat{a} \, O$, S^{-2}, RS^-, R_2NH, R_3N, PO_3^{-4}, $SO_2^{-2}4$, etc, which have high charge density or good electron density in the form of lone pairs (Singh *et al* 2023). This negative charge helps in the attraction of the positively charged metal ions on their surface. The higher the affinity of the cells for metals, the higher the concentrations of metal accumulated on the electrode surface. This significantly affects the analytical performance of the sensor. Thus, the functional groups can be chemically modified to enhance the electrode performance. The process of metal sensing using the adsorption-based sensor can be understood in three steps:

> Step 1: Preconcentration stage: Electrode exposure to the sample in open circuit or at a fixed potential.

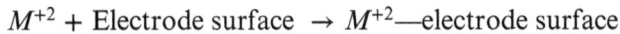

$$M^{+2} + \text{Electrode surface} \rightarrow M^{+2}\text{—electrode surface}$$

> Step 2: Deposition stage or the reduction step.

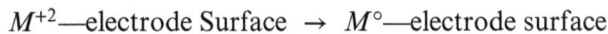

$$M^{+2}\text{—electrode Surface} \rightarrow M^{\circ}\text{—electrode surface}$$

> Step 3: Stripping the voltammetry of the electrode in stripping solution.

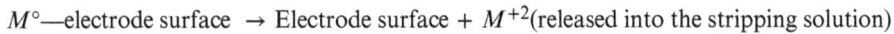

$$M^{\circ}\text{—electrode surface} \rightarrow \text{Electrode surface} + M^{+2}(\text{released into the stripping solution})$$

Electrochemical sensors function on the three-electrode system principle as described in section 2.2. The most frequently used technique is stripping voltammetry, which provides larger peak heights, thus affecting the sensitivity of the sensors. Table 9.1 provides a comparison of the analytical performance of different sensors developed for metal detection.

In recent studies, electrochemical sensors are the most preferred choice for infield application because they can easily be miniaturized, and are more robust and stable when compared to live whole-cell-based optical sensors. Several platforms of handheld devices for heavy metal detection have evolved, such as the uMED

Table 9.1. Performance comparison of different fungi-based electrochemical sensors for heavy metals.

Sl. No.	Fungi sp.	Electrode	Technique	Metal	Range of detection	Limit of detection	References
1.	*Penicillium* sp.	CPE	DPV	Cu^{+2}	0.1–20 µM	0.1 µM	Singh *et al* (2023)
2.	*Trichoderma aspergillum*	GCE	DPASV	Cd^{+2} Pb^{+2}	—	µM 0.01 µM	Dali *et al* (2018)
3.	*Rhodotorula mucilaginosa*	CPE	DPV	Cu^{+2}	0.1–10 µM	—	Yüce *et al* (2010a)
4.	*Rhizopus arrhizus*	CPE	DPASV	Pb^{+2}	0.1–0.125 µM	5 nM	Yüce *et al* (2010b)
5.	*Circinella sp.*	CPE	DPCSV	Cu^{+2}	50–0.5 µM	0.54 nM	Alpat *et al* (2008)
6.	*Tetraselmis chuii*	CPE	DPV	Cu^{+2}	0.05–10 µM	0.46 nM	Alpat *et al* (2007)

CPE: carbon paste electrode; DPASV: differential pulse anode stripping voltammetry; DPCS: differential pulse cathode stripping voltammetry; GCE: glassy carbon electrode.

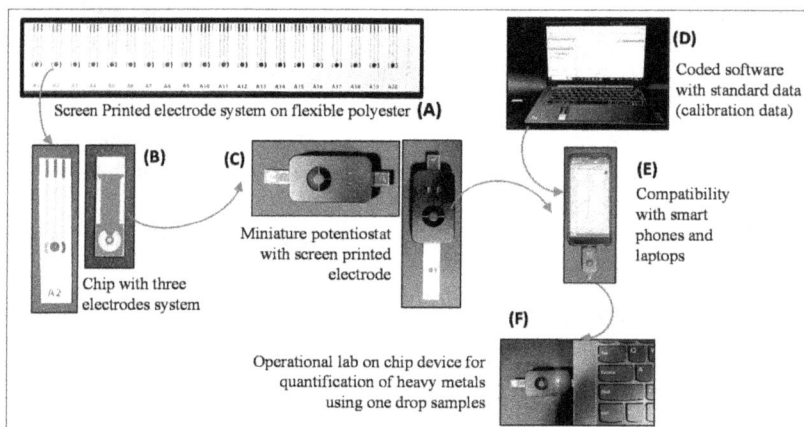

Figuree 9.2. A miniaturized system for on-spot detection of heavy metals, showing the components and their assembly. The screen-printed system allows one-drop analysis of the samples, even in field conditions, using lap-on-chip technology.

(SWASV for Cd, Zn, and Pb), DEP-Chip (DPV for Cd, Pb, As, Cu, Zn), AppliTrace (ASV for Cd, Pb, Cu, As, Zn, Fe), and IME-SJB-801 (ASV for Cd, As, and Hg) (Mukherjee *et al* 2021). However, optical sensing of heavy metals is also available in the market. For example, ANDalyze, a fluorescence-based sensing device that has been approved by USEPA (Lake *et al* 2019). It is important to note that electrochemical sensors are mostly affinity-based sensors for heavy metals. The miniaturized system for the function often consists of three components: a detachable chip consisting of printed electrodes (three-electrode system, with a modified working electrode), a transducer (a miniaturized potentiostat), and a display unit (this can be a provided screen, phone, or laptop), see figure 9.2. These miniaturized forms often require just one drop of sample for analysis, which itself functions as an electrochemical cell and is held directly on the printed chip. This is often referred to as lap-on-chip technology.

9.3.4 Fungi-derived enzymes for biosensing of heavy metals

Enzyme-based biosensors have been the basis of classical biosensing applications, including heavy metals detection (Aloisi *et al* 2019). These enzymes can be derived from plant or microbial sources, or even animal sources. However, biosensors based exclusively on fungi-derived enzymes have not yet been reported. Nevertheless, the commonly used enzymes for heavy metal sensors may also be derived from a wide range of fungal sources. Catabolic enzymes such as hydrolase, protease, esterase, lipase, pectinase, and, phenoloxidase derived from fungi can have better thermal stability when compared to other microbial enzymes (Gupta and Chaturvedi 2015). Enzymes produced extracellularly by the fungi can be easier to harvest and can significantly reduce the cost of production, lowering the cost of the sensors (Khanam *et al* 2020). Thus, fungi-derived enzyme-based sensors can provide a more economical and robust sensing platform.

9.3.5 Fungi-derived macromolecules and nanoparticles for biosensing

Fungi are a source of a variety of macromolecules, including several proteins and polymeric chitin. The latter being the second most abundant polymer on earth and a source of commercial chitosan, which finds industrial applications. A variety of chitosan composite-based electrochemical sensors for heavy metals have been reported, which have improved stability and robustness when compared to biomass (Raja 2020). The controlled composition of the chitosan-based sensors can improve the selectivity of the sensor by allowing interactions of the surface with metals of specific charge density. Similarly, several proteins have been demonstrated to function as turn-on luminescence sensors, which can be deployed for heavy metal detection (Iftikhar *et al* 2022). These proteins can be extracted from fungal sources and used for biosensing applications. Figure 9.3 shows the possible strategies for using fungi as a resource in the development of biosensors for heavy metals.

Fungal mycelium is well known for its ability to synthesize nanoparticles of different elements or composites extracellularly and intracellularly (Siddiqi and Husen 2016). These nanoparticles and quantum dots possess physicochemical characteristics (i.e., electronic, fluorescence, and surface plasmon resonance) that can be used to develop sensors for heavy metals (Khanam *et al* 2020). Uddandarao *et al* (2019) showed that endophytic species of fungi *Aspergillus flavus* can be used to synthesize nanoparticles of Gd-doped ZnS, which can be used as an optical biosensor due to its fluorescent nature. The detection of different metals was carried out on the basis of the quenching/enhancement of fluorescence of the particles. In another case, C-quantum dots derived from *Pleurotus species* were used for the detection of Pb(II) in an aqueous sample. The dots were also demonstrated to possess antimicrobial and anticancer activity (Boobalan *et al* 2020). This demonstrates the potential use of fungi to synthesize nanoparticles, which can be used in the sensing of heavy metals. However, this area remains relatively less explored.

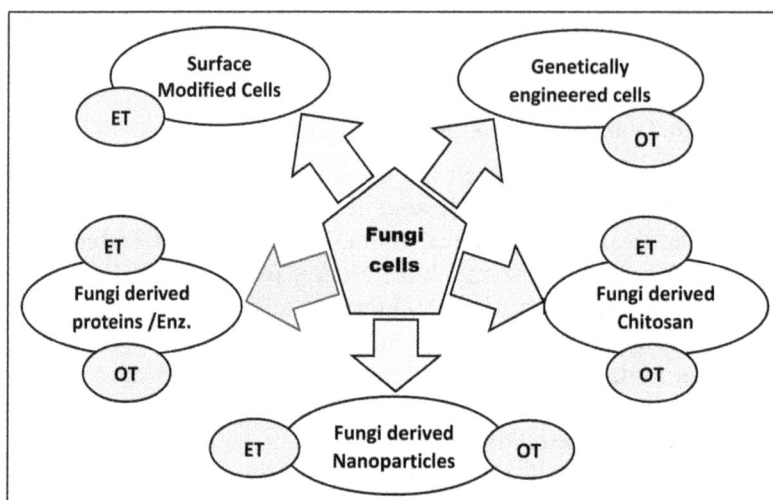

Figure 9.3. Possible strategies for using fungi for the development of biosensors for heavy metals in environmental samples. OT: optical transduction and ET: electrochemical transduction.

9.4 Conclusions

This chapter addresses the general aspects of biosensors and their applications in the field of environmental monitoring. A focus has been paid to fungi-based sensors, both optical and electrochemical sensors, for heavy metal detection, a topic that has received less attention in the biosensing world. Optical sensors using eukaryotic fungal cells are the lesser studied among the two. Multicellular fungi-based sensors have used the electrochemical platform, while unicellular yeast-based sensors are more prominently developed on the optical platform. The possibility of other macromolecule-based (derived from fungi) sensors has also been discussed. The evidence supporting the possibility of fungi-derived nanoparticles and quantum dots-based sensors, chitosan-based sensors, etc, is promising and this topic is expected to be explored in more detail in future studies from the scientific community.

References

Abu-Ali H, Nabok A and Smith T J 2019 Development of novel and highly specific ssDNA-aptamer-based electrochemical biosensor for rapid detection of mercury (II) and lead (II) ions in water *Chemosensors* **7** 27

Adeniran A, Sherer M and Tyo K E 2015 Yeast-based biosensors: Design and applications *FEMS Yeast Res.* **15** 1–15

Aloisi A, Della Torre A, De Benedetto A and Rinaldi R 2019 Bio-recognition in spectroscopy-based biosensors for* heavy metals-water and waterborne contamination analysis *Biosensors* **9** 96

Alpat S K, Alpat Ş, Kutlu B, Özbayrak Ö and Büyükışık H B 2007 Development of biosorption-based algal biosensor for Cu (II) using *Tetraselmis chuii Sens. Actuators* B **128** 273–8

Alpat Ş, Alpat S K, Çadırcı B H, Yaşa İ and Telefoncu A 2008 A novel microbial biosensor based on *Circinella* sp. modified carbon paste electrode and its voltammetric application *Sens. Actuators* B **134** 175–81

Amin N U, Siddiqi H M, Kun Lin Y, Hussain Z and Majeed N 2020 Bovine serum albumin protein-based liquid crystal biosensors for optical detection of toxic heavy metals in water *Sensors* **20** 298

Ashrafi A M, Sýs M, Sedláčková E, Shaaban Farag A, Adam V, Přibyl J and Richtera L 2019 Application of the enzymatic electrochemical biosensors for monitoring non-competitive inhibition of enzyme activity by heavy metals *Sensors* **19** 2939

Ayele A, Haile S, Alemu D and Kamaraj M 2021 Comparative utilization of dead and live fungal biomass for the removal of heavy metal: a concise review *Sci. World J.* **2021**

Boobalan T, Sethupathi M, Sengottuvelan N, Kumar P, Balaji P, Gulyás B and Arun A 2020 Mushroom-derived carbon dots for toxic metal ion detection and as antibacterial and anticancer agents *ACS Appl. Nano Mater.* **3** 5910–9

Buenger D, Topuz F and Groll J 2012 Hydrogels in sensing applications *Prog. Polym. Sci.* **37** 1678–719

Chen Z, Xie M, Zhao F and Han S 2022 Application of nanomaterial modified aptamer-based electrochemical sensor in detection of heavy metal ions *Foods* **11** 1404

Dali M, Zinoubi K, Chrouda A, Abderrahmane S, Cherrad S and Jaffrezic-Renault N 2018 A biosensor based on fungal soil biomass for electrochemical detection of lead (II) and cadmium (II) by differential pulse anodic stripping voltammetry *J. Electroanal. Chem.* **813** 9–19

Do M H, Ngo H H, Guo W, Chang S W, Nguyen D D, Pandey A and Hoang N B 2022 A dual chamber microbial fuel cell based biosensor for monitoring copper and arsenic in municipal wastewater *Sci. Total Environ.* **811** 152261

Fan C, Zhang D, Mo Q and Yuan J 2021 High-performance *Saccharomyces cerevisiae*-based biosensor for heavy metal detection *bioRxiv* 2021–12

Fan C, Zhang D, Mo Q and Yuan J 2022 Engineering *Saccharomyces cerevisiae*-based biosensors for copper detection *Microb. Biotechnol.* **15** 2854–60

Fu C, Xu W, Wang H, Ding H, Liang L, Cong M and Xu S 2014 DNAzyme-based plasmonic nanomachine for ultrasensitive selective surface-enhanced Raman scattering detection of lead ions via a particle-on-a-film hot spot construction *Anal. Chem.* **86** 11494–7

Gavrilaş S, Ursachi C Ş, Perţa-Crişan S and Munteanu F D 2022 Recent trends in biosensors for environmental quality monitoring *Sensors* **22** 1513

Gupta S and Chaturvedi P 2015 Phytochemical screening and extracellular enzymatic enumeration of foliar endophytic fungal isolates of *Centella asiatica* (L.) urban *Int. J. Pharm. Sci. Rev. Res.* **35** 21–4

Gutiérrez J C, Amaro F and Martín-González A 2015 Heavy metal whole-cell biosensors using eukaryotic microorganisms: an updated critical review *Front. Microbiol.* **6** 48

Heller J, Meyer A J and Tudzynski P 2012 Redox-sensitive GFP2: use of the genetically encoded biosensor of the redox status in the filamentous fungus *Botrytis cinerea Mol. Plant Pathol.* **13** 935–47

Iftikhar R, Parveen I, Mazhar A, Iqbal M S, Kamal G M, Hafeez F and Ahmadipour M 2022 Small organic molecules as fluorescent sensors for the detection of highly toxic heavy metal cations in portable water *J. Environ. Chem. Eng.* **11** 109030

IUPAC. Compendium of Chemical Terminology 2nd edn (The 'Gold Book') 1997. Compiled by A. D. McNaught and A. Wilkinson (Oxford: Blackwell Scientific Publications). Online version (2019–) created by S. J. Chalk.

Jarque S, Bittner M, Blaha L and Hilscherova K 2016 Yeast biosensors for detection of environmental pollutants: current state and limitations *Trends Biotechnol.* **34** 408–19

Khanam Z, Gupta S and Verma A 2020 Endophytic fungi-based biosensors for environmental contaminants-a perspective *S. Afr. J. Bot.* **134** 401–6

Kumar T, Naik S and Jujjavarappu S E 2022 A critical review on early-warning electrochemical system on microbial fuel cell-based biosensor for on-site water quality monitoring *Chemosphere* **291** 133098

Lake R J, Yang Z, Zhang J and Lu Y 2019 DNAzymes as activity-based sensors for metal ions: recent applications, demonstrated advantages, current challenges, and future directions *Acc. Chem. Res.* **52** 3275–86

Liu C, Yu H, Zhang B, Liu S, Liu C G, Li F and Song H 2022 Engineering whole-cell microbial biosensors: design principles and applications in monitoring and treatment of heavy metals and organic pollutants *Biotechnol. Adv.* **60** 108019

McConnell E M, Nguyen J and Li Y 2020 Aptamer-based biosensors for environmental monitoring *Front. Chem.* **8** 434

Mukherjee S, Bhattacharyya S, Ghosh K, Pal S, Halder A, Naseri M and Bhattacharyya N 2021 Sensory development for heavy metal detection: a review on translation from conventional analysis to field-portable sensor *Trends Food Sci. Technol.* **109** 674–89

Park J N, Sohn M J, Oh D B, Kwon O, Rhee S K, Hur C G and Kang H A 2007 Identification of the cadmium-inducible Hansenula polymorpha SEO1 gene promoter by transcriptome analysis and its application to whole-cell heavy-metal detection systems *Appl. Environ. Microbiol.* **73** 5990–6000

Raja A N 2020 Recent development in chitosan-based electrochemical sensors and its sensing application *Int. J. Biol. Macromol.* **164** 4231–44

Ramdzan N S M, Fen Y W, Anas N A A, Omar N A S and Saleviter S 2020 Development of biopolymer and conducting polymer-based optical sensors for heavy metal ion detection *Molecules* **25** 2548

Roda A, Roda B, Cevenini L, Michelini E, Mezzanotte L, Reschiglian P and Virta M 2011 Analytical strategies for improving the robustness and reproducibility of bioluminescent microbial bioreporters *Anal. Bioanal. Chem.* **401** 201–11

Shah Maulin P 2020 *Microbial Bioremediation and Biodegradation* (Berlin: Springer)

Shetty R S, Deo S K, Liu Y and Daunert S 2004 Fluorescence-based sensing system for copper using genetically engineered living yeast cells *Biotechnol. Bioeng.* **88** 664–70

Siddiqi K S and Husen A 2016 Fabrication of metal nanoparticles from fungi and metal salts: scope and application *Nanoscale Res. Lett.* **11** 1–15

Shah M P 2021 *Removal of Refractory Pollutants from Wastewater Treatment Plants* (Boca Raton, FL: CRC Press)

Singh A and Kumar V 2021 Recent advances in synthetic biology–enabled and natural whole-cell optical biosensing of heavy metals *Anal. Bioanal. Chem.* **413** 73–82

Singh A, Kumar V, Singh S and Ray M 2023 Electrochemical detection of copper (II) in environmental samples using *Penicillium* sp. IITISM_ANK1 based biosensor *Chemosphere* **313** 137294

Siontorou C G and Georgopoulos K N 2021 Boosting the advantages of biosensors: Niche applicability and fitness for environmental purpose *Trends Environ. Anal. Chem.* **32** e00146

Tripathi M K, Nickhil C, Kate A, Srivastva R M, Mohapatra D, Jadam R S and Modhera B 2023 Biosensor: fundamentals, biomolecular component, and applications *Advances in Biomedical Polymers and Composites* pp 617–33 (Amsterdam: Elsevier)

Uddandarao P, Balakrishnan R M, Ashok A, Swarup S and Sinha P 2019 Bioinspired ZnS: Gd nanoparticles synthesized from an endophytic fungi *Aspergillus flavus* for fluorescence-based metal detection *Biomimetics* **4** 11

Velusamy K, Periyasamy S, Kumar P S, Rangasamy G, Pauline J M N, Ramaraju P and Vo D V N 2022 Biosensor for heavy metals detection in wastewater: a review *Food Chem. Toxicol.* **168** 113307

Wan X, Marsafari M and Xu P 2019 Engineering metabolite-responsive transcriptional factors to sense small molecules in eukaryotes: current state and perspectives *Microbial Cell Fact.* **18** 1–13

Weitz H J, Campbell C D and Killham K 2002 Development of a novel, bioluminescence-based, fungal bioassay for toxicity testing *Environ. Microbiol.* **4** 422–9

Yüce M, Nazır H and Dönmez G 2010a A voltammetric *Rhodotorula mucilaginosa* modified microbial biosensor for Cu (II) determination *Bioelectrochemistry* **79** 66–70

Yüce M, Nazır H and Dönmez G 2010b Using of *Rhizopus arrhizus* as a sensor modifying component for determination of Pb (II) in aqueous media by voltammetry *Bioresour. Technol.* **101** 7551–5

Zhao C, Liu G, Tan Q, Gao M, Chen G, Huang X and Xu D 2022 Polysaccharide-based biopolymer hydrogels for heavy metal detection and adsorption *J. Adv. Res.* **44** 53–70

IOP Publishing

Trends in Biological Processes in Industrial Wastewater Treatment

Maulin P Shah

Chapter 10

Emerging global technologies for removal of contaminants from wastewater

Sarita Khaturia, Saloni Sahal and Harlal Singh

Due to an increase in the industrial release of emerging contaminants, environmental pollution from man-made harmful chemicals discharged into wastewater has become a big problem worldwide. The majority of emerging contaminants (ECs) are synthetic organic compounds such as pharmaceuticals, antibacterials, hormones, synthetic dyes, and flame retardants that are released into the environment by hospitals, agriculture, industry, and other sources and pose serious risks to human health and biota. Their presence has deleterious effects on both marine life and human health, even at low quantities. Therefore, there is an urgent need for the development of effective treatment technologies that can be used on industrial scales and in effluent treatment plants. ECs cannot be successfully eliminated by conventional wastewater treatment methods. The use of membrane technology, coagulation-flocculation, solvent extraction, adsorption, and sophisticated oxidation processes are only a few of the treatment options that have been studied. These methods have both benefits and drawbacks. Nanotechnology is a promising strategy to get beyond these restrictions. To completely and effectively remove ECs from polluted water, more thorough research on wastewater treatment systems that are technically and financially practical is needed. Hybrid methods have been found to be more effective than separate techniques at removing EC, but they have problems with cost, time, and energy. In this chapter, we will discuss developing pollutants, including their sources, toxicity, and treatment options. We will also give an overview of the behavior, biomonitoring, toxicity, and methods for removing ECs.

10.1 Introduction

Emerging contaminants (ECs) can relate to a variety of chemicals, including fire retardants, medicines, personal care, or household cleaning goods, lawn care products, and agricultural products. These substances can negatively impact the ecosystem and human health by bioaccumulating in the food chain. Due to the lack of comprehensive knowledge regarding their interactions or true toxicological effects, there are no strict regulations in place to control their concentration in drinking water, agricultural water, the air, or the environment in general [1]. When a chemical meets the requirements listed below, it might be categorized as an emerging contaminant [2, 3]:

1. The substance may have harmful impacts on public health.
2. Both the chemical's beneficial and harmful effects have a solid track record.
3. There is typically no regulation of these contaminants.

According to environmental monitoring data, a variety of ECs, such as anti-biotics, x-ray contrast media, plasticizers, UV filters, lipid-regulating medications, anti-microbial agents, stimulants, insect-repellents, hormones, anti-inflammatory medications, artificial sweeteners, anti-itching medications, anti-depressants, and anticonvulsants, are frequently found in water bodies as a result of unregulated or only partially regulated disposal procedures (figure 10.1) [4]. Wastewater contains a number of common household compounds, such as sunscreen, sprays, lipsticks, beauty products, and shampoos. Several of these substances mix with water totally or partially [5]. The physio-chemical properties of these compounds, such as evaporation, solubility, boiling point, chemical structure, existence of particular chemical functionalities, melting point, and complexation/sorption ability of the materials, were discovered to have a significant impact on their fate and mobility. These ECs are absorbed by plants and animals as a result of their existence in water systems. Finally, they will become part of the human food chain, which might have catastrophic repercussions on the environment and represent a major threat to human health [6, 7].

Figure 10.1. Water contaminants. This image has been obtained by the author(s) from the Pixabay website where it was made available under the Pixabay License. It is included within this article on that basis.

10.1.1 Types of emerging contaminants (ECs)

ECs can be broadly categorized into the following types according to where they come from:

1. Pharmaceuticals.
2. Products for personal care.
3. Pesticides.
4. Substances that alter hormones (EDCs).

10.1.1.1 Pharmaceuticals

Pharmaceuticals include ECs with pharmacological activities, which can be divided into the following categories: Analgesics and anti-inflammatory drugs include ibuprofen, acetylsalicylic acid, diclofenac sodium, and paracetamol. Benzodiazepines are antidepressants. Carbamazepine is an antiepileptic. Fibrates are a type of lipid-lowering medicine. Ranitidine and famotidine are examples of antihistamines. Tetracyclines, macrolides, other Medicinal ECs with pharmacological effect and degradation resistance will survive in the aquatic environment, potentially harming the ecology of fish, plants, and other aquatic living organisms. The health of people, aquatic plants, and animals who drink the contaminated water is negatively impacted as a result of this pollution (figure 10.2) [8]. The following are the main characteristics that set ECs from pharmaceutical sources apart from those obtained from other industrial sources:

1. The molecular weight, structure, functionality, form, and chemical make-up of pharmaceutical ECs vary greatly.
2. Pharmaceutical ECs are polar lipophilic compounds with numerous ionizable groups that are somewhat water soluble [9], and the degree of ionization is greatly controlled by the surroundings' chemical composition.

Figure 10.2. Testing of pharmaceutical contaminants. This image has been obtained by the author(s) from the Pixabay website where it was made available under the Pixabay License. It is included within this article on that basis.

3. Due to accumulation in humans, pharmaceutical ECs, such as erythromycin, cyclophosphamide, naproxen, and sulfamethoxazole, can persist for up to a year or longer.

4. After delivery, pharmaceutical ECs are subject to metabolic interactions, absorption, and chemical modification that may result in completely different chemical and biological consequences.

These materials can be eliminated without undergoing any chemical changes [10]. If they are subjected to biochemical processes such as hydrolysis, oxidation/reduction, alkylation, etc, or if they are metabolically altered in the body, then sulfate conjugates or glucuronide may also be created. Polar and hydrophilic substances can reach the environment in two ways: by being dumped in regular trash bins, and by passing through the feces or urine of people and other animals after ingesting them. Analgesic diclofenac has been linked to a significant death rate from renal failure in vultures in Pakistan and India [10]. Moreover, renal failure in fish and other aquatic species has been linked to this painkiller. Ibuprofen was discovered to increase *Oreochromis niloticus* micronuclei prevalence. Tetracyclines and quinolone antibiotics were shown to have many coordinating sites and have a propensity to accumulate heavy metals such as Zn, Cu, and Cd, making them even more toxic to living things. In addition, long-term use of antibiotics can result in bacterial strains that are resistant to them. Furthermore, it has been demonstrated that 5-fluorouracil and cisplatin are hazardous to aquatic life, including the alga *Pseudokirchneriella subcapitata*.

10.1.1.2 Products for personal care

Urbanization has led to the development of products for personal care. The majority of them are cosmetics (e.g., lipstick, nail polish, talcum powder, sunscreen, lotion, make-up kits, etc), as well as synthetic hormones, steroids, perfumes, shampoos, etc (figure 10.3) [12]. They undergo no metabolic biochemical alteration because their primary use is on the surface of the human body. These ECs are typically discovered

Figure 10.3. Personal care cosmetic products. This image has been obtained by the author(s) from the Pixabay website where it was made available under the Pixabay License. It is included within this article on that basis.

Figure 10.4. Spraying of pesticides. This image has been obtained by the author(s) from the Pixabay website where it was made available under the Pixabay License. It is included within this article on that basis.

in aqueous wastewater streams or urban surface water bodies that come from the relevant businesses. UV filters were reported to display estrogenic action. Personal care products can be hydrolyzed or undergo oxidation reduction reactions inside water bodies, or they can be adsorbed on sludge or biosolids, notably on biologically moderated or transformed lipophilic compounds [13].

10.1.1.3 Pesticides

Pesticides are used to safeguard crops against invasive microorganisms, which develop microbial resistance with prolonged exposure and are susceptible to transformation by plant metabolites. They are also capable of moving up aquatic plant and animal food chains [14]. Examples include chlorinated phenoxy acid, which is a frequent ingredient in pesticides used in agriculture, herbicides used on lawns, algicides used in paints and coatings, and agents used in sealants to protect roofs (figure 10.4). High polarity is a characteristic of these substances [15].

10.1.1.4 Substances that alter hormones (EDCs)

Endocrinogenic substance EDCs possess hormonal activity. Cancerous tumors, birth defects, developmental disorders, issues with the brain, the immune system, and other issues can all be brought on by endocrine disrupting substances (figure 10.5) [16]. However, the link between exposure and negative health effects is complicated. It is not always easy to connect a specific EDC with a specific health problem. EDC exposure is particularly dangerous for foetuses and embryos, whose growth and development are more closely regulated by the endocrine system. Adult diseases and lasting changes can result from prenatal exposure. Diethylstilbestrol (DES) exposure in the womb has been related to a number of malignancies and female uterine dysfunction or deformation.

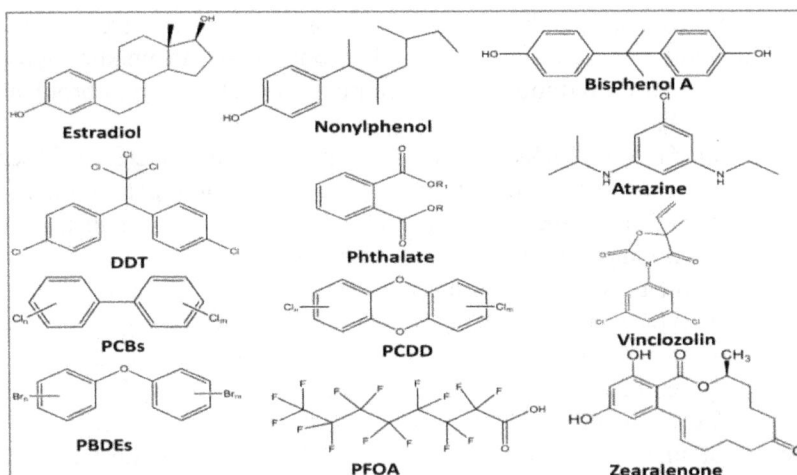

Figure 10.5. Chemical structures of some EDCs. This image has been obtained by the author(s) from the Pixabay website where it was made available under the Pixabay License. It is included within this article on that basis.

The shorter more female-like anogenital distance and smaller scrotum and penis of male babies have been linked to phthalates in the urine of pregnant women. Very low and very high levels have more impact than mid-level exposure for the majority of endocrine disruptors, which show a U-shaped dosage response curve [17]. It has been discovered that exposure to endocrine disruptors affects not only humans but also other species. For example, the thyroid gland and reproductive system of female rats are both impacted by the flame-retardant BDE-47. The synthesis, secretion, transport, binding, action, and elimination of the body's endogenous hormones—which are necessary for growth, behavior, fertility, and the maintenance of normal cell metabolism—can be affected by these substances.

Xenoestrogens—This EDC is an estrogen-mimicking xenohormone. Polychlorinated biphenyls (PCBs), bisphenol A (BPA), and phthalates are synthetic xenoestrogens that have estrogenic effects on living things. Alkylphenols (APs) are one instance. Some of the sources of APs include detergents, additives, lubricants, polymers, phenolic resins, thermoplastic elastomers, antioxidants, oil field chemicals, and flame retardants. BPA and bisphenol S (BPS) are some of the other EDCs having hydroxyl functions [18]. BPS has been found in plastics and household dust, exhibiting strong endocrine disruption activity, while BPA was found in plastic bottles, plastic food containers, dental materials, and the linings of metal food and infant formula cans. BPA has been linked to increased rates of diabetes, mammary and prostate cancer, decreased sperm count, reproductive issues, early puberty, obesity, and neurological problems.

Dichlorodiphenyltrichloroethane (DDT)—This is one of the most well-known herbicides with endocrine disruptive properties. Dichlorodiphenyltrichloroethane (DDT) exposure has harmful effects on the human reproductive system and can cause obesity in children, male infertility, and poor reproductive system development [19].

Polychlorinated biphenyls (PCBs)—This chlorinated EDC is present in industrial coolants, lubricants, and gasoline refinery byproducts. It can impair thyroid and liver function, increase childhood obesity, cause reproductive system problems, and impair fertility [20].

Polybrominated diphenyl ethers (PBDEs)—This neurotoxic EDC was discovered in foam cushions, electronics, carpets, lighting, bedding, clothing, and other fabrics. A thyroid hormone imbalance brought on by PBDEs can result in a variety of neurological, developmental, and learning problems.

Phthalates—Soft toys, carpeting, medical devices, cosmetics, and air fresheners are a few of the typical sources where phthalates exposure to people is reasonably likely. The sexual development of male newborns may suffer as a result of the medical tubing, catheters, and blood bag additive bis(2-ethylhexyl) phthalate (DEHP) [21]. Exposure to phthalates may potentially interfere with the neurological development of men. Some well-known endocrine disruptive substances include perfluorooctanoic acids (PFOAs), polychlorinated dibenzo-dioxins (PCDDs), poly-chlorinated furans (PCFs), polycyclic aromatic hydrocarbons (PAHs), phenol derivatives, atrazine, vinclozolin, 17-ethinylestradiol, and zearalenone.

10.2 Traditional wastewater treatment methods

Physical, chemical, and biological processes are used in traditional wastewater treatment methods to remove both soluble and insoluble contaminants. Although inexpensive and straightforward, biological therapy is ineffective for manufactured pollutants such as dyes because they are resistant to aerobic biodegradation [22]. Physical treatment is typically more successful than chemical treatment, which results in harmful byproducts and is less effective. The wastewater treatment process has several stages, including preliminary, primary, secondary, and tertiary.

10.2.1 Preliminary treatment

The removal of suspended items from wastewater, such as dead animals, papers, oils, and grease, is facilitated by preliminary treatment. The preliminary treatment process makes use of a variety of parts, including screening, accumulation, and flotation tanks, as well as skimming reservoirs. Sand and grit are removed using an accumulation tank, while oils and grease are removed using flotation devices and skimming tanks.

10.2.2 Primary treatment

Flotation and sedimentation techniques enable organic and inorganic components to be eliminated during primary treatment. Untreated nitrogen, unrefined phosphorus, and heavy metals associated with suspended contaminants are drained off after this treatment [23]. This approach lowers biochemical oxygen demand levels by 5%–40%, the total amount of floating particles by 50%–70%, and effluent oil and grease by up to 65%. In various developed countries, a primary treatment is required for the reuse of wastewater to irrigate crops that are not consumed by humans.

10.2.3 Secondary treatment

Organic effluent that escapes from the first treatment is removed by secondary or biological treatment. Using oxidation or nitrification, this process changes organic materials and converts them into a stable state. Filtration and activated sludge techniques are the two divisions of this sewage treatment technique. This treatment includes contact beds, uneven sand, and trickling filters.

10.2.4 Tertiary treatment

Certain effluents that cannot be entirely eliminated by secondary methods are subjected to tertiary treatment. Almost 99% of all pollutants are removed through this procedure. With the removal of inorganic elements such as nitrogen and phosphorus, wastewater may be used for agriculture and drinking purposes, and has no negative environmental effects [24].

10.3 Emerging global technologies for wastewater treatment

Typically, developing pollutants are not eliminated in traditional wastewater treatment facilities. The presence of ECs in the environment has an impact on human health and marine life, and creates cancers, neurotoxins, and resistant microorganisms. Methods such as membrane filtration, coagulation-flocculation, solvent extraction, ion exchange, catalytic oxidation, electrochemical oxidation, and precipitation, etc, have been tried and tested to remove these organic contaminants from water. However, the researchers' task is made more difficult by the fact that these procedures are less efficient, expensive, and do not totally remove toxins from polluted water. In addition to these methods, photocatalytic degradation and adsorption are thought to be viable methods for removing pollutants from wastewater [25].

10.3.1 Membrane filtration

Membrane technology is a physical technique to remove emerging pollutants from aquatic systems. To filter out suspended pollutants, materials with appropriate surface charges, small pores, and hydrophobicity are used to create membranes. Ultra-filtration (UF), nanofiltration (NF), microfiltration (MF), forward osmosis (FO), and reverse osmosis are several types of membrane filtration (RO) (figure 10.6) [26]. FO, membrane refining, and electro-dialysis of the membrane are important membrane techniques that can eliminate emergent pollutants by up to 99% but are still not widely used. Depending on the membrane and the kind of pollutant, the ultra-filtration process removes colloidal, suspended, or dissolved pollutants at low pressure [27]. Because the pore size of UF is bigger than that of dissolved, hydrated metal ions (0.001–0.1 m), it can easily pass through them. The polyelectrolyte enhanced ultra-filtration (PEUF) and micellar-enhanced ultra-filtration (MEUF) processes have been investigated to increase the removal efficiency of metal ions, including copper, zinc, chromate, arsenate, cadmium, nickel, and serinium, as well as organics like phenol and o-cresol. Microfiltration, which typically operates at

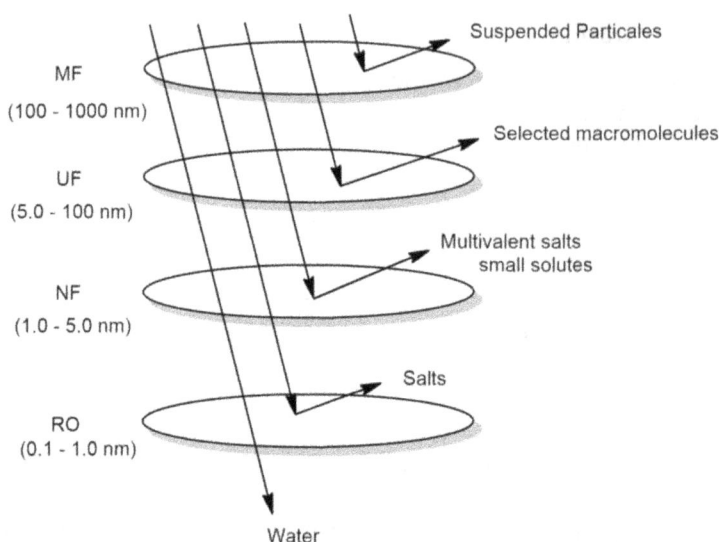

Figure 10.6. Various types of membrane [26]. This image has been obtained by the author(s) from the Pixabay website where it was made available under the Pixabay License. It is included within this article on that basis.

atmospheric pressure and has pore sizes between 0.1 and 10 m, is ineffective at removing pollutants larger than 1 m. Both forward and reverse osmosis rely on osmotic pressure gradients and a semipermeable membrane to effectively remove dissolved particles from water that are up to 1 nm in size. Depending on the kind of membrane and the pollutant, nanofiltration membranes feature tiny pores with sizes ranging from 1 to 10 nm and are highly competent at removing ECs. Anti-inflammatory medications, sulfonamide and fluoroquinolone antibiotics, as well as testosterone, estrogen, and progesterone can all be removed by NF [28].

10.3.2 Coagulation-flocculation

The coagulation-flocculation technique is useful for removing bigger colloidal or suspended particles from wastewater colored with dispersed colors. In the process of flocculation, aggregated flocs are connected to form larger agglomerates that fall to the ground as a result of gravity, whereas in the process of coagulation dye solution systems are distributed to form flocs and agglomerates. In the textile industry, flocculation and coagulation are frequently employed to clean wastewater because they are easily operated and economically viable. By combining with the contaminants and removing them by electrostatic interactions or sorption, coagulants such as lime $Ca(OH)_2$, ferric sulfate $(Fe_2(SO_4)_3 \cdot 7H_2O)$, aluminum sulfate $(Al_2(SO_4)_3 \cdot 18H_2O)$, and ferric chloride $(FeCl_3 \cdot 7H_2O)$ are used by this approach [29]. There have been reports of the use of aluminum sulfate $(Al_2(SO_4)_3)$ for the coagulation-flocculation-based clearance of medications such as betaxolol, chlordiazepoxide, bromazepam, warfarin, and hydrochlorothiazide. This method reduces the amount of suspended matter, soluble dyes, colloidal particles, and coloring agents in wastewater [30].

10.3.3 Solvent extraction

Solvent extraction is frequently used to get rid of organic and inorganic pollutants that are dumped into wastewater from various businesses. Three main operations serve as its foundation. The first step is the extraction or transfer of solute particles from water to solvent. The next stage is the separation of the solute from the solvent. The solvent recovery stage comes next. Solvent extraction is typically used to remove phenols, creosols, and other phenolic acids from contaminated water that contains only a small amount of solute and is produced by coke-oven facilities, steel mills, and plastics factories [31].

10.3.4 Adsorption

Adsorption is one of the most effective methods for treating wastewater due to its straightforward design, high competency and ease of operation, capital cost, ease of recovery, adaptability, and technical viability without producing dangerous byproducts. Although this method is not new, it is well known due to its ability to remove materials and regenerate adsorbents. This method has been widely used to remove organic and inorganic pollutants from domestic and commercial wastewater [36]. To increase the effectiveness of adsorption, numerous research projects have been committed to finding low-cost adsorbents with a wide surface area and excellent binding capacity. To increase the effectiveness of adsorption, numerous research projects have been committed to finding low-cost adsorbents with a wide surface area and excellent binding capacity. Many types of adsorbents, including peat, bamboo dust, chitosan, silica gel, activated carbon, fly ash, zeolites, and metal organic framework nano-adsorbents such as carbon nanotubes and graphene have been used to remove developing pollutants. Because of its highly porous surface area, practical pore composition, and thermostability, activated carbon is frequently used as a traditional adsorbent for the removal of dyes and pharmaceutical products from wastewater, such as 17-estradiol, 17-ethynylestradiol, bisphenol A, and fluoroquinolonic caffeine.

10.3.5 Advanced oxidation process

Advanced oxidation processes (AOPs) have been introduced as effective wastewater treatment techniques. AOPs work by producing hydroxyl (OH) or sulfate radicals to oxidize ECs, while they occasionally also use ozone and UV light to increase removal effectiveness. Instead of moving these substances to a different phase, AOPs technologies effectively remove biologically harmful or non-degradable substances such pesticides, aromatics, petroleum constituents, and volatile organic compounds. Since OH is reduced to generate H_2O as a byproduct, AOPs are appropriate for the simultaneous elimination of several organic pollutants without creating any dangerous substances in water. Ozone (O_3), hydrogen peroxide (H_2O_2), electrochemical oxidation, the Fenton process, UV radiation, and photocatalytic processes are examples of AOPs.

(A) **Non-photochemical process**

(i) Ozonation—Due to its high oxidizing power, ozone has the ability to remove both organic and inorganic substances from industrial effluents. This is a sophisticated oxidation process. While post-oxidation procedures enhance effluent quality, pre-oxidation processes significantly advance biological decomposition. Low solubility, instability, and short half-life are the limitations of ozonation [37]. The generation of hydroxyl radicals, which effectively removed organic pollutants such as antibiotics, antiphlogistics, beta blockers, lipid regulators and their metabolites, natural oestrogen estrone, antiepileptic drug carbamazepine, and musk fragrances in wastewater effluents, has been studied using O_3/H_2O_2 and catalytic ozonation.

(ii) Electrochemical method—Toxic pollutants from textile effluents are typically removed using an electrochemical technique by direct or indirect oxidation. This method is frequently used to get rid of ECs such as dye by utilizing SS304 as the cathode and either mercury electrode, graphite rod, platinum foil, boron doped diamond electrode, or titanium/platinum as the anode. Since only a small amount of chemical is needed for this process, stability can be easily achieved by adjusting the electric current and it is cost-effective.

(iii) Fenton process—Fenton's reaction is the name given to the interaction between ferrous iron and hydrogen peroxide. Organic contaminants such as phenols, reactive dyes, and insecticides are removed using the Fenton procedure. The Fenton process is inexpensive because it uses no energy to activate H_2O_2, is safe for the environment, simple to operate, and effective at getting rid of organic contaminants.

(B) **Photolytic chemical process**

(i) **Ultraviolet lamp**

In this method, an oxidizing agent such as H_2O_2 is produced by the UV process to create OH, which has the ability to efficiently destroy micropollutants depending on a number of factors, including pH, dye structure, effluent composition, and UV radiation intensity [32]. UV processes typically take place at low pressure and a standard wavelength of 254 nm. Mecoprop and diclofenac were successfully removed 98% from wastewater in a pilot plant using UV/H_2O_2 generated hydroxyl radicals. $O_3/H_2O_2/UV$ methods have been investigated to completely degrade textile effluent.

(ii) **Photo-Fenton process**

In this procedure, UV radiation enhances the production of the hydroxyl radical in the presence of iron, effectively degrading wastewater effluents. Fenton and photo-Fenton processes are comparable, but mineralization is significantly better in the latter. By using the photo-Fenton technique, the removal of many ECs, such as medicines, beta blockers, and pesticides, is greatly improved (95%–100%). Different UV wavelength ranges, such as UVA (315–400 nm), UVB (285–315 nm), and UVC (285 nm), are used as a source of light energy in the photo-Fenton process. It has been discovered that the elimination of organic contaminants is significantly influenced by the UV radiation's strength and wavelength. The solar photo-Fenton process, which is utilized for the photocatalytic decontamination of waterways, is a method that uses sunlight (at wavelengths > 300 nm) as a renewable energy source [AOP]. It is a cost-effective and environmentally sound technique.

10.3.6 Nanotechnology

Materials with at least one dimension less than 100 nm are referred to as nanomaterials. The greater density and surface area of nanomaterials lead to greater adsorption effectiveness, surface reactivity, and resolution mobility. Adsorption, AOPs, and filtration are three techniques for using nanotechnology in wastewater treatment that are currently being investigated. According to reports, wastewater can be successfully cleaned of developing pollutants using nanomaterials. Numerous nanomaterials have been described for wastewater treatment, including carbon nanomaterials, metal-oxide nanoparticles, zerovalent metal nanoparticles, and nanocomposites [33].

10.3.6.1 Nanomaterials made of zerovalent metal

Due to its small size and large surface area, zerovalent metal is an important nanomaterial for wastewater treatment. Researchers have recently focused on a number of zerovalent metal nanoparticles for pollutant removal, including silver, zinc, iron, aluminum, and nickel. Silver nanoparticles are typically employed as disinfectants to get rid of a lot of germs, SUCH AS viruses, bacteria, and fungi [34]. They may also have antibacterial characteristics. They have numerous channels for wastewater treatment and are very responsive, economical, and environmentally benign. By using the adsorption, redox reaction, and co-precipitation techniques, iron nanomaterial can successfully remove pollutants from wastewater, including cadmium nitrates, colorants, and antibiotics. Zerovalent metal nanoparticles covered in silica and polydopamine ($nZVI/SiO_2/PDA$) were created in two steps, according to Li et al [35], and they can be used as sorbents with good capacity, selectivity, and reusability (up to 10 times).

10.3.6.2 Metal-oxide nanomaterials

Arsenic, uranium, phosphate, and organic waste have all been successfully removed using metal-oxide nanoparticles such ferric oxides, manganese oxides, aluminum oxides, and titanium oxides. Titanium oxide nanomaterial is a competent photocatalyst with a band gap of 3.2 eV, good photostability, cheap cost, and exceptional photocatalytic behavior. Pollutants such as organic chlorine, PAHs, pigments, phenols, pesticides, and heavy metals can all be degraded using TiO_2 nanoparticles

[36]. ZnO nanomaterial, which has excellent photocatalytic capabilities and a broad wavelength range, is an effective material for treating wastewater. Comparing ZnO nanomaterial to other metal oxides with semiconducting characteristics, ZnO nanomaterial is more environmentally friendly and absorbs more light. Iron oxide nanoparticles are accessible as effective sorbent materials to remove heavy metals from wastewater and have a variety of uses [37].

10.3.6.3 Nanomaterials made of carbon

Due to their unique structural and electrical characteristics, carbon nanostructures are extremely useful in adsorption [38]. They have a large surface area, high surface area adsorption capability, and aromatic selectivity. These nanomaterials include carbon beads, nonporous carbon, carbon nanotubes (CNTs), and carbon fibers. CNTs feature strong physicochemical interactions, well-defined cylindrical structures, porosity, wide surface areas, adaptive hydrophobic sides, and good adsorption capacities for dyes, dichlorobenzene, ethylbenzene, Pb^{2+}, Zn^{2+}, Cd^{2+}, and Cu^{2+} [39]. The single carbon atom layer known as graphene, which has a honeycomb-like structure, is a member of another class of nanomaterials [40]. Heavy metals such as lead, zinc, copper, cadmium, mercury, and arsenic can be removed using graphene oxide, a graphene layer made up of hydroxyl, epoxy, carboxyl, and carbonyl groups. It is possible to effectively remove Pb(II), As(III), and As(V) from contaminated water using graphene hybrid with manganese ferrite nanoparticles. Rajabi et al [41] varied the experimental conditions, such as pH, periods, and temperatures, to compare the adsorption effectiveness of multi-walled carbon nanotubes (MWCNTs) and functionalized CNTs. Their results showed that f-CNTs have a greater capacity for removal than pure CNTs. In comparison to MWCNTs, which have a maximum removal capacity of 100 mg g^{-1}, methylene blue with functionalized multi-walled carbon nanotubes (f-MWCNTs) had a greater maximum removal capacity of 166.7 mg g^{-1}.

10.4 Upcoming challenges

It is very difficult for humans to control ECs in water sources. Every year, the global population grows significantly, which increases the need for freshwater for home use and produces waste that depletes water supplies. In addition, urbanization, industry, and water demand from agriculture are key contributors to water shortage, which has a negative impact on the environment. Therefore, the need for highly effective and affordable wastewater treatment techniques, as well as increased societal awareness, exists. Currently, wastewater treatment is a difficult problem that depends on socioeconomic conditions and has a significant impact on the bio-physical environment and living things. However, it is difficult to develop a generic method that completely removes all pollutants from wastewaters. Although the elimination of emerging pollutants has been studied using a variety of biological, physical, and chemical wastewater management technologies, the best approach to address operational complexity, ecological impact, effectiveness, feasibility, and cost-efficiency has not been found. Two or more removal methods are combined for increased removal to provide desirable water quality at a reasonable price.

Some suggested future directions that may help to overcome these obstacles follow:

- To use cutting-edge ideas like nanotechnology and genetic engineering to provide safe and non-hazardous methods for creating nanoparticles that will degrade pollutants.
- Effective treatment evaluation to determine the best treatment method based on a variety of factors, including water quality, environmental compatibility, consistency, flexibility, usability, and cost-effectiveness.
- To use industrial-scale green technologies such as membrane filtration nanotechnology and microbial fuel cells as effective and low-maintenance solutions.
- To design the best model, it is necessary to investigate cross-treatment systems, such as combinations of photo-versus-electro-Fenton, UV photolysis, ozonation, and biological treatment technologies.

10.5 Conclusions

Emerging contaminants are harmful substances that are created by humans and released into wastewater. The origins of developing pollutants, their toxicity, and treatment methods are covered in this chapter. The primary sources of ECs are medications, personal care items, and fertilizers. Their presence has deleterious effects on both marine life and human health, even at low quantities. They cannot be effectively removed using standard wastewater treatment techniques. There have been discussions on a number of treatment techniques, including membrane technology, coagulation-flocculation, solvent extraction, adsorption, enhanced oxidation processes, and nanotechnology. These methods have both benefits and drawbacks. Although hybrid systems are more efficient than individual procedures for removing EC, they have drawbacks in terms of time, energy, and cost. Nanotechnology is a promising solution to get beyond these restrictions. To completely and effectively remove ECs from polluted water, extensive research on wastewater treatment technologies that are technically and economically viable is necessary.

Disclaimer

References

[1] Lee C S, Robinson J and Chong M F 2014 A review on application of flocculants in wastewater treatment *Process Saf. Environ. Protect.* **92** 489–508
[2] Abdelbasir S M and Shalan A E 2019 An overview of nanomaterials for industrial wastewater treatment *Korean J. Chem. Eng.* **36** 1209–25

[3] Raouf M E A, Maysour N E and Farag R K 2019 Wastewater treatment methodologies, review article *Int. J. Environ. Agric. Sci.* **3** 18

[4] Karimi-Maleh H, Ranjbari S, Tanhaei B, Ayati A, Orooji Y and Alizadeh M *et al* 2021 Novel 1-butyl-3-methylimidazolium bromide impregnated chitosan hydrogel beads nano-structure as an efficient nanobio-adsorbent for cationic dye removal: kinetic study *Environ. Res.* **195** 110809

[5] Taheran M, Naghdi M, Brar S K, Verma M and Surampalli R Y 2018 Emerging contaminants: here today, there tomorrow! *Environ. Nanotechnol., Monitor. Manag.* **10** 122–6

[6] Petrie B, Barden R and Kasprzyk-Hordern B 2015 A review on emerging contaminants in wastewaters and the environment: current knowledge, understudied areas and recommen-dations for future monitoring *Water Res.* **72** 3–27

[7] Hemavathy R V, Kumar P S, Kanmani K and Jahnavi N 2020 Adsorptive separation of Cu (II) ions from aqueous medium using thermally/chemically treated *Cassia fistula* based biochar *J. Clean. Prod.* **249** 119390

[8] Varsha M, Senthil Kumar P and Senthil Rathi B 2022 A review on recent trends in the removal of emerging contaminants from aquatic environment using low-cost adsorbents *Chemosphere* **287** 132270

[9] de Oliveira M, Frihling B E F, Velasques J, Magalhães Filho F J C, Cavalheri P S and Migliolo L 2020 Pharmaceuticals residues and xenobiotics contaminants: occurrence, analytical techniques and sustainable alternatives for wastewater treatment *Sci. Total Environ.* **705** 135568

[10] Rout P R, Zhang T C, Bhunia P and Surampalli R Y 2021 Treatment technologies for emerging contaminants in wastewater treatment plants: a review *Sci. Total Environ.* **753** 141990

[11] Calderon A G, Duan H, Seo K Y, Macintosh C, Astals S and Li K *et al* 2021 The origin of waste activated sludge affects the enhancement of anaerobic digestion by free nitrous acid pre-treatment *Sci. Total Environ.* **795** 148831

[12] Poonia T, Singh N and Garg M C 2021 Contamination of arsenic, chromium and fluoride in the Indian groundwater: a review, meta-analysis and cancer risk assessment *Int. J. Environ. Sci. Technol.* **18** 1–12

[13] Chinnaiyan P, Thampi S G, Kumar M and Mini K 2018 Pharmaceutical products as emerging contaminant in water: relevance for developing nations and identification of critical compounds for Indian environment *Environ. Monit. Assess.* **190** 1–13

[14] Richardson S D and Kimura S Y 2017 Emerging environmental contaminants: challenges facing our next generation and potential engineering solutions *Environ. Technol. Innov.* **8** 40–56

[15] Kim E, Jung C, Han J, Her N, Park C M and Jang M *et al* 2016 Sorptive removal of selected emerging contaminants using biochar in aqueous solution *J. Ind. Eng. Chem.* **36** 364–71

[16] Juliano C and Magrini G A 2017 Cosmetic ingredients as emerging pollutants of environmental and health concern. A mini-review *Cosmetics* **4** 11

[17] Mandaric L, Celic M, Marcé R and Petrovic M 2015 Introduction on emerging contaminants in rivers and their environmental risk *Emerging Contaminants in River Ecosystems* (Cham: Springer) pp 3–25

[18] Tollefsen K E, Nizzetto L and Huggett D B 2012 Presence, fate and effects of the intense sweetener sucralose in the aquatic environment *Sci. Total Environ.* **438** 510–6

[19] Venier M, Dove A, Romanak K, Backus S and Hites R 2014 Flame retardants and legacy chemicals in Great Lakes' water *Environ. Sci. Technol.* **48** 9563–72

[20] Tijani J O, Fatoba O O, Babajide O O and Petrik L F 2016 Pharmaceuticals, endocrine disruptors, personal care products, nanomaterials and perfluorinated pollutants: a review *Environ. Chem. Lett.* **14** 27–49

[21] Rathi B S, Kumar P S and Show P L 2021 A review on effective removal of emerging contaminants from aquatic systems: current trends and scope for further research *J. Hazard. Mater.* **409** 124413

[22] Munthe J, BrorströmLundén E, Rahmberg M, Posthuma L, Altenburger R and Brack W *et al* 2017 An expanded conceptual framework for solution-focused management of chemical pollution in European waters *Environ. Sci. Eur.* **29** 1–16

[23] Stuart M, Lapworth D, Crane E and Hart A 2012 Review of risk from potential emerging contaminants in UK groundwater *Sci. Total Environ.* **416** 1–21

[24] Crini G and Lichtfouse E 2019 Advantages and disadvantages of techniques used for wastewater treatment *Environ. Chem. Lett.* **17** 145–55

[25] Ungureanu N, Vlăduţ V and Voicu G 2020 Water scarcity and wastewater reuse in crop irrigation *Sustainability.* **12** 9055

[26] Singh H L, Khaturia S, Chahar M and Bishnoi A 2023 Membrane and membrane-based processes for wastewater treatment *Membrane and Membrane-Based Processes for Wastewater Treatment* pp 51–65 (Boca Raton, FL: CRC Press)

[27] Vieira W T, de Farias M B, Spaolonzi M P, da Silva M G C and Vieira M G A 2020 Removal of endocrine disruptors in waters by adsorption, membrane filtration and biodegradation. A review *Environ. Chem. Lett.* **18** 1113–43

[28] Nghiem L D and Fujioka T 2016 Removal of emerging contaminants for water reuse by membrane technology *Emerging Membrane Technology for Sustainable Water Treatment* (Amsterdam: Elsevier) pp 217–47

[29] Dhangar K and Kumar M 2020 Tricks and tracks in removal of emerging contaminants from the wastewater through hybrid treatment systems: a review *Sci. Total Environ.* **738** 140320

[30] Teh C Y, Budiman P M, Shak K P Y and Wu T Y 2016 Recent advancement of coagulation–flocculation and its application in wastewater treatment *Ind. Eng. Chem. Res.* **55** 4363–89

[31] Mohamed Noor M H, Wong S, Ngadi N, Mohammed Inuwa I and Opotu L A 2021 Assessing the effectiveness of magnetic nanoparticles coagulation/ flocculation in water treatment: a systematic literature review *Int. J. Environ. Sci. Technol.* **28** 1–22

[32] Deng Y and Zhao R 2015 Advanced oxidation processes (AOPs) in wastewater treatment *Curr. Pollut. Rep.* **1** 167–76

[33] https://alfaauv.com/blog/uv-disinfection-system-water-treatment

[34] Borrego B, Lorenzo G, Mota—Morales J D, Almanza-Reyes H, Mateos F and López-Gil E *et al* 2016 Potential application of silver nanoparticles to control the infectivity of Rift Valley fever virus *in vitro* and *in vivo Nanomed. Nanotechnol. Biol. Med.* **12** 1185–92

[35] Li J, Zhou Q, Liu Y and Lei M 2017 Recyclable nanoscale zero-valent iron-based magnetic polydopamine coated nanomaterials for the adsorption and removal of phenanthrene and anthracene *Sci. Technol. Adv. Mater.* **18** 3–16

[36] Fagan R, McCormack D E, Dionysiou D D and Pillai S C 2016 A review of solar and visible light active TiO_2 photocatalysis for treating bacteria, cyanotoxins and contaminants of emerging concern *Mater. Sci. Semicond. Process.* **42** 2–14

[37] Lu H, Wang J, Stoller M, Wang T, Bao Y and Hao H 2016 An overview of nanomaterials for water and wastewater treatment *Adv. Mater. Sci. Eng.* **2016** 4964828

[38] Nasrollahzadeh M, Sajjadi M, Iravani S and Varma R S 2021 Carbon-based sustainable nanomaterials for water treatment: state-of-art and future perspectives *Chemosphere* **263** 128005

[39] Ouni L, Ramazani A and Fardood S T 2019 An overview of carbon nanotubes role in heavy metals removal from wastewater *Front. Chem. Sci. Eng.* **13** 1–22

[40] Al-Wafi R, Ahmed M K and Mansour S F 2020 Tuning the synthetic conditions of graphene oxide/magnetite/hydroxyapatite/cellulose acetate nanofibrous membranes for removing Cr (VI), Se (IV) and methylene blue from aqueous solutions *J. Water Process Eng.* **38** 101543

[41] Rajabi M, Mahanpoor K and Moradi O 2017 Removal of dye molecules from aqueous solution by carbon nanotubes and carbon nanotube functional groups: critical review *RSC Adv.* **7** 47083–90

IOP Publishing

Trends in Biological Processes in Industrial Wastewater
Treatment

Maulin P Shah

Chapter 11

Bioremediation, phytoremediation, and mycoremediation of wastewater

Joydeep Das, Divyajeet Kumar and Soma Nag

The availability of fresh water is one of the major issues in the present world. Pollutants present in wastewater are causing danger to human health and the environment. Sustainable wastewater treatment is a major challenge to both researchers and policy makers. The treatment method varies based on the sources and characteristics of the contaminants present. Effluents produced by domestic, hospital, agricultural, or industrial sources need proper treatment to avoid their hazardous impacts. In spite of conventional methods, the recent trend is to follow biological methods that are cheap, eco-friendly, and easy to operate. Bioremediation, phytoremediation, and mycoremediation are recent biological techniques for treating hazardous pollutants in a very economic and environmentally-friendly way. Bioremediation uses natural biological methods that are simple and helpful in converting organic or inorganic micropollutants into less harmful and relatively eco-friendly substances. This method is considered to be the soundest way of treating sewage. Bioremediation systems are capable of eliminating heavy metals, dyes, total solids, ammonia and other nutrients, and chemical and biochemical oxygen demand and pathogens from the wastewater. Phytoremediation is a novel bioremediation technique that uses plants and microbes for the degradation of heavy metals or metal contaminated soils and wastewater in an eco-friendly, efficient, and solar driven path. Meanwhile, mycoremediation makes use of fungi to get rid of the pollutants. Some fungi are able to bear extreme conditions of severe metal concentration, nutrient availability, pH, or temperature. For mycoremediation, land-based species, particularly mushroom, white-rot fungus, and brown rot fungi are thoroughly explored in this chapter. The principles, mechanisms, and role of process parameters are discussed for each process. Finally, the advantages and disadvantages of each method are highlighted.

11.1 Introduction

The fast growth of the population has led to speedy industrialization, thereby producing large quantities of wastewater that contains residual chemicals, metals, and metalloids. However, contaminated water is unsafe for people, aquatic animals, and soils. Crops can be contaminated when in contact with this water, which is a risk for food safety. Sustainable wastewater treatment is a major challenge to both researchers and policy makers. Treatment methods vary based on the sources and characteristics of the contaminants present in the wastewater [1]. Apart from industrial wastewater, municipal and household wastewater contains a large number of pesticides, unused or expired medicines, hospital wastes, traces of personal care products, and many other contaminants used in our daily lives. They are not easily traceable, and consequently require special attention and removal techniques.

The operation and maintenance of conventional techniques are costly. Moreover, there is an additional burden of secondary pollution by producing sludge. The treatment of wastewater is one of the most significant biotechnological procedures employed globally and is a major source of concern. The microbial communities from various settings have a significant impact on a variety of substances present in wastewater from both natural and manmade systems. In bioremediation, contaminated water is treated using helpful microbiological agents, such as plants and their roots, yeast, fungi, bacteria etc. Most treatments typically involve seeding contaminated wastewater with competent microflora that can degrade hazardous waste to hasten the bioremediation process [2–4]. In bioremediation, microorganisms are used to modify and utilize toxic pollutants to get energy and biomass, and extinguish the environmental contaminants into less toxic forms [5, 6]. Since fossil fuel supplies are few and could run out soon, necessitating the development of alternate energy sources. Microbial treatment of wastewater can be used as a green source of energy [13].

By 2025, it is estimated that half of the world's population will live in water stressed areas [7]. The bioremediation technique makes use of plant interactions in contaminated areas to lessen the hazardous effects of contaminants. The phytoremediation procedure is a profitable and sustainable plant-based technique. It takes advantage of the capacity of plants to concentrate substances from their environment and metabolize different molecules in their tissues.

Mycoremediation is a fungi-based method of treating polluted soil or harmful effluents. It is possible to perform mycoremediation while filamentous fungus (molds) and macrofungi are present, e.g., mushrooms [8–10]. In the bioremediation process, many indigenous naturally occurring organisms are utilized as remediating agents. Various types of microorganisms are used for efficient remediation because contaminated water is likely to include a variety of pollutants [11]. Different types of catabolic-capable microbes utilize organic pollutants as a source of energy and transform these impurities into non-toxic intermediate molecules [12]. These toxins need to be removed from water otherwise the crops would be contaminated, which can be a risk to human health [14, 15].

Diverse contaminants may be eliminated by microbes using a number of biological wastewater remediation techniques. The utilization of bacteria and filamentous fungus is gaining attention due to their high efficacy among the various groups of microorganisms utilized for wastewater treatment. It is quite advantageous to use fungus to remediate wastewater. The use of fungi in wastewater treatment results in the production of important biochemicals that are highly resistant to inhibitory compounds, as well as high-value prospective fungal protein and biomass for fodder and human diet [16]. Bioremediation has various benefits, including low cost, high efficiency, high precision, reduced sludge or chemical production, lack of additional nutrient needs, ability to regenerate biosorbents, and environmental friendliness [23].

11.2 Principles of bioremediation

Physical, chemical, or biological procedures are employed to treat wastewater. In this section, the biological treatment procedures are discussed in detail. Biological unit procedures are the name given to treatment techniques when biological activity is used to remove pollutants.

In the bioremediation process, contaminated soil and water are cleaned up using helpful microbiological agents such as yeast, fungi, or bacteria. This is described as the removal, attenuation, or modification of contaminated or polluting chemicals using biological operations. Bioremediation relies on biological processes to reduce the concentration of pollutants and transform them into harmless forms. Enzymes that are naturally present in microorganisms or plants and which are used to damage or reduce toxins and unwanted environmental pollutants are known as 'enzyme-mediated bioremediation'. Microorganisms attack impurities enzymatically and then convert them into harmless products for effective bioremediation. Enzymes are biocatalysts that enable rapid and full collapse of substrates by lowering the activation energy. Enzymes such as laccases, hydrolases, oxidoreductases, and others play a key role in the bioremediation process. There are two types of enzyme-mediated bioremediation, which can be distinguished as intracellular and extracellular [6].

Thus, bioremediation uses biological organisms to address environmental issues such contaminated soil or, with the use of technical advances, contaminated groundwater. Complex hazardous substances are changed by living microbes into innocuous byproducts of cellular metabolism, such as carbon dioxide and water. Nitrogen and phosphorus are two nutrients that are removed from wastewater through biological treatment. Most wastewater can be treated biologically with the right environmental controls. The general goals of biologically treating industrial wastewater are to produce acceptable end products by oxidizing dissolved and particulate biodegradable components, capture and incorporate colloidal soil that is suspended and unable to settle into biological floc or biofilm, transform and eliminate nutrients such as phosphorus and nitrogen. It can also remove trace organic components and chemicals in some circumstances. Microorganisms play a significant part in the bioremediation process. As single-celled organisms reach a particular size, they multiply and split into two. Biological oxygen demand (BOD) and the chemical oxygen demand (COD) are two vital guides for water quality [17].

Wastewater comprising nitrates and phosphorus in many forms causes eutrophication, which decreases the dissolved oxygen (DO) and raises the BOD level of water [18]. This process eliminates (colloidal or dissolved) biodegradable organic materials from wastewater.

The extraordinary ability of microorginisms such as bacteria and fungi to destroy a diverse xenobiotics and organic pollutants, including long-chain and aromatic hydrocarbons, is a well-known fact [18]. Through the enhancement of specific living organisms that can remove, degrade, or change dangerous organic substances into safe or less toxic metabolic products, bioremediation is utilized to repair environmental damage [20]. The use of fungi in wastewater treatment is popular because they seem to have faster rates of organic matter breakdown. In addition to producing extracellular enzymes, the components of fungal cell walls are essential for the biosorption of toxic substances during wastewater treatment [21]. Bioremediation can happen naturally or only effectively when fertilizers, oxygen, and other factors that promote the growth of pollution-eating bacteria inside the medium are added [22].

Aquatic ecosystems are the earliest and most severely impacted ecosystems, whether due to pollution from a single point cause or multiple sources. Point sources of pollution happen when the pollutant is released into the stream directly. Municipal sewage, industrial effluents, runoff from solid waste disposal and industrial sites, and release from vessels are the typical point sources of pollution. However, the water flow from farming is an example of a non-point source. Water contamination has severe consequences on both land animals and birds, in addition to aquatic organisms. More seriously, polluted water kills aquatic life and hinders their ability to reproduce. In the end, the water becomes unsuited for household or human use, and in extreme circumstances it can even pose a risk to human health. The application of bioremediating agents can lower the financial and environmental costs associated with waste disposal. Most treatments typically involve seeding contaminated wastewater with competent microflora that can degrade hazardous material to hasten the bioremediation process. The injected microorganisms may be specially bred in the laboratory to attack the target waste or they may be naturally occurring varieties.

Hydro-geologic circumstances, the contaminant, microbial ecology, and financial considerations all affect how successful and cost-effective a microbial bioremediation program is when compared to other highly variable geographical and temporal elements. Each bioremediation procedure involves the introduction of microorganisms that utilize the pollutants as food or fuel [5]. The most important factor influencing the procedure for removing pollutants is the type of pollutant, such as agrochemicals, dyes, heavy metals, hydrocarbons, plastics, sewage, or chlorinated chemicals. Aside from this, some of the primary selection variables considered while selecting any bioremediation technology are the depth and degree of pollution, type of site and environment, treatment cost, and green policies.

A well-known and widely acknowledged alternative to environmental cleanup that does not hurt the ecosystem is bioremediation. Moreover, it can be utilized for *in situ* circumstances with cost-effectiveness because its on-site application lowers the cost of excavation, transportation, treatment, etc. However, this method has also drawbacks. For example, it is slower than the alternatives and less effective at breaking down non-biodegradable substances. It also requires the donor organism to have a particular metabolic profile. Nonetheless, it has nonetheless established itself as a strong contender in this situation. With bioremediation, pollutants are either completely removed or significantly reduced in concentration.

The most significant contributors to this process are fungi, bacteria, and plants in the categories of remediation, mycoremediation, and phytoremediation. Moreover, mycoremediation and microremediation approaches include a few subprocesses, such as biostimulation, bioaugmentation, and biosparging, to improve the on-site removal of toxins. By providing extra nutrients and growth factors, the biostimulation procedure revitalizes the local microorganisms, whereas bioaugmentation introduces exogenous microorganisms into an ecosystem to digest toxins. Meanwhile, biosparging uses pressured air to give oxygen and/or nutrients to a specific zone to promote microbial activity.

The topic of bioremediation has undergone tremendous research and development in recent years. Various technologies are employed to carry out and track the nanotechnology. Some examples of bioremediation methods include genetic engineering, biochemical engineering, spectroscopy, spectrometry, and chromatography. Science is now paying attention to bioluminescence-based assays for the eco-toxicological evaluation of contaminants. Microbial biosensors have also been investigated as a sensitive and trustworthy method for locating and monitoring the pollutants within an ecosystem. Enzymatically, microbes fight the contaminants and turn them into harmless compounds. Microbes rely on the participation of several enzymes to cleanup persistent and organic contaminants. The use of microbial enzymes in bioremediation is regarded as a new, promising, and cost-effective method. Thus, using microbial enzymes in waste management for environmental safety and general welfare may be taken as an important option [16].

Bioremediation retains microbial activity and growth, which may require changing environmental parameters based on different applications to promote microbial growth and breakdown at a faster rate or quicker pace. Thus, the microorganism counts or the strengths of nutrients and variation of temperature are the key controls of bioremediation. Using biological organisms to address environmental issues such as polluted soil or water is known as bioremediation. In other words, it is a technology for cleaning up the environment and regaining the original balance and preserving the environment and halting additional pollution. Using bioremediation, specific contaminants can be removed, such as chlorinated pesticides that are broken down by bacteria, or a broader strategy can be used, such as with wastewater that has been contaminated with oil and is broken down using a variety of methods, including the addition of biostimulation to speed up the breakdown of oil by bacteria.

A wide variety of bacteria and fungi can be utilized for bioremediation. The first recyclers in nature were microorganisms and they are able to digest pollutants, both natural and manmade, into energy and basic materials for their own survival. Consequently, the use of costly treatment methods might be reduced with less expensive and more environmentally-friendly biological processes. For novel environmental biotechnologies, microorganisms constitute a valuable and mostly untapped resource.

11.3 Factors affecting bioremediation

Bioremediation techniques are controlled by a complicated system of different variables. The parameters include nutrients, pH, temperature, and oxygen concentration.

Nutrients: Since carbon (C) is the simplest element known to exist in living things, it is more highly sought-after than other elements, such as nitrogen (N) and phosphorus (P). By improving the bacterial C: N: P ratio, nutrient balancing, particularly the supply of essential nutrients like N and P, can increase the efficiency of biodegradation. Nutrients such as carbon, nitrogen, and phosphorous are essential by microbes for survival and continuous microbial activities.

In marine environments, nutrients such as N, P, and iron (Fe) are extensively more vital than oxygen (O) in controlling the speed of biodegradation. The biodegradation of alkane and polyaromatic hydrocarbons are accelerated by the addition of nitrogen, whereas adding phosphorus accelerated alkane biodegradation but not that of polyaromatic hydrocarbons (PAHs). Minerals are frequently lacking in wastewater ecosystems because phytoplankton and other non-oil-degrading microbes compete with oil-degrading types for their resources. At high pH, phosphorus precipitates as calcium phosphate. The most likely elements to restrict biodegradation are N and P; however, the absence of Fe or other trace minerals can sometimes matter. These nutrients enable bacteria to emit the enzymes required to degrade the pollutants. In addition, these microbes require carbon in high proportions and the other essential components are P and N.

pH: The optimum pH condition for the effective bioremediation must be in the range of 6.5 to 8.0 because pH plays the major role in affecting the solubility and biological availability of nutrients and other constituents.

Temperature: Temperature directly impacts how quickly microorganisms breakdown food, which in turn determines how active they are in the environment. In some cases, the rate of biodegradation increases with rising temperatures and decreases with lowering temperatures. Temperature ranges between 25 and 45 °C are ideal for microbial development.

The metabolism of oils decelerates at cold temperatures, resulting in the lighter portions of petroleum becoming less volatile, allowing the oil's hazardous components to remain in the water for longer and also inhibiting microbial activity. At higher temperatures, biodegradation occurs more quickly due to higher solubility and diffusion of lipids. As a result, a drop in viscosity occurs.

Concentration of oxygen: Anaerobic microbes do not require oxygen for the biodegradation process. However, some microorganisms, such as aerobic microbes, require oxygen to improve their biodegradation rate. In many circumstances, oxygen can speed up the metabolism of hydrocarbons [6]. Oxygen is a vital factor for the microbial breakdown of oil. Microbes make use of enzymes that incorporate oxygen begin attacking oil. Anaerobic breakdown is a slow process that happens in the absence of oxygen.

The initial breakdown of oil typically requires oxygen, and at later stages reactions may also require it directly. In most cases, the abundance of oxygen at or near the ocean's surface, where oil has spread out to an open and broad surface area, does not limit the rate of biodegradation. The rates of biodegradation slow down when oxygen availability is reduced. As a result, the degradation of oil that has sunk to the sea floor and been buried by sediment takes much longer. The height of the water column, sediment depth, and the turbulence all affect how much oxygen is available [22].

11.4 Degradation techniques

Aerobic biodegradation: According to the idea of aerobic biodegradation, oxygen is necessary for the destruction of contaminants by degradable species. Oxygen from air is utilized to breakdown the organic matters, reduce pathogens, and alter P and N from the the effluents. Bacteria and fungi can produce the enzymes oxygenase and peroxidase. Aerobic biodegradation reactors are included in membrane bioreactors and activated sludge reactors [23]. This treatment technique is more effective and faster for the decay of organic solids and pathogens than conventional methods and the treated water is pure and suitable for long-term storage without smell. However, though it is a widely used method, it generates large amounts of sludge and can effectively degrade the strength of pollutants. These disadvantages and high operation costs are directing people towards anaerobic treatments [3].

Anaerobic degradation: This process is most suitable for organic pollutants and a variety of bacteria are used to decompose natural polymers, proteins, and lipids, especially in warm ambiences. Anaerobic degradation is a mechanism through which microorganisms breakdown biodegradable material in the absence of oxygen. In this technique, air is not required from outside. Controlled anaerobic digestion of the organic pollutants, produce CO_2 and CH_4. The following processes are part of the anaerobic degradation concept:

Insoluble organic pollutants are first split into soluble substances to get access to other bacteria. Acidic bacteria then convert sugars and amino acids into CO_2, H_2, NH_3, and organic acids. Finally, organic acids are altered into acetic acid, H_2, and NH_3 [23]. In contrast to aerobic degradation, anaerobic degradation processes are slow and ineffective. They can effectively manage wastewater with high organic pollutant loads. Wastewater from the sugar industry, slaughterhouses, food and paper industry, and tanneries can be treated by this method. Enzymes with azoreductase activity have been found in many types of aerobic and anaerobic microorganisms, including bacteria, fungi, and algae [27]. Anaerobic and aerobic bacteria seem to have resemblances in their degradation of aromatic hydrocarbons: the different structure is degraded in some key structures [28].

11.5 Mechanisms of bioremediation

Microorganisms that are extensively employed for bioremediation include yeast, fungus, microalgae, and bacteria. Bioremediation involves two mechanisms, such as biosorption and bioaccumulation.

Biosorption: Biosorption is a rapid and reversible passive adsorption mechanism. Both living and dead biomass may be utilized. This is an inexpensive method because the biomass used in this process is generally agricultural or industrial waste and can be repeatedly used after regeneration. The method uses microbial biomass as biosorbent to detoxify water sources that have been contaminated with heavy metals.

Biosorption is dependent on factors such as pH, temperature, the amount of dye present, and the strength of biomass. Several microorganisms, including bacteria, yeast, algae, protozoa, and fungus, are useful for extracting heavy metals, such as, As, Hg, Pb, Cu, Cd, and their cyanide complexes. Actinobacteria biomass for heavy metal biosorption is a promising tool for the creation of green technology. Actinomycetes can stay alive at high metal concentrations and serve a special function in heavy metal degradation at high concentrations. They can change the toxicity of heavy metals and they can also many physiologically active substances. Metal-binding proteins on the cell wall are involved in the biosorption of metal ions. Heavy metals can be remediated with a variety of processes, including spontaneous physiochemical routes, the utilization of energy (ATP), and inactive microbial biomass having metal-binding capacity. Microbes are referred to as 'bioremediators' because they can degrade almost all types of organic contaminants [29].

Bioaccumulation: By absorbing poisons from their surroundings, marine animals can increase their concentration or accumulation of toxins through a process known as bioaccumulation. These are both intracellular and external processes. The only kind of biomass that can happen is living biomass. However, this method is expensive because it takes place in living cells and recycling is restricted.

Bioremediation in aquatic life: The optimization of nitrification and denitrification rates is an essential step to keep the ammonia content low and eliminate overloading of nitrogen in ponds for a successful bioremediation. As nitrogen gas increases sulfide oxidation to limit hydrogen sulfide buildup, and also promotes carbon dioxide production and reduces sludge build up, ultimately it enhances primary productivity, thus boosting secondary crop production and prawn production. It was found that ponds pre-inoculated with nitrifying bacteria had lower levels of ammonia and nitrite in the rearing water and higher prawn survival rates. It was also shown that the introduction of Bacillus near pond aerators decreased the need for chemical oxygen and enhanced prawn yield. To diminish or eradicate specific pathogenic bacteria, and usually enhance the growth and survival of the targeted species, bioremediation chemicals modify or manipulate the microbial communities in water and sediment.

About 15 different species of Bacillus make up the majority of commercial probiotic (bioremediation) products for pond aquaculture. Bioremediation agents are an important management tool in prawn cultivation, but their effectiveness depends on an understanding of the nature of competition between different species or bacterial strains. The results of microcosm trials to determine how well commercial shrimp farm bioremediators removed ammonia show that when the NH3 level was high, commercial bioremediators were unable to remove most of the total ammonia from the environment [19]. Bioremediation is a sustainable method to reduce the number of harmful bacteria, probiotics, and biodegrading micro-organisms in pond water and soil.

The effect of aquaculture on the environment: Probiotics play a crucial role in preserving good water quality, ensuring that helpful bacteria outnumber harmful bacteria in aquaculture ponds. Throughout the culture phase, they are crucial in maintaining the best water quality indicators, including dissolved oxygen, ammonia, nitrite, nitrate, and phosphates. They can also stop bacterial illnesses from developing in fish ponds.

11.6 Wastewater remediating agents

Bacteria: Bacteria have a wide spectrum of bioremediation potential. They are advantageous from an economic and environmental standpoint. Heavy metal emissions from industrial use cause a serious environmental problem [23]. The majority of bacteria in activated sludge are facultative, which means they can survive either with or without oxygen [24].

Algae: Algae are crucial to the process of naturally purifying water. They have wonderful biological characteristics of minimal structure, high photosynthesis rate and ability to grow in extreme toxic conditions. They can be utilized to recover expensive metal ions such as gold and silver, as well as poisonous and radioactive metal ion sorption. They aid in the remediation of nutrients by growing swiftly and assimilating the C, N, and P nutrients found in wastewater. This is a new technique for the environmentally-friendly and cost-effective treatment of sewage effluent [23]. Wastewater treatment with microalgae is a more environmentally-sound approach to decrease nitrogen and phosphorus, and to eliminate heavy metals from wastewater [25].

Microalgae: Microalgae are able to withstand high salinity, stress for nutrients, the presence of heavy metals, or very high or low temperatures. They have high surface area, high metal ions, and good dye binding ability, which reflects in elevated removal efficiency [55]. Textile wastewater includes organic dyes as well as the nutrients needed to grow algae (phosphate, nitrate, micronutrients, etc). Microalgae use the nutrients and colors in wastewater for their growth. Both living and dead algae have been used to remove color from dyes and wastewater.

Bioconversion and bioaccumulation can be used to biodegrade textiles using microalgae. During the bioconversion process, microalgae consume these colors as a source of carbon and convert them into metabolites. Moreover, they can also serve as a biosorbent by adsorbing these dyes on their surface. Both phenomena are possible in textile wastewater bioremediation. The aggregation of microalgae may be caused by enzyme breakdown, adsorption, or both. Large surface area and strong binding affinity for azo dyes have made microalgae a potential biosorbent. Metal ions are biosorbed on the surface of the cell membrane of dead algae. An eco-friendly, less expensive, and more efficient way to remove metal ions from wastewater is through algae-based biosorption of heavy metal ions. Spirogyra has been demonstrated as a beneficial biosorbent for removing reactive dye from textile wastewater. Certain algae can bio convert colors into simpler chemicals. Color is extracted from textile dyes using immobilized algae. Spirulina and chlorella are both good at handling wastewater cleanup.

Fungi: Filamentous fungi eat heavy metals. Fungal biomass may be used as biosorbents to remove radionuclides and heavy metals from polluted streams. Straw, sawdust, or corn cobs are examples of dangerous impurities that white-rot fungi can eliminate [23]. The use of fungus in wastewater treatment systems has several benefits, such as their ability to create extracellular enzymes to solubilize insoluble substrates and their improved cell-to-surface ratio, which allows them to have better enzymatic interaction with their environment [26].

Yeast: Yeast may reduce COD levels and can absorb heavy metals, such as Zn, Cd, and Pb. They are also effective in getting rid of mono- and polyphenols. They are employed in the processing of textile wastewater as a result of their capacity to absorb and degrade harmful chromophores into less dangerous chemicals [23]. They could possess enzymes for the breakdown of dyes and be employed as biosorbents for dye biosorption.

11.7 Types of bioremediations

There are primarily two approaches based on trash removal and transportation for treatment: (i) *ex situ* and (ii) *in situ* bioremediation.

In the case of *ex situ* bioremediation, contaminated materials are removed from the site and treated elsewhere, whereas *in situ* technique purifies the contaminated substances on the spot. The choice of which method to use is based on how well a region is aerated and saturated.

***Ex situ* bioremediation:** *Ex situ* techniques are helpful for treating soil and groundwater by excavating the soil (for soil treatment) or pumping the water (for water treatment) from polluted areas. *Ex situ* bioremediation techniques typically depend on the following factors: the cost of treatment, the level or severity of pollution, the type of contaminant, and (most importantly) the location of the contaminated area. It is possible to treat the contaminated material on or off-site if it is dug, which is frequently a quicker way to decontaminate the area. Composting, bioreactors, biopiles, and land farming are some of the methods that can be applied.

Land farming: Land farming is a straightforward process in which contaminated soil is excavated, deposited on a bed that has been prepared, and repeatedly tilled until impurities are destroyed. The intention is to encourage naturally occurring biodegradative microorganisms and facilitate their aerobic breakdown of pollutants. This procedure is often restricted to treating the top 10–35 cm of soil. Land farming has drawn a lot of interest as a disposal alternative since it requires low cost and minimum monitoring and maintenance.

Composting: Composting mixes polluted soil with non-hazardous organic materials such as manure or agricultural waste. These organic components encourage the development of a diverse microbial population and the high temperature necessary for composting.

Biopiles: Composting and land cultivation are combined in biopiles. Engineered cells are basically made of ventilated compost piles. They are a more sophisticated kind of land farming and are frequently employed for the treatment of surface pollution from petroleum hydrocarbons because they tend to control physical losses of the pollutants through leaching and volatilization. Original aerobic and anaerobic microorganisms thrive in the favorable environment provided by biopiles.

Bioreactors: Bioreactors are used in case of an *ex situ* treatment procedure. When dirty water is pumped from a polluted area, slurry reactors or aqueous reactors are used to process the water or sludge in a controlled manner. A slurry bioreactor is a containment vessel and device used to produce a three-phase (i.e., solid, liquid, and gas) mixing condition to speed up the bioremediation of contaminants that are bound to soil and those that are water-soluble in the form of a polluted soil and biomass water slurry that is capable of degrading target pollutants. In general, a bioreactor system has a higher rate and scope of biodegradation than *in situ* systems because the enclosed environment is easier to control [30]. Although reactor systems have several benefits, they also have some drawbacks. Before being placed in a bioreactor, the contaminated soil needs to be pre-treated (excavated), or alternatively the contaminant can be physically extracted (by vacuum extraction) or washed from the soil.

***In situ* bioremediation:** *In situ* procedures cause the minimum disturbance to the soil and groundwater of the polluted location. These technologies are used 'in place', i.e., without removal of the polluted matrix. *In situ* remediation includes techniques such as bioventing, biosparging along with physical, chemical, and thermal processes. Both intrinsic and engineered bioremediation technologies can be used *in situ*. Techniques for *in situ* bioremediation have proved effective in cleaning up sites that have been contaminated by heavy metals, hydrocarbons, dyes, and chlorinated solvents [6].

In situ bioremediation is carried out with the aid of external dosing of bacteria in flowing sewage. The microbial consortiums that are applied are not genetically altered. To treat wastewater, they must be activated by creating favorable environmental circumstances for multiplication. Enzymes are occasionally introduced in order to get the microorganisms going. The ability of the microbial consortia to remain active in a variety of environments, including facultative, anaerobic, and aerobic, is a crucial component in bioremediation.[23]

Sewage from running battery flow may be decomposed by applying microbial consortia in aerobic environment without displacing it. This process leads to the generation of carbon dioxide and various other gases. In essence, *in situ* bioremediation is characterized by the application of microbial consortia in water, for reducing odours of the sewage and the natural breakdown of contaminants. This technology is user-friendly, less expensive than traditional methods, and does not require trained manpower. The process involves activating the bacteria, allowing them to proliferate in the presence of oxygen, and making food in the form of organic waste and sewage breakdown. The heavy metals and hazardous compounds are decreased during the treatment of contaminants. The hazardous pathogenic bacteria are either inhibited or removed from the treated water as a result of the dominating microbial consortia at work.

The choice of which bioremediation technique to use will be determined by the polluted area, the characteristics of the compounds involved, the level of contaminants present, and the amount of time needed to accomplish the bioremediation. *In situ* procedures entail bioventing, biosparging, bioaugmentation, and biostimulation [22]. The most important land treatments are:

Bioventing: The most typical *in situ* remedy is bioventing, which involves providing air and nutrients through wells to dirty soil to encourage the local bacteria. Low air flow rates and just providing the quantity of oxygen required for biodegradation are used in bioventing to reduce volatilization and the release of contaminants into the atmosphere. When the contamination is deep down, it can be employed and is effective for simple hydrocarbons.

Biosparging: To improve the rate of biological breakdown of pollutants by naturally occurring microorganisms, biosparging involves injecting air under pressure beneath the water table. By increasing the saturation zone's mixing, biosparging improves the interaction between soil and groundwater. Small diameter air injection points are simple to instal and inexpensive, which gives designers and system builders a great deal of freedom.

Bioaugmentation: The addition of microorganisms, either native or exogenous to the polluted locations, is a common step in bioremediation. The utilization of additional microbial cultures in a land treatment unit is constrained by two factors. First, it is uncommon for non-indigenous civilizations to successfully coexist with an indigenous population at useful population levels. Second, if the land treatment unit is properly run, then most soils exposed to biodegradable waste include native microbes that can decompose materials [30].

In the case of bioaugmentation, the microorganisms breakdown the oil at the location of the oil spill. The microbes that breakdown oil are collected from various sites and industrial areas. They are chosen to survive challenging environmental factors, including high salt content and fluctuating temperature, paired with an exceptional capacity to use nutrients such as oxygen, nitrogen, and phosphorus that are readily available. They can easily decompose the area because they can out compete native microbes. According to bioaugmentation advocators, once the oil is depleted, these organisms lose their competitive edge over the local microbes due to shortage of carbon source and finally vanish.

Biostimulation: Biostimulation is a technique in which substrates, vitamins, oxygen, and other substances are added to stimulate microbe activity so that they can breakdown the waste more quickly. In biostimulation of bacteria, the availability of important inorganic nutrients such as N and P tends to quickly deplete when substantial amounts of carbon sources are introduced. Mixing fertilizer with oil effluent is one instance of this. This works by delivering nutrients such as nitrogen and phosphorus, which prevent bacteria in wastewater tainted with oil from growing. This addition will enable microbes to breakdown the oil quickly because it makes the oil a carbon source and fertilizer a nitrogen and phosphorus source [22].

11.8 Advantages and disadvantages of bioremediation

Advantages: Bioremediation is a natural process for suitably treating contaminated wastewater. Carbon dioxide, water, and cell biomass are harmless byproducts of the treatment process. A wide range of pollutants can be eliminated with the help of bioremediation and can be converted into safe goods. In this way, the possibility of future liability of treatment and disposal of polluted materials is largely eliminated. Target pollutants may be destroyed rather than being transferred from one environmental medium to another, i.e., from land to water or air.

Bioremediation can often be done on-site, with little or no disruption to regular operations. Additionally, this removes the need to move large amounts of waste off-site and thus avoids potential risks to public health and the environment. So, compared to the other methods, bioremediation can be a more affordable technique for cleaning up hazardous wastes [30].

Disadvantages: Only biodegradable substances may be subject to bioremediation. Not all chemicals have the tendency to degrade quickly and completely. There are chances that the byproducts formed from biodegradation are more harmful or persistent than the original pollutant. In addition, some of the processes are very specialized. The presence of proper microbial populations that are suitable environmental growth conditions, and the correct quantities of nutrients and pollutants are vital for success. Another disadvantage is that it is difficult to scale-up. Moreover, bioremediation typically requires a longer time in comparison to alternative treatment techniques such as soil excavation and removal or cremation [30]. Some wastewater pollutants, such as heavy metals, might prevent native microorganisms from growing and producing toxins, which reduces the effectiveness of bioremediations [31].

11.9 Phytoremediation

The Latin words for plant and remedy are combined to form the term phytoremediation. Phytoremediation is a practical, cost-effective, and durable approach for eliminating pollutants. Because it employs plants to filter the air, it is not only environmentally benign but also has no adverse impacts on those who live nearby or work at the location. The broad and evolving word 'phytoremediation' has been used in recent decades to describe a collection of green environmentally-friendly solutions that are mostly based on plants. Another emerging technology that is effective at treating waste is phytoremediation. This technology should be supported so that it can be used in real-world situations to restore the water and soil resources on-site. Nowadays, these technologies are frequently utilized in addition to microbial degradation to remove contaminants from polluted soil, sludge, sediment, groundwater, surface water, and wastewater. Because natural plants or transgenic plants can bioaccumulate these toxins as they grow, phytoremediation is sometimes a more effective treatment method for biotreatment of water that has been contaminated with heavy metals such as cadmium and lead. Marine plants are exceptional in reducing the levels of dangerous metals, BOD, and total solids in effluent. This approach encompasses a number of methods that can be used to cleanse wastewater (surface and groundwater), remove unwanted nutrients from water reservoirs, and more [6].

The most commonly used plants for phytoremediation of wastewaters are aquatic macrophytes [32]. Fast biomass growth, low sensitivity to variations in the composition of treated fluids or soils, and a high amount of salts, phosphates, and nitrogen should all be present in phytoremediation plants [33]. Phytoremediation plants are raised using hydroponic or aeroponic methods in a bed of inert granular substrate, such as sand or pea gravel, with the purpose of cleaning polluted wastewater [34]. This is an emerging technology to remove damaging or toxic pollutants using green plants [8, 35]. Additionally, potential targets for phytoremediation technologies include heavy metals, aromatic, chlorinated solvents, petroleum hydrocarbons, pesticides, crude oil, and other organic and inorganic pollutants. To lessen the toxicity of pollutants, this strategy focuses on using a variety of plant interactions at polluted locations, including physical, biochemical, biological, chemical, and microbiological ones [36]. Phytoremediation is an environmentally-friendly way of treating wastewater. Wastewater pollutants and nutrients may be removed via floating aquatic macrophyte-based treatment systems [37]. Phytoremediation is a broad and developing term that has been used in recent decades to describe a group of environmentally-friendly green technologies that are primarily based on plants (aquatic, semi-aquatic, and terrestrial) and related associated enzymes, microorganisms, and water consumption, uptake, remove, retain, transform, degrade, or immobilize contamination (organic and/or inorganic) with different origins, from soil, sediment, aquatic media, or atmosphere.

To recover contaminated water, soil, and sludge, phytoremediation employs plants and the accompanying microbes in the rhizosphere. This biotechnology can be used *in situ* and has neither high costs nor side effects. Thus, this technique has positive environmental and ecological effects. Phytoremediation is the practise of using plants and microorganisms to remove or immobilize contaminants from soil or water through physiological and biochemical processes. These processes include phytoextraction, phytodegradation, or phytovolatilization, as well as phytostabilization and rhizofiltration. Fast-growing plants with high biomass production and tolerance that collect metals are recommended for phytoremediation. Local species that are typical of the ecosystem and are also simple to harvest are preferred. As a result, plants have specialized defences against the presence of metals in their environment. Some plants rely on the existence of effective mechanisms for metal exclusion that limit the transport of metals to aerial parts, whereas others base their resistance on other factors. Others allow harmless chemical species to accumulate metals in the aerial portions. Because plants can absorb and collect these inorganic pollutants, phytoremediation of heavy metals has proved successful. Nevertheless, different plants have different capacities for absorbing and accumulating harmful metals from water, soil, and sludge [20].

A modest level of persistent pollutants may be accumulated, immobilised, and transformed by vegetation-based remediation. Plants serve as filters and digest chemicals produced by nature in natural ecosystems. Phytoremediation is a new technique that employs plants to filter toxins out of water and soil. However, more research is necessary to determine whether it can promote the biodegradation of organic pollutants, but it may be a fruitful topic in the future [22]. The term 'phytoremediation' refers to a range of technologies that use plants to diminish, eliminate, degrade, or immobilize environmental toxins, mostly those with anthropogenic origins, with the goal of rehabilitating the area's sites so that they can be used for either private or public purposes. To remove toxins from the environment, phytoremediation uses plant-based systems and microbial activities. This is a passive wastewater treatment approach.

The physiochemical processes of plants and related microbes, such as photosynthesis, metabolism, and mineral nutrition, are the foundation of phytoremediation. This is a bioremediation-based facility. The phytorid system has advantages in that it is inexpensive, requires little electricity, and is highly effective at recycling and reusing water. Plants detoxify or stabilize pollutants as part of the phytoremediation process using a variety of techniques. In the process of phytoremediation, contaminants are primarily removed by six different mechanisms, i.e., phytoextraction, phytodegradation, rhizosphere degradation, rhizofiltration, phytostabilization, and phytovolatization. Both plant species and microbial consortia unique to the used plant system are required for wastewater treatment. Thus, the biota's consolidation effect is what ultimately removes the organic and inorganic burden from the wastewater [38].

There are several phytoremediation procedures that can be used for cleaning wastewater, purifying surface, and groundwater, removing excess nutrients from water reservoirs, and reclaiming soil that has been contaminated by environmental disasters.

11.10 Aquatic plants used for phytoremediation

Macrophytes that grow in water are frequently employed in phytoremediation. These plants are typically present in Poland and are an important part of wetlands. Additionally, the environmental properties of macrophyte populations are frequently used to evaluate the purity of water. Macrophytes are crucial components of both natural and artificial filtering systems in the biochemical processes of water purification. They have a beneficial impact on the environment. Macrophytes help to stabilize sludge, ensure favorable conditions for water filtration, and offer a place for the development of microorganisms.

Macrophytes have evolved to live permanently in contact with both groundwater and surface water. They have delicate exterior tissues and aerenchyma, which is a specialized tissue that creates a network of channels and gaps via which air is transported to a plant's submerged portions. Macrophytes are naturally occurring, which lowers the cost of acquisition. By incorporating various pollutants into the design of their cells, macrophytes gather a lot of pollutants. In addition, when adapting to difficult conditions in contaminated environments, these plants have a natural ability to take in and metabolize xenobiotics [39].

There is a good chance that different approaches with effectiveness and financial benefits will be used. Phytoremediation has recently become more significant due to its affordability, long-term applicability, and ecological component. This technology is based on plants' ability to remove a significant quantity of metal pollutants from water or groundwater by absorbing and accumulating them in their tissues. Metal must be absorbed by roots and moved to shoots and leaves for phytoremediation to take place. Aquatic plants are used in a bio-removal process that involves two absorption processes: biosorption, which is a quick and reversible metal-binding step, and bioaccumulation, which is a delayed and irreversible ion-sequestration step.

11.11 Mechanism of phytoremediation

Polluted places can be cleaned up or remedied via plant metabolism. Roots, which have multiple detoxifying mechanisms, provide surface area for adsorption and the collecting of water and nutrients that aid in growth, and are the main entry point for contaminants into the plant. When absorbed by the plant, these contaminants may be stored in the stems, leaves, or roots; converted internally into less harmful chemicals; or transformed internally into gases that are released into the atmosphere as the plant transpires. Therefore, phytoremediation may occur in one or more of the following ways: the impurity may enter through the root system and be stored in the plant's harvestable portion (phytoextraction), become volatile (phytovolatization), be metabolized (phytodegradation), or any combination of these may occur. In addition, impurities such as heavy metals may occasionally be either adsorbed or absorbed by the root systems of plants, helping to stabilize the impurity, which means reducing its bioavailability, rather than cleaning the environment. Rhizofiltration is a technique used to remediate waste streams or marine bodies (phytostabilization). However, the rate of phytoremediation is determined by the kind of soil, bioavailability, and contaminants [23]. Because they can be employed for phytoremediation by rhizofiltration, phytoextraction, phytovolatilization, phytodegradation, or phytotransformation processes, aquatic plants are essential in biological wastewater treatment systems [40].

Phytoextraction: This method is used by plants to buildup contaminants in their leaves and roots above ground. Here, the plant roots absorb heavy metal ions and transfer to shoots or other parts above ground. As plants deteriorate, they must be harvested and kept to prevent contaminants from being recycled. This technique allows the elimination of pollutants from the soil, groundwater, or surface water by plants that have a high capacity for buildup of toxic substances [39].

By using this strategy, plants with a high capacity for toxin accumulation can remove toxins from the soil, groundwater, or surface water. The plants utilized in this process may withstand high levels of organic or heavy metals and should have fast growth rate. Phytoextraction falls into two categories: triggered processes and continuous processes. Plants that acquire large quantities of harmful pollutants over the course of their whole lives are used in continuous phytoextraction. Utilizing chelators during a specific stage of plant growth is known as 'induced phytoextraction,' which causes a greater buildup of toxins in plant tissues.

Phytodegradation: In this process, organic substances are decomposed into more dependable, immobile, and safe forms. For example, the toxic Cr(VI) can be changed to Cr(III), which is less mobile and non-carcinogenic. This technique eliminates organic contaminants such as nitroaromatics and aliphatic organic molecules. They are phyto degraded inside the plant.

In this technique, xenobiotic degradation processes are catalyzed by enzymes produced by plants. Whenever a plant develops enzymes that are secreted into the soil of the root zone, phytodegradation may take place inside the plant or outside of it. This method is used to treat ground and surface water, as well as soil, river sediments, and sludges [39].

Phytovolatilization: For this technique to work, specific metals such as mercury or selenium must be absorbed in groundwater, transform into volatile chemical species, and then be released into the atmosphere. The release of contaminants as gas from plants to the atmosphere is known as phytovolatilization. This can be utilized for a few inorganics that can exist in volatile form, while it works well for organics [41].

Rhizofiltration: This is a method for water remediation that reduces pollution from flowing water. This involves the sorption and filtration of wastewater pollutants by plant roots and by microorganisms found in the rhizosphere. The main pollutants that may be removed with this method include metals, radio-nuclides, and hydrophobic organic molecules. This technique may involve the usage of specific plants, such as duckweed. Industrial wastewater treatment facilities have used duckweed in the past [23].

Rhizofiltration uses plant roots to absorb and adsorb the heavy metals form wastewater [42]. In a process called rhizofiltration, which is related to phytoextraction, wastewater is filtered via a dense network of hydroponically grown plant roots, causing any dissolved toxins to be absorbed or adsorb or accumulated [20].

Surface wastewater produced by industry and agriculture is treated using this technology. Plants are either submerged in purified water or have wastewater sprayed on top of their roots. Toxic chemicals and low oxygen concentration should therefore not be a problem for the plants employed in this process, and they should also have vast roots that develop quickly and create a lot of biomasses. In particular, lead (Pb) and radioactive materials are removed using rhizofiltration [38].

11.12 Phycoremediation

Phycoremediation is the process of bio-transforming or removing contaminants from wastewater, as well as CO_2 as air pollution from waste air, utilizing macro- and microalgae. Algae is a suitable plant for the nutrient removal from various wastewaters, the sanitization of metallic elements, and the consumption of carbon dioxide from exhausts. Microalgae are effective plants for the aerobic and anaerobic treatment of several types of wastewater, including home, industrial, and solid wastes [40]. It is possible to lessen the negative effects of different contaminants on the environment by promoting phycoremediation, which is a natural process [43]. Phycoremediation is the process of removing nutrients and xenobiotics from wastewater using macroalgae, microalgae, and cyanobacteria [44].

Phytostabilization: This technique is used to lessen the bioavailability of pollutants in the environment and stabilize them rather than having plants remove them (often metallic elements; hydraulic control). The bioavailability of metallic elements decreases as appropriate soil modification by plants increases, whereas plant cover decreased leaching and improved environmental protection. Plants can absorb pollutants in their root systems or uptake them in adsorption systems to help stabilize them. The choice of plants for phytostabilization is a critical issue, and it is proposed that perennial species with high biomass output and strong resistance to pollution should be used [45].

Phytostabilization involves the absorption and precipitation of pollutants, primarily metals, by plants. This reduces the mobility of the pollutants and prevents them from moving into groundwater (leaching), the atmosphere (wind transport), or the food chain [46]. Toxicants, primarily metals, are absorbed and precipitated by plants, which reduces their mobility and stops them from moving to groundwater (leaching), the air (wind transmission), or the food chain. Plant roots are utilized in this procedure to remediate the soil. Through phytostabilization, pollutants are kept from migrating to groundwater, surface soil, and farther with precipitation runoff. The root systems of plants employed for phytostabilization should be highly developed to allow for the adsorption, absorption, accumulation, and conversion of pollutants into less soluble chemicals in the rhizosphere. Additionally, the plants should be very resilient to changes in pH, salinity, and soil moisture, as well as having a low capacity to accumulate contaminants in their above ground sections.

Phytovolatization: In this method, pollutants are taken up by plants, processed through their metabolism, and then released into the atmosphere in a volatile and less hazardous form. This technique is primarily used to cleanup water and soil that have been contaminated with organic substances such as trichloroethylene, benzene, nitrobenzene, and phenol as well as selenium (Se), mercury (Hg), or arsenic. This type of phytoremediation uses plants to absorb pollutants from contaminated media, convert them into volatile form, and then transpire them into the atmosphere.

Plants typically engage in phytovolatilization to absorb water, organic substances, and inorganic substances. As a result, some pollution can move through the plant's components to the leaves and, at low concentrations, volatilize into the atmosphere. In brief, plant species are used to volatilize toxins from the leaves, which can lead to soil and sediment pollution, air pollution, and water pollution. Several studies exemplify the application of phytovolatilization in different contexts. For instance, some research focused on the phytovolatilization of radionuclides like Tritium (3H) from soil and some other study demonstrated the significant uptake of selenium compounds, including dimethyldiselenide and dimethylselenide, by Brassica species. However, it's important to note that there are drawbacks of phytovolatilization, particularly in the case of mercury. One disadvantage is the potential for recycling of contaminants by rain, allowing them to re-enter the ecosystem [45].

Phytomining: A green technology known as phytomining, or bio-ore, has the potential to generate income from the saleable metallic elements that accumulate in plant biomass ash. *In situ* phytomining typically takes place at polluted mine areas or low-value material, sub-economic ore, or contaminated mine areas. Cropping, harvesting, drying, are the primary processes involved in phytoextraction, also known as bioextraction, which is the extraction of metals for financial gain [45].

11.13 Advantages and disadvantages of phytoremediation

Advantages: Techniques for phytoremediation are widely employed because they have a number of benefits. One of these is the direct application of this approach *in situ* to the reclamation of the polluted environment. Because biological approaches do not result in secondary contamination, they may be more effective than conventional methods based on chemical extraction of xenobiotics. The large root systems of plants can prevent soil erosioBy strengthening the soil's structure, permeating deep layers of the soil, and promoting soil productivity and aeration.

Phytoremediation is less expensive when compared to more traditional methods, which is another benefit. Techniques for phytoremediation are acknowledged locally and do not call for specialized equipment. It is a totally free process that happens naturally. It provides *in situ* remediation, can preserve soil and water resources, originates energy from the sun, and plants are the agents for remediation. It prevents groundwater pollution by monitoring seepage, enhances soil quality, and promotes productivity

Disadvantages: There are several restrictions to phytoremediation. Because aquatic plants have shallow root systems, the depth to which the rhizosphere may treat the soil during phytoremediation is also constrained. Another drawback is the lengthy environmental clean up process, which could take up to 10 years. Additionally, phytoremediation's efficiency decreases in the winter because of the slower plant growth. When compared to traditional approaches, the rate of remediation is slower. This remediation is restricted to top layers only. In addition, it is not allowed to work in developed areas with high land costs [39].

The primary limiting elements are the plants, their availability, their capacity to thrive in contaminated soils, and the limitations placed on their development by the climate, the soil, and management techniques. The pollution and plant species have an impact on the remediation rate. However, it is appropriate for big agricultural and dirty sites. The effectiveness of phytoremediation will be impacted by the poor development rate and quantity of phytoremediation plants, the low transfer rate of heavy metals, and the high or low concentration of heavy metal pollution, which restricts the practical application of phytoremediation [47].

11.14 Mycoremediation

Mycoremediation is a phrase used to describe the employment of fungi in the degradation of impurities and pollutants from the environment. The words my (Greek for fungus) and remediation (reversing environmental harm) are the ancestors of this term. Using techniques related to bioremediation to cleanup the environment produces amazing outcomes. There are many different bioremediation techniques, and mycoremediation is one of them. It is crucial in the treatment of contaminated soils and wastewaters to restore naturally occurring ecosystems.

This is a naturally occurring and non-toxic process. Maintenance cost is almost nil and it is a less labor-intensive process, and is therefore an eco-friendly and economic process. The oyster mushroom is one of the fungi that can be utilized both as food and as a mycoremediator in industry. By allowing dead wood to rot, oyster mushrooms benefit the ecosystem. Mycoremediation is a method of pollution mitigation that employs biological systems to catalyze the breakdown or conversion of certain polluted substances into less hazardous forms. The treatment of several industrial wastes, such as sewage water, dairy industrial sludge, waste from tanneries, and waste from the paper and pulp sector, can be done via mycoremediation. Mycoremediation is a simple and effective decontamination technique that is gaining popularity nowadays for clearing environmental contaminants. Mycoremediation is naturally occurring, non-toxic, requires no maintenance, is recyclable, is less intrusive, and is safe compared to other cleaning techniques. The oyster mushroom is one of the edible mushrooms that can be utilized commercially for mycoremediation [48]. Another strategy for addressing the growing issue of water pollution is mycoremediation, which involves using fungi or their derivatives to remove water contaminants [49]. Mycoremediation is the term used to describe the use of fungus in the remediation process. This is a technique that uses fungi to transform polluted areas into clean areas [50]. Some fungal species have the capacity to exchange soil ions, modify their permeability, and eliminate hazardous materials from a polluted environment. Insecticides, petroleum hydrocarbons, polycyclic aromatic hydrocarbons, phenolic compounds, metals, chlorides, and biopolymers can also be broken down and degraded by some fungi.

Being able to produce several extracellular enzymes that target the metals in industrial waste and turn them into soluble substrates is one of the most crucial characteristics of fungal mycelia minerals that contain metals or low-grade materials. Their enzymes can assimilate different carbohydrate complexes without the need for hydrolysis, which speeds up the destruction of a variety of contaminants. A rich network of filaments that occur in a significant volume and occupy the top layer of the soil allows different fungal species to mineralize, release, and store numerous elements, ions, and accumulate many hazardous chemicals.

Moreover, fungi play a significant part in the breakdown of diverse organic compounds and the recycling of components that result in the secondary metabolites of multiple essential chemicals, including antibiotics, mycotoxins, different organic acids, vitamins, and many other essential compounds. Fungi are also known to thrive in a variety of different environmental conditions, such as low pH and low nutrient concentrations. The goal of this study is to discover the fungal species that thrived in industrial wastewater and demonstrate their effectiveness in reducing the danger posed by wastewater toxins [25].

Numerous scientific studies have shown that different fungus species can assist in the detoxification of a wide range of environmental pollutants, including heavy metals, pesticides, effluents, and some accidental petroleum spills such as the Exxon-Valdes disaster. Mycoremediation is a term that is used to describe the use of fungi in remediation. The approach involves employing fungi to transform dirty areas into clean ones. By using the fruiting bodies of fungi, this technique also aids in the removal of heavy substances from the surrounding area and the land. The appropriate species of fungi must be chosen to remove a specific pollution, from which an easy screening method has been evaluated, in order to achieve successful mycoremediation. Laboratory research demonstrates that the molecular structure and composition of soil contaminants and other nutrient sources may be altered by the fungal mycelial networks of all fungi. During their metabolic breakdown, certain toxins are excreted, which serves as an appropriate output to ease their removal from afflicted areas with the least amount of environmental impact possible [50].

Another effective strategy for addressing the ever-growing issue of water pollution is mycoremediation, which uses fungus or their derivatives to remove water pollutants. Its strong growth, extensive hyphal network, and production of a variety of fungi appear to be the best options for the purification of wastewater and landfill leachate due to their extracellular ligninolytic enzymes, high surface area to volume ratio, tolerance to heavy metals, adaptability to changing pH and temperature, and presence of metal-binding proteins. Fungi use a variety of methods, such as biodegradation, biosorption, and bioconversion of contaminants, to cleanup contaminated areas and invigorate the ecosystem. The use of fungal biomass as a biosorbent for hazardous metal remediation has drawn considerable interest because of its ample availability, quick biosorption and desorption efficiency, and cost competitiveness [49].

11.15 Advantage and disadvantage of mycoremediation

Advantages: Mycoremediation is rapid and economic. Because it does not require new homes, structures, buildings, or other items and resources, it is a less expensive technique than other bioremediation techniques. It is a low-cost technique because fungal spores are widely available and reasonably priced. In a short period of time, fungi will grow within the soil using their own capacity for reproduction. No additional costs are required to begin the mycoremediation process. Only a minimal amount of fungus or fungal spores are needed to completely cleanup the polluted surface. Since this method of repair does not involve excavating and disposal, it is safer than other remediation methods. Moreover, this method does not secondary waste streams; therefore, additional cleanup is not necessary. No equipment is needed. Thus, it is a simple process and more secure than other technologies. Animals and people are unaffected because the results are not toxic. The products can be reused. By converting poisonous molecules to non-toxic ones, this process creates goods that are safe to reuse for both humans and animals. It yields rapid results, is less noisy, and is eco-friendly. This method is quieter than other options because it does not require any machinery, structures, or other sounds. This approach does not bother surrounding habitations, making it more advantageous for the environment and nearby ecosystems. This method is all-natural and does not introduce any corrosives or other chemicals into the environment. When the fungal system advances to treat hazardous compounds caused by pollution, and eventually restore the system's normal function and equilibrium, one must bring the pollution into balance with the ecosystem [50].

Disadvantages: Bio systems are slower and are not fully capable, which makes it difficult for users to recognize and choose the best approach for their needs. Mycoremediation is still in its infancy and it is difficult to convince organizations to use it [42]. The use of standard procedure can result in problems because of the hostile environment in diverse places or due to infrequent efficiency in intense habitats.

11.16 Role of mushrooms in mycoremediation

Wastewater can be treated using fungi and mushrooms. Enzymes that are released by mushrooms and other fungi breakdown enormous amounts of trash. Fruiting bodies are a source of protein in mushrooms. The effectiveness of mushrooms is in the provision of food protein via biomass in the form of various wastes present to reduce waste and emission of various hydrolase and oxidoreductase types. However, basidiomycetous mushrooms have recently become more well-known for the remediation process because they play a crucial role in mycoremediation. Many researchers have been drawn to the remediation process, and they are faming mushrooms. Some have highlighted the biosorption and biodegradation functions of mushrooms.

Mushrooms can produce extracellular enzymes such as oxidases and peroxidases. These enzymes could oxidize waste in a laboratory. These chemicals traditionally supported these enzymes. These enzymes have been eliminating stubborn non-polymeric contaminants such nitrotoluenes, organic and synthetic PAHs, and artificial colorants in the environmental control situation. Currently, it is known that mushrooms can decompose polymers, including plastics. Biodegradation is a very complicated process. However, growing mushrooms will simultaneously address two important global issues, namely garbage collection and the need for protein-rich food [51].

The fleshy fungi, or mushrooms, make up a significant portion of the lower plant world. A typical fungus fruiting body, the mushroom produces basidiospores at the tips of club-like structures called basidia that are arranged along the gills of the mushroom. Water is a valuable natural resource and the only universal solvent. Many of the aquatic bodies have recently been contaminated by sewage, industrial waste, and artificial chemicals. The main factor contributing to the lack of appropriate water for irrigation is water contamination. A high degree of the main source of contaminants in river water is organic matter, which raises the levels of total dissolved solids, BOD, and COD. They render water unsafe for consumption, irrigating crops, or any other purpose. Ecosystem characteristics are changed when effluent is released into the environment without being properly treated. Farmers have found that using these raw effluents for irrigation reduces growth, yield, and soil health.

Mycoremediation can be used to cleanse sewage water, dairy industrial sludge, effluents from tanneries, and effluents from the paper and pulp sector, among other industrial effluents. Mycoremediation is a low cost and effective decontamination technique that is currently widely used to eliminate environmental population [53].

11.17 Role of white root fungi in mycoremediation

The white root fungus, which can effectively mineralize lignin, exhibit positive results in mycoremediation. Additional cellular enzymes, including lignin peroxidase, are released, such as laccase and manganese peroxidase. These enzymes can reduce organic waste products with similar molecular configurations. As a result of plant and animal tissues decay, humic substances (HS) are produced. White root fungi are mostly employed to cleanup water pollution and to raise awareness of enzyme-related mechanisms. Researchers have studied the growth of fungi and subsequent activity, such as the production of enzymes, in germ-free conditions. In this technique, several biological components such as carbohydrates, proteins, lipids, fiber, ash, and vitamins have been discovered. Sorghum is used in this technique in place of defined media because it is believed that fungi will benefit from sorghum's access to nutrients more than bacteria will. Early mycoremediation trials were successful under sterile conditions to promote the capacity of Sorghum for fungus growth and enzyme activity while removing the humic acid (HA) from simulated contaminated water. Various experiments have been carried out in contaminated settings to evaluate the capacity of weak white root fungi and to remove the HA from polluted water [51].

White-rot fungi are known for the capability to decay lignin, the highly recalcitrant 'wood-polymer' that provides rigidity to trees and other woody plants [53]. The fungus species that causes white rot is found in various Ascomycota groups, specifically the Xylariaceae and mainly the Basidiomycota in groups. The utilization of this method of remediation by fungi was predicted after the lignolytic enzymes of the white-rot fungus were discovered. This species has been thoroughly investigated and proved to be the most important fungus associated with the degradation of organic contaminants.

11.18 Role of brown rot fungi in mycoremediation

The class Agaricomycetes includes Brown Rot Fungi. Moreover, Basidiomycetes are entirely to blame for the wood degradation. Just 6% of all wood-decaying fungus can breakdown cellulose and hemicellulose after some modification of lignin. After being altered by polymer degradation, decayed wood loses its ability to grow. Heavy metals have been found in many kinds of fungi. Lead, silver, cobalt, and copper are just a few of the heavy metals that they can all absorb. According to recent studies, *Saccharomyces cerevisiae* can absorb 92% of lead Pb, 100% of copper, and 68% of Mo ions from the solution in just one hour. Moreover, it has the capacity to take up Cd and Co ions. Their effectiveness as an adsorbent is influenced by their affinity, specificity, and ability, which includes their physical and chemical makeup. Penicillium strains that grow in soil exhibit the ability to produce extracellular enzymes. Several of these strains can breakdown hydrocarbons, including phenol, polyaromatic hydrocarbons, and halogenated chemicals, in addition to not being able to oxidize them.

11.19 Applications of mycoremediation

Mycoremediation and household electronic wastes: The rapid expansion of home electronics has increased the need for additional production, and the use of simple and inexpensive energy sources power. These portable power sources are referred to as cells or batteries, and the electrochemical cell and alkaline AA battery are the most popular types of cells used in homes. However, the ecology is harmed by the disposal of these batteries. These batteries end up in landfills after being thrown away in the trash. This severe environment causes the protective coating of batteries to fracture, which causes leaks. Leacheate is a hazardous substance that might leak out of the battery. Zinc fumes and potassium hydroxide are the other two most dangerous components. As mentioned earlier, these cells contain very little mercury. Batteries could potentially release more chemicals over a longer length of time in landfills, allowing them to contaminate subterranean water. The application of the mycoremediation technique is one workable solution to overcome this challenge [50].

Mycoremediation and soil pollution: Crude oil, which is created as a result of the removal and distribution of the oil in industry, is another significant contaminant of soil. The use of items containing petroleum and diesel has contaminated the environment through a variety of channels, including unintentional spills, landfill leaching, incorrect waste disposal, leakage from underground storage tanks, and pipeline leaks. Certain fuel ingredients have the potential to induce neurological, renal, respiratory, and hepatic issues when they meet people or animals by ingestion, inhalation, or skin contact. Two crucial stages in the growth of plants that are susceptible to environmental contamination are root elongation and germination. In mushrooms, there are enzymes that can breakdown the complex organic chemicals found in agricultural waste and industrial byproducts. The importance of non-toxic mushrooms has grown as a result of advances in farming technology, which make the use of industrial and agricultural wastes as substrates for agriculture attractive, and lead to low-cost production and a large market.

Mycoremediation and sewage sludge: Sludge contamination and its management is a major issue. Sun drying, dumping on land, use in agriculture fields, and other methods are recognized experiments and disposal techniques for wastewater sludge land filling, incineration, and dumping at sea. Nothing had previously worked to completely solve this issue because it has already been a problem for so long, emerging with toxic chemicals that affect both the groundwater and the atmosphere. Membrane filtration, flocculation, and other highly developed chemical treatments are currently used for their disposal, settling, and dewatering. These methods are useful for removing hazardous substances from the body. A few of the expensive and sophisticated advanced techniques are not appropriate for removing harmful chemicals.

Further microbial processing is less expensive and causes less site interruption. The microorganisms undergo biological processes and breakdown textile waste-water and solids. According to recent studies, filamentous fungi can be grown in liquid culture utilizing a process known as liquid state bioconversion (LSB). Yet, this method is effective at treating wastewater when used with pre-sterilized domestic wastewater sludge. Presterilization processes modify the structure and properties of the treated plants, making them unsuitable for usage at a commercial level. So, a relatively novel method is required and suitable for situations involving raw or uncooked wastewater sludge. Mycoremediation is both affordable and environmentally friendly, and it functions best at commercial scale.

Mycoremediation and paper pulp industry waste: Paper manufacturing is generally renowned for being the most polluting business in the world. Due to its environmental pollution, the paper and pulp industry is involved in environmental research on a local and worldwide scale. Effluents from the paper industry have a very complex nature because they are made up of several organic materials that are difficult to degrade, such as hemicelluloses, cellulose, lignin, and resin. Microorganisms provide a more straightforward, affordable, and environmentally-friendly way to both lessen ecological damage and aid in the breakdown of toxic substances. Mycoremediation is crucial in the degradation of numerous toxic substances, including pesticides, polychlorinated biphenyls, and petroleum, heavy metals, phenol derivatives, hydrocarbons, and so forth. These toxic substances are used as food and digested into simplified form by the fungal organisms. Fungus is an inexpensive remediation method that can be used in place of costly chemicals [50].

Many types of waste, such as wastepaper, sawdust, wood chips, bagasse, black liquor from pulp and paper industry effluents, etc, are easily colonized and degraded by fungi. Some of these poisonous substances are consumed by fungi as a source of nutrients. They also breakdown the contaminants into less harmful and simpler forms. Cellulases are used to prepare cardboard that is easily biodegradable, as well as to remove ink, coating, and toner from paper. Hemicellulase enzymes are suggested as an efficient bio-reagent for bio-bleaching in place of the toxic chlorine chemicals that are currently employed to achieve pulp brightness in the production of high-quality paper goods. The growth of fungus on medium with Cellulose and Xylan as substrates, respectively, has been used to evaluate the ability of fungi to degrade materials. The more fungus growth there is, the more space there is around the fungus colony, and the better it can use cellulose and xylan and create the enzymes cellulase and hemicellulase. By utilizing enzymes, a wide range of bacteria and fungi may breakdown lignocellulosic macromolecules into simpler shapes. The absence of any negative side effects in the fungus's enzyme combination is essential for many applications. For instance, xylanases are utilized for enzymatic pulp bleaching and the enzyme preparation should not have any processes that degrade the cellulose fiber's quality.

The incredible adaptability of microbes provides a more straightforward, affordable, and environmentally beneficial approach to reduce environmental pollutants and aid in hazardous chemicals biodegrade. Recently, the pulp and paper sector has been forced to explore affordable and environmentally responsible alternatives due to the high cost of input energy and growing environmental concerns. The decomposition of several harmful compounds, including petroleum hydrocarbons, polychlorinated biphenyls, heavy metals (via biosorption), phenolic derivatives, persistent pesticides, etc, is greatly aided by mycoremediation. Some of these harmful substances are used by fungi as a source of nutrition, and they are transformed into more straightforward fragmentary forms [54].

11.20 Conclusions

Bioremediation is an effective and significant alternative technique for remediating, cleaning, and recovering pollutants to solve ecological problems involving contamination. With this method, locations can be effectively and cheaply disinfected. Large amounts of wastewater can be treated using the low-cost and simple infrastructure method of bioremediation. Bioremediation is widely regarded as an environmentally responsible technology because it is inherently clean, green, and sustainable. Because of inaction on the manufacturing, use, and disposal of dangerous chemical compounds, there are now more contaminated sites. Academics, governments, and scientists from all around the world strongly support addressing such toxins that are endangering the ecosystem. Hence, bioremediation is a crucial tactic for the sustainable development of our society with the least possible negative effects on the environment.

Another cutting-edge technology, phytoremediation, is effective at treating waste and should be supported so that it can be used practically to restore water. The energy requirements for phytoremediation are lower, it is easier to operate, and there are no sludge disposal issues. Moreover, it can be maintained by untrained workers. It is a green technology that relies on plants for remediation, making it a safe method for reclaiming the atmosphere. Phytoremediation, the use of plants for ecological restoration, is a developing clean up technology for the remediation of polluted water.

Mycoremediation is a recent technique to remediate wastewater. It has bioaccumulation capacity towards heavy metals, and can play a vital role in breaking down various poisonous substances into smaller and safe chemicals. It also helps in cleaning the atmosphere. Mycoremediation uses fungi for cleaning up raw residential wastewater sludge and is described as a successful bio-separation and bio-filtration technique.

However, due to the complex mechanisms of these processes, special efforts are required in the field of genetic engineering, pre-treatment techniques, and others to explore the full potential of these biological methods.

References

[1] Das J, Mondal A, Biswas S and Nag S 2022 The eco-friendly treatment of rubber industry effluent by using adsorbent derived from *Moringa oleifera* bark and *Pseudomonas* sp, cultured from effluent *Water Sci. Technol.* **86** 2808–19

[2] Saha R and Nag S 2022 Microbial remediation of petroleum hydrocarbons in liquid wastes ed M P Shah and S Rodriguez-Couto *Development in Wastewater Treatment Research and Processes: Microbial Degradation of Xenobiotics through Bacterial and Fungal Approach* (Elsevier) pp 117–29

[3] Biswas S and Nag S 2022 Removal of emerging contaminants through bionanotechnology *Development in Wastewater Treatment Research and Processes* (Elsevier) pp 291–310

[4] Deb S and Nag S 2022 Use of microalgae for the removal of emerging contaminants from wastewater *Biodegradation and Detoxification of Micro pollutants in Industrial Wastewater* (Elsevier) pp 193–210

[5] Divya M, Aanand S, Srinivasan A and Ahilan B 2015 Bioremediation—an eco-friendly tool for effluent treatment: a review *Int. J. Appl. Res.* **1** 530–7

[6] Saxena A, Gupta V and Saxena S 2020 Bioremediation: a green approach towards the treatment of sewage waste *J. Phytol. Res.* **33** 171–87

[7] Das J, Debnath C, Nath H and Saxena R 2019 Antibacterial effect of activated carbons prepared from some biomasses available in North East India *Energy Sources Part* A 1–11

[8] Nag S and Biswas S 2021 Accumulation and detoxification of metals by plants and microbes *Removal of Emerging Contaminants Through Microbial Processes* (Springer) pp 359–72

[9] Nag S and Biswas S 2021 Cellulose-based adsorbents for heavy metal removal *Green Adsorbents to Remove Metals, Dyes and Boron from Polluted Water* (Springer) pp 113–42

[10] Biswas S and Nag S 2021 Biomass-based absorbents for heavy metal removal *Green Adsorbents to Remove Metals, Dyes and Boron from Polluted Water* (Springer) pp 351–76

[11] Sciences B, Sreedevi S, Vineetha A, Manogna K B and Sravani V N 2019 Bioremediation of wastewater photosynthetic bacterial isolate using *Int. J. Pharm. Biol. Sci.* A **9** 60–3

[12] AJAO A and AWE S 2018 Bioremediation of wastewaters from local textile industries *Nat. Appl. Sci. J.* **1** 16–25

[13] Bhatia R K, Sakhuja D, Mundhe S and Walia A 2020 Renewable energy products through bioremediation of wastewater *Sustainability (Switzerland)* **12** 1–24

[14] Das J, Saha R, Nath H, Mondal A and Nag S 2022 An eco-friendly removal of Cd (II) utilizing banana pseudo-fibre and moringa bark as indigenous green adsorbent and modelling of adsorption by artificial neural network *Environ. Sci. Pollut. Res.* **29** 86528–49

[15] Jain B, Akolkar A B and Choudhary S K 2013 In-situ bioremediation for treatment of sewage flowing in natural drains *Int. J. Biotechnol. Food Sci.* **1** 56–64

[16] Yadav A N, Suyal D C, Kour D, Rajput V D, Rastegari A A and Singh J 2022 Bioremediation and waste management for environmental sustainability *J. Appl. Biol. Biotechnol.* **10** 1–5

[17] Das C, Naseera K, Ram A, Meena R M and Ramaiah N 2017 Bioremediation of tannery wastewater by a salt-tolerant strain of *Chlorella vulgaris J. Appl. Phycol.* **29** 235–43

[18] Deepthi, Lakshmi V S and Babu M C 2020 Bioremediation of wastewater using invasive bivalves *Int. J. Innov. Technol. Explor. Eng.* **9** 3077–9

[19] Ghosh S, Sasmal D and Jawahar Abraham T 2006 Efficacy of commercial shrimp farm bioremediators in removing ammonia in microcosm experiments *Indian J. of Fisher.* **53** 469–73

[20] Mora-Ravelo S G, Alarcón A, Rocandio-Rodríguez M and Vanoye-Eligio V 2017 Bioremediation of wastewater for reutilization in agricultural systems: a review *Appl. Ecol. Environ. Res.* **15** 33–50

[21] Kshirsagar A D 2013 Bioremediation of wastewater by using microalgae: an experimental study *Int. J. Life Sci. Bt. Pharm. Res.* **2** 339–46

[22] Basharudin H B 2008 Bioremediation of oil contaminated wastewater using mixed culture *PhD Dissertation* (Husni Bin Basharudin Universiti Malaysia Pahang)

[23] Ojha N, Karn R, Abbas S and Bhugra S 2021 Bioremediation of industrial wastewater: a review *IOP Conf. Ser.: Earth Environ. Sci.* **796** 012012

[24] Akpor O B and Muchie M 2010 Bioremediation of polluted wastewater influent: phosphorus and nitrogen removal *Sci. Res. Essays* **5** 3222–30

[25] Kadhim N F, Mohammed W J, Al Hussaini I M, Al-Saily H M N and Ali R N 2021 The efficiency of some fungi species in wastewater treatment *J. Water Land Develop.* **50** 248–54

[26] Author C 2018 Bioremediation of dyes in textile wastewater *Turk. J. Sci. Rev.* **11** 24–8

[27] Hossain K, Quaik S, Ismail N, Rafatullah M, Avasan M and Shaik R 2016 Bioremediation and detoxification of the textile wastewater with membrane bioreactor using the white-rot fungus and reuse of wastewater *Iran. J. Biotechnol.* **14** 94–102

[28] Shah M P 2018 Bioremediation-waste water treatment *J. Bioremed. Biodegrad.* **09** 1–10

[29] Ng Y S and Chan D J C 2017 Wastewater phytoremediation by *Salvinia molesta J. Water Process Eng.* **15** 107–15

[30] Amin A, Naik A T R, Azhar M and Nayak H 2013 Bioremediation of different waste waters —a review *Continent. J Fish. Aquat. Sci.* **7** 7–17

[31] Nyika J and Dinka M O 2022 A mini-review on wastewater treatment through bioremediation towards enhanced field applications of the technology *AIMS Environ. Sci.* **9** 403–31

[32] Oliveira L W 2021 Phytoremediation in Portugal : a comparison between plants and different wastewaters https://sigarra.up.pt/faup/pt/pub_geral.show_file?pi_doc_id=335050

[33] Polińska W, Kotowska U, Kiejza D and Karpińska J 2021 Insights into the use of phytoremediation processes for the removal of organic micropollutants from water and wastewater; a review *Water (Switzerland)* **13** 2065

[34] Patel S, Maurya R and Solanki H 2017 Phytoremediation of treated industrial effluent collected from Ahmedabad mega pipe line *J. Ind. Pollut. Control* **33** 1202–8

[35] Garad A D 2022 Phytoremediation of domestic wastewater *Int. J. Rec. Technol. Eng. (IJRTE)* **10** 73–5

[36] Bhavsar S R, Pujari V R and Diwan V V 2010 Potential of phytoremediation for dairy wastewater treatment *J. Mech. Civil Eng.* **16** 23

[37] Alam A R and Hoque S 2018 Phytoremediation of industrial wastewater by culturing aquatic macrophytes, *Trapa natans* L. and *Salvinia cucullata* Roxb *Jahangirnagar University J. Biol. Sci.* **6** 19–27

[38] SKNH P 2018 Feasibility study of phytoremediation in wastewater treatment *Int. J. Sci. Res.* **7** 1019–26 https://ijsr.net/archive/v7i5/ART20182544.pdf

[39] Materac M, Wyrwicka A and Sobiecka E 2015 Phytoremediation techniques in wastewater treatment *Environ. Biotechnol.* **11** 10–3

[40] Mustafa H M and Hayder G 2021 Recent studies on applications of aquatic weed plants in phytoremediation of wastewater: a review article *Ain. Shams Eng. J.* **12** 355–65

[41] Hooda V 2007 Phytoremediation of toxic metals from soil and waste water *J. Environ. Biol.* **28** 367–76

[42] Meena A L, Meena L K and Kumar A 2004 Phytoremediation of wastewater and effluents using aquatic weeds *Sci. Agric. Allied Sect. A Mon. e-Newsletter* **2** 1–12

[43] Sharma G K and Khan S A 2013 Bioremediation of sewage wastewater using selective algae for manure production *Int. J. Environ. Eng. Manag.* **4** 573–80 http://ripublication.com/ijeem_spl/ijeemv4n6_10.pdf

[44] Brar A, Kumar M, Vivekanand V and Pareek N 2017 Photoautotrophic microorganisms and bioremediation of industrial effluents: current status and future prospects *3 Biotech.* **7** 1–8

[45] Farraji H and Sains U 2016 Wastewater treatment by phytoremediation methods wastewater engineering: types, characteristics and treatment technologies *Int. J. Sci. Res* **7** 205–18

[46] Ralinda R and Miller P G 1996 Phytoremediation: technology overview report *GWRTAC Ser. TO-96-03* 1–26

[47] Liu Z, Lin H, Cai T, Chen K, Lin Y and Xi Y *et al* 2019 Effects of Phytoremediation on industrial wastewater *IOP Conf. Ser.: Earth Environ. Sci.* **371** 032011

[48] Bhatnagar A, Tamboli E and Mishra A 2021 Wastewater treatment and Mycoremediation by *P. ostreatus* mycelium *IOP Conf. Ser.: Earth Environ. Sci.* **775** 012003

[49] Mrozik A 2021 Microbial action in wastewater and sludge *Water (Switzerland)* **13** 846

[50] Rehman S, Mahboob L, Atta A, Ahmad W, Saqib K A and Zeeshan S *et al* 2018 Mycoremediation: a green technology approach *J. Adv. Biotechnol. Res.* **9** 672–86

[51] Barrech D 2018 A review on mycoremediation—the fungal bioremediation *Pure Appl. Biol.* **7** 343–8

[52] Shekhar S, Maurya C and Srivastava J N 2017 Remediation of wastewater using mushroom: *Pleurotus ostreatus* (Oyster Gill Mushroom) *Int. J. Sci. Eng. Res.* **8** 352–63 https://www.ijser.org/researchpaper/Remediation-ofWastewater-using-Mushroom-Pleurotus-ostreatus-Oyster-Gill-Mushroom.pdf

[53] Stiebeling D A and Labes A 2022 Mycoremediation of sewage sludge and manure with marine fungi for the removal of organic pollutants *Front. Mar. Sci.* **9** 1–11

[54] Singh A and Sharma R 2013 Mycoremediation an eco-friendly approach for the degradation of cellulosic wastes from paper industry with the help of cellulases and hemicellulase activity to minimize the industrial pollution *Int. J. Environ. Eng. Manag.* **4** 199–206 http://ripublication.com/ijeem.htm

[55] Leong Y K and Chang J S 2020 Bioremediation of heavy metals using microalgae: recent advances and mechanisms *Bioresour. Technol.* **303** 122886

Chapter 12

Myco-remediation in industrial wastewater treatment

Ashraf El-Baz, Yousseria Shetaia, Dina Y Abdelghani and Amera Adel Abaza

Myco-remediation (myco-enzymes, myco-degradation, and myco-sorption) is a technique that uses fungi, such as macro-fungi mushrooms, and micro-fungi, to remediate or biodegrade contaminants through a variety of processes, including biosorption, bioaccumulation, bioconversion, and biodegradation. The ability to break down pollutants, sustainability, low cost, source of edible protein, and adaptation to harsh environments are just a few of the benefits of myco-remediation. Fungal enzymes are versatile and effective in treating a wide range of contaminants because of their broad specificity. Industrial wastewater is produced from industrial activities such as manufacturing, mining, and power generation. Effluent often contains a wide range of contaminants, including heavy metals, organic compounds, and other contaminants that cause harm to public health and the environment, and must be treated before discharge. Myco-remediation offers promising solutions with the potential to provide a cost-effective, sustainable, and effective treatment of a wide range of pollutants produced from industrial wastewater effluent.

12.1 Introduction

Industrialization contributes to worldwide pollution through effluents. The dyes, paints, and related hydrocarbons that are used, among other things, in fabrics, printing, pharmaceuticals, food, toys, cardboard, plastic, and cosmetics are manufactured and discharged into the environment without adequate treatment, posing a threat to our lives. Pollutants that accumulate in dangerous amounts damage the environment. Depending on the business and the manufacturing process being used, industrial effluent can contain a complex mixture of contaminants. Common contaminants include heavy metals, organic compounds, suspended matter, nutrients, strong acids, and bases. Heavy metals accumulate in soil and water, and are harmful to humans and the ecosystem. Organic compounds, including solvents, oils, and pesticides, can be toxic and produce hazardous byproducts when released into the environment. Suspended matter such as sediment, silt, and debris can clog streams and harm aquatic life. High levels of nutrients such as nitrogen and phosphorus in industrial wastewater can contribute to an overgrowth of harmful algal blooms. Strong acids and bases found in industrial wastewater can be corrosive and harmful to the environment if not properly treated.

Myco-remediation is a type of bioremediation that uses the fungus 'mycelia' to *in situ* and *ex situ* clean up polluted areas and has been extensively explored since the early-1990s. Fungi are fast-growing organisms with a broad hyphal network. Fungal mycelia can adapt to harsh environmental conditions. To maintain ecological balance, fungi thrive in soils of all climates, including extreme conditions, and spread by airborne spores. Additionally, marine, freshwater, and complex soil matrices all support the growth of fungi. Due to their strong morphology and capacity to remove recalcitrant materials through mycelia or enzymes, fungi are crucial to repair soil and industrial effluents in an inexpensive and environmentally beneficial manner. How fungi detoxify and biodegrade such contaminants using extracellular and intracellular enzyme systems such as peroxidases, laccases, and cytochrome P450 has been widely researched. Similar techniques such as myco-filtration employ mycelia as filters to get rid of germs and dangerous substances from contaminated water. Paul Edward Stamets, a leading proponent of myco-remediation, has endorsed the idea of the creation of 'mycological response teams' that would utilize fungi to recycle and recover healthy soil in the area after any pollution event (Stamets 2005).

Fungi are the only known species on Earth capable of digesting wood. Mycelium secretes strong extracellular enzymes and acids that can break down lignin and cellulose, the two main building blocks of plant fiber (Rhodes 2014). The goal of bioremediation is to mineralize contaminants, which means that they are changed into harmless substances such as carbon dioxide, water, nitrogen gas, and other byproducts. Materials that have not been degraded have been transformed into less hazardous forms of low solubility. In phytoremediation, phycoremediation, and myco-remediation, heavy metals and radioactive cations are taken out of a site by harvesting plants, algae, or fungi. Biostimulation, in which adds nutrients to the soil to promote native fungal development, and bioaugmentation, in which introduces allochthonous fungi to contaminated locations, are two ways to remediate vast amounts of pollution with minimum effort and money.

Fungi remove polyaromatic hydrocarbons (PAHs) with ligninolytic enzymes and Cytochrome P450 monooxygenase. In addition, fungi produce laccase, manganese peroxidase, and lignin peroxidase as key enzymes used in myco-remediation. Due to their vigorous growth and biomass synthesis, basidiomycetes secrete strong and low-specificity catabolic extracellular enzymes from their mycelium to contact contaminants. They also produce dietary protein in mycelium, or fruiting bodies, from waste. Basidiomycete white rot accounts for at least 30% of myco-remediation studies.

Oyster mushrooms, *Pleurotus ostreatus*, can tolerate and clear a wide range of heavy metals from coal washery effluents. The ability of fungi to tolerate toxic metals is dependent on the production of antioxidant enzymes. Otherwise, surfactants have been used to reclaim manganese from industrial effluents. Surfactants allow high surface area and binding locations on fungal hyphae for high manganese bioaccumulation from wastewater. Mushroom species bioconvert industrial lingocellulosic waste into high-protein mushrooms and remove toxins. Thus, this is a useful multifunctional utility.

Aspergillus sp. and *Penicillium* sp. were isolated from fields contaminated with herbicides able to degrade xenobiotic substances. In addition, fungi have algicidal action, such as *Trichoderma citrinoviride* and *Trametes versicolor*, which inhibit the development of microalgae. Fungal enzymes, such as hydrolase, protease, cellulase, laccase, and manganese peroxidase, are crucial in the inhibition of the development of algal blooms and the biodegradation of the cyanotoxins that they generate (table 12.1). The main objective of this chapter is to explain how myco-remediation provides promising solutions for industrial wastewater treatment.

12.2 Methods applied in the myco-remediation processes

12.2.1 Direct injection

This approach involves introducing the fungus right away to contaminated soil or water so that the mycelia can interact with the pollutants as much as possible. The efficient degradation of petroleum compounds by fungi such as *Trichoderma* spp. and *Pleurotus* spp. in polluted soil or water makes them useful for cleaning up areas that have been contaminated with gasoline, diesel, or other petroleum products. It has also been demonstrated that *P. ostreatus* and *Phanerochaete chrysosporium* break down PAHs, while *P. chrysosporium* breaks down chlorinated solvents. In addition, *Aspergillus niger* and *Rhizopus* spp. can biosorb and remove heavy metals. The site conditions and nature of contaminants will determine the species of fungi that are used and the parameters they require. Several techniques can be used to accomplish direct injection, including the examples listed below.

    I. *In situ* cultivation: In this technique, nutrients are introduced to polluted soil or water to encourage the growth of indigenous fungi. The toxins are then consumed by the fungi, which transform them into less hazardous substances.

    II. Liquid inoculation: In this technique, a mix of fungi is put into the contaminated soil or water. As they develop and distribute throughout the environment, the fungi degrade the contaminants.

    III. Solid-state injection: In this technique, after being mixed with a solid carrier substance like sand or vermiculite, the fungi are then injected into the polluted area. In addition to materials serving as a source of energy for the fungus to develop, the solid carrier aids in spreading the fungi throughout the environment.

Table 12.1. Summary of fungi involved in myco-remediation across different business industries.

Industrial wastewater	Contaminants	Fungal enzymes	Fungal species	References
Battery manufacturing	Heavy metals (Cd, Cr, Co, Cu, Fe, Pb, Mg, Hg, Ni, Ag, Zn,) cyanide, O&G.	Ligninolytic enzymes and antioxidants enzymes.	Micro-fungal species *Aspergillus niger, Aspergillus fumigatus, Penicillium rubens, Mucor indicus, Trichoderma atroviride, Trichoderma harzianum, Trichoderma ghanense, Rhizomucor sp., Fusarium sp., Emericella sp., Funneliformis geosporum, Penicillium rubens* Macro-fungal species *Galerina vitiformis, Hypholoma capnoides, Marasmius oreades, Pleurotus ostreatus*	(Akhtar and Mannan 2020, Thakare *et al* 2021)
Chemical manufacturing	Organic chemicals: polymers (polyester, nylon, polypropylene, acrylics), plastics (PE, PVC, PP, PS), petrochemicals (ethylene, propylene, benzene, toluene, xylenes, methanol, VCM, styrene, butadiene, ethylene oxide, chloroform, naphthalene, phenols), PAHs, naphthalene, anthracene, phenanthrene, fluorene, pyrene, chrysene, and B[a]P. Inorganic chemicals: salt, chlorine, caustic soda, soda ash, acids (nitric acid, phosphoric acid, hydrofluoric acid and sulfuric acid),	Ligninolytic enzymes, cytochrome P-450 monooxygenase, dioxygenase, dehydrogenases, FAD-dependent monooxygenases, glutathione transferase, epoxide hydrolases, cutinase, and esterase.	Micro-fungal species *Aspergillus strains, Trichoderma lixii, Fusarium oxysporum, Cochliobolus lunatus Fusarium oxysporum, Fusarium subglutinans, Penicillium brocae, Purpureocillium lilacinum* Macro-fungal species *Pleurotus tuber-regium, P. pulmonaris, P. ostreatus, Pycnoporus sanguineus, Marasmiellus sp., Dentipellis sp., Trametes versicolor, Pleurotus ostreatus, Pleurotus eryngii, Phanerochaete chrysosporium*	(Akhtar and Mannan 2020, Mishra and Srivastava 2022, Stiebeling and Labes 2022)

Industry	Pollutants	Mechanism/Enzymes	Species	References
	titanium dioxide, hydrogen peroxide, aluminum compounds, borax; chrome, fluorine-based compounds, and heavy metals (Cd, Cr, Co, Cu, Fe, Pb, Hg, Ni, Zn,)			
Electric power plants	Heavy metals (Cd, Cr, Pb, Hg, As, Se). Nitrogen compounds (nitrates, nitrites) and sulfur dioxide.	Ligninolytic enzymes, and antioxidants enzymes.	Micro-fungal species: *Aspergillus* sp., *Rhizomucor* sp., *Fusarium* sp., *Emericella* sp., *Funneliformis geosporum*, *Trichoderma harzianum*, *Trichoderma ghanense*, *Penicillium rubens*. Macro-fungal species: *Pleurotus ostreatus*	(Akhtar and Mannan 2020)
Food industry	Fish farms (nitrogen, phosphorus, drugs), dairy processing plants (BOD, TSS, fats), animal slaughter (blood, gut contents, BOD, TSS, coliform bacteria, O&G, organic nitrogen, ammonia),	Ligninolytic enzymes, esterification, deoxygenation, dehydrogenation, dechlorination, demethylation, Proteases, Lipase, and esterase.	Micro-fungal species: *A. tamarii*, *Aspergillus flavus*, *Verticilium* sp., *Trichoderma hamatum*, *Rhizopus arrhizus*, *Botryosphaeria laricina*, *Aspergillus glaucus*, *Penicillium spiculisporus*, *Penicillium verruculosum*. Macro-fungal species	(Saglam *et al* 2018, Akhtar and Mannan 2020, Elhussiny *et al* 2020, Chaurasia *et al* 2023)

(*Continued*)

Trends in Biological Processes in Industrial Wastewater Treatment

Table 12.1. (*Continued*)

Industrial wastewater	Contaminants	Fungal enzymes	Fungal species	References
	processing food plants (salt, flavorings, coloring, FOG, BOD, sugarcane vinasse, surfactants, pesticides), and frying oil waste.		Pycnoporus sanguineus, Metacordyceps sp., P. ostreatus, Trametes pavonia	
Iron and steel industry	Inorganic chemicals: iron, ammonia, cyanide, highly acid ferrous sulfate, ferrous chloride, fluoride, cyanide. Heavy metals: Co, Cu, As, Fe, Hg, and Sb. Organic chemicals: PAHs, benzene, naphthalene, anthracene, phenols, creosols, hydrochloric acid, sulfuric acid, and hydraulic oils.	Ligninolytic enzymes, cytochrome P-450 monooxygenase, dioxygenase, dehydrogenases, FAD-dependent monooxygenases, glutathione transferase, and epoxide hydrolases.	Micro-fungal species Aureobasidium pullulans, Cladosporium resinae, Penicillium spp., Aspergillus niger Macro-fungal species Ganoderma lucidum, Funalia trogii, Trametes versicolor, Marasmiellus sp., Agaricus bisporus, Agrocybe praecox Yeast cell Candida lipolytica, Starmerella (Candida) bombicola	(Vishwakarma 2019, Akhtar and Mannan 2020, Kajla et al 2021, Stiebeling and Labes 2022)
Metal finishing and metallurgical industries	Inorganic chemicals: cyanide. Heavy metals: Cd, Cr, Cu, Pb, Ni, Ag, and Zn. Organic chemicals, various organic chemicals, solvents; O&G.	Ligninolytic enzymes and antioxidants enzymes.	Micro-fungal species Rhizopus nigricans, Aspergillus niger, Penicillium rubens Macro-fungal species Agaricus macrosporus, Agaricus bisporus, Pleurotus ostreatus, Galerina vittiformis, Amanita rubescens, Clitocybe nebularis	(Vishwakarma 2019, Akhtar and Mannan 2020)

Industry	Pollutants	Organisms	References
		Yeast cell *Saccharomyces cervisiae, Candida pelliculosa*	
Mines and quarries waste	Inorganic chemicals: hematite, hydraulic oils, sulfuric acid, AMD, sulfate, lime. Heavy metals: Au, Ag, Fe, Pb, Ni, Ag, and Zn. Organic chemicals: surfactants and oils.	Micro-fungal species *Mycorrhizal fungus, Rhizophagus intraradices, Arbuscular mycorrhiza fungi (AMF), Ectomycorrhizal fungi* Macro-fungal species *Agaricus sp., Lentinula edodes*	(Kumar et al 2018, Kumar and Dwivedi 2021)
Nuclear industry (radioactive waste.)	Radionuclides: 3H, 32P, 63Ni, 241Am, 90 Sr, 60 Co, 85 Kr, 137 Cs, 235 U, 234 U, and 239Pu.	Micro-fungal species *Alternaria alternata, Fusarium verticillioides, Aspergillus pulverulents, Rhizopus sp., Penicillium digitatum* Macro-fungal species *Pleurotus ostreatus, Amanita muscaria, Hebeloma cylindrosporum, Pleurotus pulmonarius* Yeast cell *Rhodotorula taiwanensis*	(Eskander et al 2012, Thakare et al 2021, Vandana et al 2021, Mahmoud and Darwesh 2022)

(Continued)

Table 12.1. (*Continued*)

Industrial wastewater	Contaminants	Fungal enzymes	Fungal species	References
Oil and gas extraction	Inorganic chemicals. Heavy metals: As, Cd, Cr, Hg, and Pb. Salts. Organic chemicals: oils, EHCO, HPAs, asphalt, organic chemicals, solids, drilling (drilling fluid), drill cuttings, and sulfonated lignite.	Lipases and proteases.	Micro-fungal species *Aspergillus flavus, Pseudallescheria ellipsoidea, Stachybotrys chartarum, Scopulariopsis brevicaulis* Macro-fungal species *Pycnoporus sanguineus, Auricularia polytricha, Pleurotus pulmonarius, Heterobasidion annosum* Yeast cell *Saccharomyces cerevisiae, Rhodotorula mucilaginosa*	(Kulshreshtha 2019, Liu *et al* 2020)
Petroleum refining and petrochemicals	BOD, O&G, TSS, NH₃, chromium, phenols, sulfides, and BTEX		Micro-fungal species *Alternaria alternata, Penicillium chrysogenum, Curvularia brachyspora, Cladosporium sphaerospermum, Scopulariopsis brevicaulis, Stemphylium botryosum, Trichoderma asperellum, Trichoderma harzianum*	(Daccò *et al* 2020, Chaurasia *et al* 2023)

Industry	Contaminants/parameters	Enzymes	Species	References
Pharmaceutical manufacturing	Antibiotics, high sulfate concentration, BOD, COD, TOC, NH3-N, total (TKN), and pH.	versatile peroxidase, laccase, manganese peroxidase, cytochrome 450 system, ligninolytic enzymes.	Micro-fungal species *Aspergillus luchuensis, Leptospaherulina sp, Mucor hiemalis* Macro-fungal species *Pleurotus ostreatus, Trametes versicolor, Phyllotopsis nidulans, Hypholoma fasciculare, Fomes fomentarius, Irpex lacteus, Lentinula edodes, Phanerochaete chrysosporium*	(Akhtar and Mannan 2020, Ortúzar *et al* 2022, Stiebeling and Labes 2022, Chaurasia *et al* 2023)
Pulp and paper industry	Lignosulphonic acid, chlorinated resin acids, chlorinated hydrocarbons, chlorinated phenols, biocides, physicochemical parameters (TSS, TDS, COD, BOD) phenols, lignin, and heavy metals, chloroform, dioxins, and furans.	Heme peroxidases (lignin peroxidase, manganese peroxidase, versatile peroxidases), and phenol oxidases (laccase).	Micro-fungal species *Aspergillus niger, Aspergillus flavus, Penicillium notatum.* Macro-fungal species *Trametes versicolor, white-rot fungi*	(Rhodes 2014, Hubbe *et al* 2016, Gupta and Gupta 2019)
Non-ferrous mining and smelting	Bauxite smelters (phenols), aluminum smelters (fluoride, benzo(a)pyrene, Sb, Ni, Al), copper smelters (Cd, Pb, Zn, As, Ni, Cu), lead smelters (Pb, Zn, Ni), cobalt smelters (NH3, Cu), and zinc smelters (As, Cd, copper, Pb, Se, Zn).	Ligninolytic enzymes and antioxidants enzymes.	Micro-fungal species *Aspergillus sp., Rhizomucor sp., Fusarium sp., Emericella sp., Funneliformis geosporum, Trichoderma harzianum, Trichoderma ghanense, Penicillium rubens* Macro-fungal species *Pleurotus ostreatus*	(Akhtar and Mannan 2020, Chaurasia *et al* 2023)
			Micro-fungal species	

(*Continued*)

12-9

Table 12.1. (*Continued*)

Industrial wastewater	Contaminants	Fungal enzymes	Fungal species	References
Textile mills and clothing industry	Synthetic dyes (azo dyes, Xylene cynol FF, Brilliant blue R, Aniline Blue, Orange G II and Crystal violet.), chloride, sulfate, total nitrogen, COD, BOD, salts, Fe, Insecticide, sulfide, O&G, phenol, TSS, and heavy metals (Cr, Zn, Pb, Cu).	laccase, manganese peroxidase, lignin peroxidase, tyrosinase, aminopyrine Ndemethylase, NADH–DCIP reductase, and azoreductase.	*Aspergillus niger, Galactomyces geotrichum, Aspergillus flavus* Macro-fungal species *Pleurotus ostreatus, Bjerkandera adusta, Pleurotus eryngii, Trametes hirsute, Marasmius cladophyllus, Phlebia acerina* Yeast cell *Candida zeylanoides, C.oleophila.*	(Vishwakarma 2019, Akhtar and Mannan 2020, Ekanayake and Manage 2022, Santhilatha et al 2022, Shukla et al 2022, Chaurasia et al 2023)
Water treatment	Cyanotoxins, algal bloom, detergents, calcium, magnesium, and carbonate.	laccase, hydrolase, protease, cellulase, and manganese peroxidase.	Micro-fungal species *Mucor hiemalis,* *Geotrichum candidium,* *Penicillium verrucosum,* *Cladosporium cladosporioides, Mucor hiemalis, Trichoderma citrinoviride.* Macro-fungal species *Trametes versicolor.*	(Akhtar and Mannan 2020, Shukla et al 2022)
Wood preserving and timber products (surface finishing)	Inorganic chemicals (As, Cd, Cr, Cu, Pb, Hg, F, Cl), COD, BOD, Phenols, O&G, TSS,	Ligninolytic enzymes, cytochrome P-450 monooxygenase, dioxygenase,	Micro-fungal species *Cochliobolus lunatus,* *Aspergillus sp., Rhizomucor sp., Fusarium sp., Emericella sp.,*	(Akhtar and Mannan 2020, Stiebeling and Labes 2022)

Industry	Pollutants/Parameters	Enzymes	Fungal species	Reference
	chlorinated phenolic compounds, PAHs, PCP, creosote, and B[a]P.	dehydrogenases, FAD-dependent monooxygenases, glutathione transferase, epoxide hydrolases and antioxidants enzymes, cellulases.	*Fumeliformis geosporum, Trichoderma harzianum, Trichoderma ghanense, Penicillium rubens,* Macro-fungal species *Trametes versicolor, Pleurotus ostreatus, Pleurotus eryngii, P. dryinus, T. hirsute, Wolfiporia cocos, Antrodia xantha, Fibroporia radiculosa, Fomitopsis palustris, Dentipellis sp., Phanerochaete chrysosporium.*	
Tannery industry	COD, BOD, sulfates, chlorides, Cr, nitrogen, sulfides, NH_3.		Micro-fungal species *Cladosporium perangustum, Penicillium commune, Paecilomyces lilacinus, Fusarium equiseti.* Macro-fungal species *Trametes versicolor*	(Boujelben *et al* 2022)
Distillery industry	COD, BOD, Nitrogen, Phosphorus, Potassium, Sulfates, Calcium,		Micro-fungal species *C. cladosporioides, Penicillium sp., P. decumbens, A. fumigatus, A. terreus, F. flavus*	(Bezuneh 2016)

(Continued)

12-11

Table 12.1. (*Continued*)

Industrial wastewater	Contaminants	Fungal enzymes	Fungal species	References
	Heavy metals (Mg, Cr, Ni, Pb, Fe)		Macro-fungal species Trametes sp., P. chrysosporium T. pubescens	
Extraction of metals (bioleaching)	Heavy metals: V, Ni, Fe, Ca, Mg, Na, and K.		Cladosporium cladosporioides	(Seddiek et al 2021)

Cobalt (Co), copper (Cu), chromium (Cr), iron (Fe), magnesium (Mg), manganese (Mn), molybdenum (Mo), nickel (Ni), selenium (Se) and zinc (Zn), cadmium (Cd), lead (Pb), mercury (Hg), silver (Ag), arsenic (As), antimony (Sb), gold (Au), aluminum (Al), ammonia (NH3), fluorine (F), chlorine (Cl), polyethylene (PE), polyvinyl chloride (PVC), polypropylene (PP), polystyrene (PS), vinyl chloride monomer (VCM), polycyclic aromatic hydrocarbons (PAHs), hydrocarbons polycyclic aromatics compounds (HPAs), benzo[a]pyrene (B[a]P), pentachlorophenol (PCP), chemical oxygen demand (COD), biological oxygen demand (BOD), total Organic Carbon (TOC), total dissolved solids (TSS), fats, oil and grease ('FOG'), acid mine drainage (AMD), tritium 3 (3H), phosphorus 32 (32P), nickel 63 (63Ni), americium 241 (241Am), strontium 90 (90 Sr)- cobalt 60 (60 Co), krypton 85 (85 Kr), cesium 137 (137 Cs)), uranium 235,(235 U) uranium 234 (234 U), plutonium (239Pu), extra-heavy crude oil (EHCO), benzene, toluene, ethyl benzene, xylene (BTEX), total ammonia-nitrogen (NH3-N), kjeldahl nitrogen (TKN), and oil and grease (O&G).

12-12

IV. Bioaugmentation: In this technique, commercial fungi cultures (allochthonous) grown in a laboratory are inserted into a polluted area to speed up the breakdown of certain contaminants. The fungi are carefully chosen to make sure they will not harm the environment, and they are chosen based on their capacity to break down particular pollutants.

V. Biostimulation: Native microorganisms can be stimulated by adding nutrients (nitrogen and phosphorus), electron acceptors (oxygen), substrates (methane, phenol, and toluene), pH, and temperature needs. Microbes with desirable enzyme functions thrive using this technique; enzymes are produced that reduce, precipitate, accumulate, and remove pollutants. Oxygen and nutrients encourage microbial growth, which results in the cleansing of polluted areas.

VI. Biosorption: This process removes xenobiotics, contaminants, and metallic ions from wastewater using either living or dead biomass. Biosorbents can be produced using fungus mycelium and compost of decayed mushrooms. One way that biosorption works is through bioaccumulation (active uptake), which depends on the metabolism and involves moving toxic metals across cell membranes. Only a living cell can experience intracellular aggregation, where contaminants concentrate inside the cytoplasm of cells that are actively growing. Another metabolism-independent method includes pollutants binding to the cellular surface of fungus mycelium (dead fungal cell). Cell surface sorption, external accumulation, and internal accumulation are the three kinds of biosorption that can involve adsorption, ion exchange, and covalent binding. Different processes, such as electrostatic interaction, ion exchange, complexation, and metal precipitation, can lead to biosorption. Chemical reactions between heavy metals and cell membranes can lead to metal precipitation. Fungi such as *Ganoderma lucidum* and *A. niger* can remove chromium from the environment through electrostatic interaction-based biosorption. In addition, ion exchange-based copper biosorption has been demonstrated in both strains. Siderophores are specific iron-chelating proteins used by fungi that are essential for removing radionuclides and toxic metals by making them more soluble. Bacteria, filamentous fungi, and even microalgae and orange peels have also been found to sequester metal ions, such as Strontium and radionuclides, through biosorption and ion-exchange mechanisms (Gopal *et al* 2022)

VII. Biomethylation: Fungi can convert toxic metal ions to less harmful forms through biomethylation, a process involving the addition of methyl groups to metal ions. Metallothioneins are proteins found in all organisms that help sequester toxic metals by binding to their thiol groups, protecting cells from reactive oxygen species.

VIII. Phytoremediation: In this technique, polluted areas are cleaned up by combining the power of fungus and plants. The fungi get their energy from the plants, while the fungi break down the toxins and aid in soil detoxification. In this procedure, contaminants in the water and soil are extracted, transformed, or stabilized using plants. The rhizosphere effect, where the root systems of the plants provide a habitat that is favorable for the development of microorganisms, including fungi, is a term used to describe the relationship between plants and fungi in phytoremediation. Symbiotic connections between the fungi and plant roots allow them to exchange carbon from the plant for nutrition and water. The fact that phytoremediation is an eco-friendly way to clear up polluted areas is one of its major advantages. Unlike conventional techniques of excavating and incinerating, phytoremediation does not generate pollution or emit toxic gases. Heavy metals such as lead, cadmium, and zinc, as well as organic pollutants such as petroleum compounds and chlorinated solvents, are a few of the pollutants that can be remedied using phytoremediation. Phytoremediation offers a powerful tool for cleaning up polluted environments in an eco-friendly and sustainable manner.

Several types of fungi are commonly used in phytoremediation, including:

- White-rot fungi (WRF): These fungi are well known for their capacity to breakdown lignin, the primary constituent of plant matter. Additionally, harmful substances that are frequently found in polluted soils, such as dioxins, polychlorinated biphenyls (PCBs), and PAHs, can be decomposed by WRF. *P. chrysosporium* and *P. ostreatus* are two common types of WRF used in phytoremediation.
- Mycorrhizal fungi: These fungi develop symbiotic relationships with plant roots, supplying nutrition and water in return for sugars made during photosynthesis by the plants. Additionally, mycorrhizal fungus can aid plants in absorbing toxic heavy metals such as lead and cadmium, which are frequently present in polluted soils. *Glomus mosseae* and *Rhizophagus irregularis* are two examples of prevalent mycorrhizal fungus used in phytoremediation.
- Endophytic fungi: The fungi that occupy plant tissues help the plants withstand stress and boost their defenses against diseases and parasites. Through the degradation of toxic materials and an increase in the absorption of heavy metals, some endophytic fungus species can also contribute to phytoremediation, such as *Fusarium oxysporum* and *Penicillium chrysogenum*. These fungi must be carefully selected and maintained to prevent them from endangering the ecosystem or human health. The particular location conditions and pollutants must also be taken into account when choosing the suitable fungi for phytoremediation. (Gopal *et al* 2022)

IX. Biofiltration: This technique removes pollutants from air or water by passing polluted air through a layer of mycorrhizal fungi (rhizofiltration), which captures and decomposes the pollutants. To help fungi develop and eliminate contaminants through biochemical processes, polluted air, or water must be passed through this biofilter, which acts as a bioreactor containing a substrate such as compost, peat, or soil. Volatile organic substances (VOCs), such as benzene and toluene, can be broken down in this process, along with fumes and toxic gases. This can be applied in a number of ways, such as the cleaning of industrial effluent, processing water, and air.

X. Leaching: This is a process in which microorganisms bio-catalyze by converting metal compounds into water-soluble forms. The microbial byproducts such as organic acids, released by filamentous fungi help dissolve the metal ions in the mineral. Through biological and chemical processes, fungi interact with metal components causing changes in the molecules and physical properties of metals. Iron from ore is dissolved by the fungi *A. niger* and *Penicillium verruculosum*. *A. niger* bioleaching of heavy metals from bauxite waste reduced the toxicity of red mud (Sharma *et al* 2021)

XI. Bioventing: This is a process that boosts aerobic microbial growth and bioremediation by adding oxygen to nutrients. This technique is applied in a low-oxygen environment where the population of native aerobic microorganisms is high. The microbial decomposition of hazardous pollutants, particularly spilt petroleum compounds, was optimized by moisture, nutrients, and ventilation. Petrol compounds, such as petroleum, gasoline, fuel oil, and kerosene, may only be remedied by microbes using this technique (Hegde *et al* 2022)

XII. Biosparging: This technique injects air into the soil subsurface below the water table to enhance groundwater oxygen concentration and microbial activity, allowing naturally occurring microorganisms to degrade contaminants faster. Injecting air into the saturated zone elevates volatile organic molecules to the unsaturated zone, which speeds up microbial breakdown. Pollutant bioavailability to microorganisms is determined by soil permeability and biodegradability where the efficacy of elimination is determined (Hegde *et al* 2022)

XIII. Composting is also a method of bioremediation to scavenge and breakdown pollutants in industrial effluents with the support of microbes in the soil. These microbes consume pollutants from water and metabolize or transform it into inert byproducts such as water, CO_2, and mineral salts. This is a conclusively proven method of bioremediation in degrading a wide range of pollutants such as wood-preserving chemicals, heavy metals, pesticides, herbicides, explosives, petroleum products, solvents, and chlorinated and nonchlorinated hydrocarbons. Unique composts are designed for specific pollutants at specific sites, though they are termed tailored composts. For example, tree leaf compost was designed for stormwater filtration and bioremediation. This method was found useful in the removal of 85% of grease and oil and 98% of heavy metals from stormwater runoff.

12.2.2 Substrate-based

This technique involves growing the fungi on a carrier medium, such as sawdust or straw, and then adding them to the polluted area. The fungi use the carrier material as a source of energy. The need for a source of a clean substrate and the potential for competition between the fungi and other environmental microbes are two constraints of substrate-based myco-remediation.

Agrocybe and *Pleurotus*, are edible fungi that can remove pollutants from soil, such as heavy metals and pesticides. Cellulosic materials, such as paper, cardboard, and wood chips, can be used for growing fungi, including *P. chrysosporium*, *Pleurotus*, *Lentinula*, and *T. versicolor*, which can remediate persistent organic pollutants, such as PCBs, furans, and PAHs, as well as VOCs, such as benzene and toluene, found in contaminated soil and groundwater. A combination of fungi and bacteria can be used to clean electronic waste, such as computer parts and circuit boards, by using the electronic components as a base to break down hazardous substances and heavy metals (Pant *et al* 2018). To cultivate fungi such as *Aspergillus* spp. and *Fusarium* spp., cotton and wool can be used as growing media for degrading azo dyes, a prevalent class of manufactured dyes found in polluted water, soil, and textile waste.

12.2.3 Bioreactor

Microbial clean up may be applied *in situ* (in location via bioaugmentation, biostimulation, biosorption, phytoremediation etc) or *ex situ* (off the site via biopiles, landfarming, composting or bioreactors, etc). *In situ* remediation is sluggish, and difficult to control and manage, hence bioreactor designs are chosen to enhance microbial activities. Bioreactors are defined as 'an engineering or manufactured device or system for managing an enclosed bio-environment.' Microorganisms or enzymes are carried out in the microbial bioreactor. Bioreactors are suitable for microorganism cultivation in bioremediation. Bioreactors vary in mode of operation from batch, continuous, and fed-batch to packed, stirred tanks, air lift, slurry phase, and partitioning phase reactors (Tekere 2019).

Bioreactors may be simple benchtop units for use in laboratories or large-scale systems for industrial applications. They fall into two main categories: attached-cell bioreactors (biofilm bioreactors) and suspended-cell bioreactors (suspended growth bioreactors). In suspended-cell bioreactors, microorganisms are suspended in a liquid media, and metabolic processes take place in the liquid, e.g., batch reactors, continuous stirred-tank (CSTR), plug-flow reactors etc. In attached-cell bioreactors, microorganisms have adhered to a solid surface, and the metabolic activity takes place at the surface, e.g., fluidized bed, packed bed, air lift, and up-flow anaerobic sludge blanket reactors. However, it is crucial to carefully manage the conditions in the bioreactor to prevent the growth of unwanted microorganisms and ensure the efficient operation of the system. Bioreactors can be used in conjunction with other treatment technologies to enhance the removal of pollutants or the production of bioproducts. A variety of contaminants in the soil, groundwater, and air can be cleaned using bioreactors. Since they may be configured for a variety of purposes and remedial applications, bioreactors offer green solutions for bioremediation. The design of a bioreactor operation should accommodate microbial cell proliferation, nourishment availability, and waste collection.

The following types of bioreactors are favorable for fungal hyphal growth and facilitate myco-remediation (Tekere 2019, Gopal *et al* 2022):

A. Slurry bioreactors: When a solid substrate is turned into a slurry, a slurry bioreactor provides an ecologically favorable method for remediating mostly soils and sediments contaminated by petrochemical hydrocarbons, tars, creosotes, chlorinated solvents, herbicides, pesticides, and explosives. Because they are hydrophobic and sorb to soil or sediments, most long-lasting chemicals are difficult to biodegrade. In a slurry reactor, the addition of water to a polluted solid matrix in the right proportions boosts interactions between microbes, pollutants, media, and oxygen. Bioavailability is increased when contaminants are dissolved in water. Bacteria and fungi clean up soil from VOCs, organochlorines, PAHs, and 2,4-Dichlorophenoxy acetic acid (2,4-D). Hexachlorocyclohexane (HCH) was degraded by *Bjerkandera adusta* after 30 days.

B. Packed bed bioreactors (tricking filters or biological towers): These bioreactors are designed for high substrate flow, in the presence of live microbes, which were used to remediate textile dyes, PAHs, and amines. The medium to which microbes are attached is solid media (e.g., polyurethane foam, sintered glass, porous ceramics, propylene agarose, agar gel beads, silicone tubing, stainless steel, and nylon web) for their respective industrial effluent used. They are commonly used for aerobic wastewater treatment. Various fungi and bacteria are used to remove organochlorine pesticides, PAHs, pharmaceuticals, amines, and textile dyes.

C. Fluidized bed bioreactors: The upward movement of gas or liquid fluidizes a bed of solid particles, such as sand grains or glass beads (biofilm carriers), resulting in the dispersion of the particles. The fluidized bed gives fungi a large surface area to attach to, enabling effective nutrient and oxygen uptake. Fluidization also creates a homogeneous environment, promoting the growth of fungi throughout the bioreactor. The constant movement of the particulates aids in preventing the development of inactive zones, which could prevent the growth of fungus. WRF have been used in fluidized bed bioreactors to clean up soil contaminated with PAHs.

D. Air-lift bioreactors: In these bioreactors, mixing is accomplished without any mechanical agitation. Air is fed into the bottom of a central draught tube through a sparger ring. The flow passes up through the draft tube to the headspace of the bioreactor, where excess air, byproduct, and CO_2 disengage. These bioreactors are used for textile dye effluent decolourization by fungi, olive mill effluent, and cellulose industry bleaching effluent.

12.3 The steps of the myco-remediation technique for detecting fungi

1. Isolation and identification of fungi: Isolating and identifying fungi isolates that can break down the particular contaminants is the first step in using them for myco-remediation. It is possible to accomplish this by using culture collections of soil or water samples, or by using advanced molecular techniques, such as:

   - Molecular assays (immune assays): Immunological fungus identification was first created at the beginning of the twentieth century. New fungal detection techniques use monoclonal and polyclonal antibodies that can attach to bioremediation enzymes such as manganese peroxidase, laccase, and lipase. Many researchers created many immune assays, immunofluorescence, ELISA, and immunoblotting tools. Polyclonal antibodies can detect the brown rot fungus, whereas fluorescent pigments (immunofluorescence stains) may be able to distinguish Basidiomycetes fungi (Saglam *et al* 2018).
   - Internal transcribed spacers (ITSs): These are used in taxonomic research to identify taxa and establish interspecific relationships. ITS repetitions of Ribosomal RNA are recognized in fungus classification as genetic bar codes. Microorganisms with the defined trait can be recognized by this code. This approach is accomplished by using PCR and DNA sequence. Following these procedures, the specific fungi can be identified by determining the target gene sequence or taking into account DNA size bands (Saglam *et al* 2018, Kumar *et al* 2021).

2. Screening for degradation ability: The capacity of the fungi to decompose the desired contaminants can be tested after the identification step. This can be accomplished through trials in the laboratory or by cultivating the fungus on media containing the contaminants and monitoring the rate of deterioration over time.

3. Optimization of degradation conditions: The ideal circumstances for decomposition can be identified after the fungi have been examined. This might involve selecting the most suitable variety of fungi for the particular pollutant, as well as adjusting environmental factors like temperature, pH, and nutrition levels.

4. Application in the environment: A four-phase investigation is necessary before using a fungus in bioremediation, i.e., bench-scale treatability, on-site pilot testing, inoculum formation, and full-scale application (Rhodes 2014). The use of the fungus in the ecosystem can begin once the ideal conditions have been identified. To remediate polluted effluent, the fungus may be injected directly into contaminated soil or water or used by bioreactor technique.

5. Monitoring and evaluation: Myco-remediation needs to be continuously observed to make sure its efficacy. To do this, it may be necessary to collect samples of polluted material, analyze them for contaminants, and keep observing the development of fungal biomass.

For maximum output efficiency, we ultimately want to keep the highest possible fungal inoculum. This is achieved by engineering mutant types.

12.4 Myco-remediation and bioengineering

Genetic engineering is a powerful tool for improving fungal capacity to decompose particular contaminants. The strategy is to alter the genetic makeup of the fungi to improve the production of genes that code for enzymes that can degrade the target contaminant. Another approach is to provide the fungi DNA from other species that can also decompose the contaminants.

However, genetically modified organisms may present environmental risks. For example, the genetically modified fungi could spread and harm non-target species, or the inserted genes could move to other organisms and cause new environmental problems. Thus, genetic engineering can aid myco-remediation, but it should be used with caution and evaluation of risks and benefits. Common genetic engineering methods in myco-remediation include:

- Transformation: Agrobacterium-mediated transformation or electroporation are used to insert foreign DNA into the fungal genome.
- Metabolic engineering: Genes that improve fungi breakdown of specific pollutants can be introduced. Fungal species can decompose lignocellulosic materials by adding genetic code for ligninases, cellulases, or other hydrolytic enzymes.
- Gene expression regulation: Regulating the amounts of particular enzymes or other factors crucial for myco-remediation involves changing gene expression patterns. For instance, to increase the fungi degradation of particular pollutants, promoters that are controlled by the presence of those pollutants can be applied.
- Synthetic biology: This involves the development of novel biological systems for particular uses, such as myco-remediation. For instance, fungal types with improved capacities for degradation or improved survival in polluted areas can be developed.

There are some examples of genetic engineering role in myco-remediation, such as the genetically modified *Pichia pastoris* that contains the yeast laccase gene (YlLac) from *Yarrowia lipolytica*, which destroys phenolic substances (Kalyani *et al* 2015). The laccase gene from *Phanerochaete favidoalba* was also expressed in *A. niger*, which was known to break down synthetic textile dyes like Acid Red and Brilliant blue R (Benghazi *et al* 2014). Additionally, it has been observed that a parasexual hybridized recombinant strain of *Fusarium solani* is efficient in DDT decomposition (Mitra *et al* 2001). Thus, fungi have been used as a bioremediation tool for a long time, both in their naturally occurring form and as genetically-engineered organisms.

12.5 Application of enzymes in myco-remediation

It is important to note that different fungi produce different combinations and amounts of enzymes called ligninases (lignin-modifying enzymes) that are capable of degrading lignin into simple united of sugars, which help some fungi to be suitable for certain types of myco-remediation. Furthermore, different environmental conditions, such as pH, temperature, and nutrient availability, can affect the activity of fungal enzymes, so the optimal conditions for myco-remediation may vary depending on the specific pollutants and fungal strains involved.

12.5.1 Classification of fungal enzymes

12.5.1.1 Oxidases

A group of enzymes play a role in receiving electrons from the substrate and use oxygen (O_2) as the electron donor to catalyze oxidation–reduction processes. The enzymes generate either water (H_2O) or hydrogen peroxide (H_2O_2) as a result of the process. These enzymes usually have metal or flavin coenzymes on their active regions. It has been demonstrated that lignin-modifying enzymes, particularly phenoloxidase enzymes such as Laccase (Lac) and Tyrosinase (Tyr), as well as peroxidases like Manganese Peroxidase (MnP) and Lignin Peroxidase (LiP), are responsible for the biodegradation of synthetic dyes by yeast cells.

Laccase (EC 1.10.3.2)

Laccase belongs to the urushiol and P-diphenol oxidase families and is an enzyme that can be found in a variety of organisms, including lichens, bacteria, and fungi. It is differentiated by the presence of multiple copper atoms in the catalytic center. This enzyme can decompose a wide range of organic substances, including polyphenols, aromatic diamines, and contaminants. It has a high-redox potential of 780 mV. WRF, which can break down lignin, are the species that have the greatest concentration of laccase. Compared to other species, fungal laccase is simpler to separate and concentrate because it is an external enzyme. Because of this, there is now more commercial interest in using it for organic synthesis, bleaching pulp and textiles, bioremediation, chemical grafting, and polymer surface modification. Additionally, it has been demonstrated that laccase, which does not need cofactors, participates in the biodegradation of synthetic dyes in yeast cells. Through non-specific free radical processes, it directly degrades azo dyes without generating harmful byproducts such as aromatic amines.

Tyrosinase (EC 1.14.18.1)

This is a phenoloxidase that contains copper and is present in a variety of species, including plants, fungi, bacteria, and yeast. It is used in a variety of industrial applications, including the creation of L-dihydroxyphenylalanine and the biosynthesis of melanin. Additionally, it contributes to environmental biotechnologies such as the purification of phenol wastewater. Tyrosinase has been shown to decolourize synthetic dyes in yeast cells in a few species, including isolates of *Candida krusei*, *Saccharomyces cerevisiae*, and *Galactomyces geotrichum MTCC*. The two stages in the enzymatic reactivity of dyes with tyrosinase are monophenol hydroxylation and monophenol oxidation to o-quinones (Danouche *et al* 2021).

Lignin peroxidase (LiP) (EC 1.11.1.14)
This is a glycosylated enzyme that reacts with peroxide, and is a member of the family of oxidoreductases and heme peroxidases. It is a nonspecific substrate enzyme that can break down various phenolic and non-phenolic substances, has a high-redox potential, and functions best at acidic pH levels. Two successive one-electron oxidations of the H_2O_2-oxidized versions of LiP have shown that LiP participates in the biodegradation of sulphonated azo pigments. Several yeast isolates from the groups Ascomycota and Basidiomycota have been found to exhibit LiP activity while degrading different pigments.

Manganese peroxidase (MnP) (EC 1.11.1.13)
This is a heme peroxidase and oxidoreductase family member glycoprotein enzyme with a molecular weight of 38 to 62.5 kDa. It is a substrate-specific enzyme that oxidizes organic pollutants or phenolic substrates such as lignin and Mn^{2+} to Mn^{3+}. However, it has also been discovered that yeast species like *Pichia occidentalis*, *Debaryomyces polymorphus*, *Candida tropicalis*, *Trichosporon multisporum*, and *Trichosporon laibachii* are engaged in the biodegradation of synthetic dyes using MnP. In addition, MnP is made by specific soil litter-decomposing and wood-decaying fungi.

Versatile peroxidases (VPs) (EC.1.11.1.16)
These are hybrid heme-containing (heme peroxidases) ligninolytic peroxidases with multiple oxidation-active sites. VP was discovered in *Pleurotus eryngii*, a WRF. Only *Pleurotus* and *Bjerkandera* genera have VP. Like MnP, VP enzymes reduce Mn^{2+} and high-redox potential non-phenolic molecules.

Cytochrome P450 (EC 1.14.14.1)
This is a group of widely distributed heme enzymes. Extracellular heme peroxidases and intracellular monooxygenases called unspecific cytochrome P450 synthesizes complicated natural products, breakdown pharmaceuticals, and bio-transform toxic materials. Cytochrome can diminish xenobiotics through chemical modifications such as aliphatic hydroxylations and dehalogenation, epoxidations, dealkylations, and other bioremediation chemistry-related processes (Mousavi *et al* 2021).

Aromatic peroxygenases (APOs) (EC 1.11.2.1)
This is a group of enzymes that belong to the family of heme-thiolate ferryl proteins. APOs have been identified in different fungi, such as *Agrocybe aegerita*, and *Marasmius rutola*. The potential biotechnological applications of APOs have attracted significant interest due to their capacity to oxidize aromatic compounds, including lignin. For example, APOs can be used in the production of high-value chemicals from lignin, such as vanillin. In addition, APOs have the potential to be utilized for bioremediation purposes in both soil and water systems. They exhibit the

ability to break down a diverse array of pollutants, such as PAHs, pesticides, and dyes.

12.5.1.2 Reductases

This is a group of enzymes that break down azo bonds in azo dyes to produce an aromatic amine that is colorless. They have been demonstrated to be active in yeast, filamentous fungi, and microalgae in addition to bacteria, which are the primary organisms that use them to biodegrade dyes.

The following are the major reductase enzymes found in yeast strains:

Azoreductase (AzoR) (EC 1.7.1.6)

This is present in various microorganisms and higher eukaryotes can reduce azo bonds in organic compounds such as azo dyes, nitroaromatic compounds, and azoic drugs. AzoR can be divided into two classes based on their secondary and tertiary structures: flavin-dependent and independent. The flavin-dependent AzoR can be further classified based on the required co-enzyme, NADH, NADPH, or both. The reduction of azo bonds by AzoR is a critical step in the biotransformation of dyes. The involvement of AzoR in dye biodegradation has been detected in yeast strains such as *S. cerevisiae MTCC 463*, *C. krusei*, *Issatchenkia occidentalis*, and *Trichosporon beigelii NCIM-3326*

NADH-DCIP reductase (EC 1.6.99.3)

This is an oxidoreductase enzyme that uses NADH as an electron donor to decrease 2,6-dichloroindophenol (DCIP). When DCIP is reduced, its color changes from blue to colorless. According to some research, yeast cells decolourize azo dyes by increasing the activity of NADH-DCIP reductase.

Malachite green reductase

This is an enzyme that uses NADH as an electron donor to convert green malachite into green leucomalachite. These enzymes, which have been found in yeast isolates of *S. cerevisiae MTCC 463*, *C. krusei*, *I. occidentalis*, *T. beigelii NCIM-3326*, and *G. geotrichum MTCC 1360* among others, are crucial for the biotransformation mechanisms of dyes (Das and Charumathi 2012).

12.6 Myco-remediation of recalcitrant contaminants in selected industrial sectors

12.6.1 Food industry

The food business is a combination of many sectors that create different products. The majority of water consumed by large-scale industries is used to cool and clean equipment and raw materials. Effluent from the food processing sector varies in its volume and composition. Biological methods that use both aerobic and anaerobic processes can treat wastewater that is loaded by a lot of organic matter. This type of wastewater can be used again because it has low amount of harmful chemicals or pathogens. The typical food sectors that produce wastewater are beverages, oil production, dairy foods, confectioneries, poultry, and meat processing. Manufacturers can use fungal remediation to eliminate polluted wastewater. For example, in WRF capsule technology hydrogel capsules include a polymeric substance that swells and retains a lot of water but does not dissolve in water then inoculated by fungal mycelium. Wastewater contamination can be reduced near food processing hubs, food firms, and others by applying hydrogel capsules in high-risk regions. Mycelium can be kept for a long time in a hydrogel capsule and then released when the capsule is exposed to wastewater (Ahmed 2015).

Pesticides are commonly used in the packing of fruit, thus the wastewater from this process must be treated before being released into the environment. Karas *et al* (2011) examine the ability of three WRF (*P. chrysosporium, T. versicolor, P. ostreatus*) and *A. niger* strain to degrade the pesticides thiabendazole (TBZ), imazalil (IMZ), and thiophanate-methyl (TM). The role of peroxidases (LiP, MnP) and laccases (Lac) in pesticide breakdown was also examined by measuring their activity. *T. versicolor* and *P. ostreatus*, were able to break down all pesticides (10 mg l^{-1}) except TBZ. For example, *T. versicolor* was able to fully degrade ortho-phenylphenol (OPP), diphenylamine (DPA) pesticide concentrations, and partially degrade TBZ and IMZ at the spillage level of 50 mg l^{-1}. For the bioremediation of fruit packing wastewaters, both strains exhibited promising results. However, achieving TBZ degradation needs further investigation. In addition, soil and water contamination from herbicides and pesticides can be remedied *in situ* using myco-remediation. Several reports suggest that some indigenous fungi found in polluted areas could be the best remediator candidates (Gajendiran and Abraham 2017)

12.6.2 Steel and iron industries

On average, 28.6 m^3 of water is used to create one ton of steel, with a discharge rate of 25.3 m^3/ton of steel (Colla *et al* 2017). Heavy metals and organic contaminants are discharged into effluents during different manufacturing stages, including coke, sinter, iron, steel, pickling, furnace blasting, shaping, cooling, and finishing. Regarding the concept of 'zero waste' and environmentally-responsible steel manufacturing, industrial effluent must be treated thoroughly before being discharged into the environment. Yeast and fungi are used in the biological treatment of steel wastewater to consume various pollutants as nutrients and energy sources. Whether they are native or recently brought to the location, they surpass traditional treatments.

Yeast is commonly used as the core for wastewater treatment in industry and home sewage water cleaning. Yeast may physically entangle their mycelia or pseudo mycelia to produce floes, which aid in the process of coagulation and flocculation. Yeast also produces proteins and vitamins, which can make it a desirable component of an animal diet. Its application in the treatment of wastewater includes organic waste, heavy metal ions, and residential sewage (Wang *et al* 2018, Qadir 2019). Yeast may detoxify heavy metals in live cells, on their surface. Adsorption can either be passive or active, and both are involved in the mechanism. The effects of adsorption time, temperature, pH, and Cr^{6+} concentration on the maximum adsorption rate of Cr^{6+} was 94.71% by beer yeast have been studied (Zhang *et al* 2020). Mercury, strontium, and arsenic ions, as well as radioactive uranium can be effectively adsorbent by yeast (Wang *et al* 2018)

In addition, Filamentous fungi are commonly utilized in industrial bioremediation and fermentation because they are simpler to remove from liquid substrates. *A. niger*, *Rhizopus oryzae*, and *P. chrysogenum* biomass transformed Cr^{6+} to Cr^{3+}. In contaminated soils and industrial toxins, *Curvularia* spp, *Acrimonium* spp, and *Pithyum* spp were Cd, Ni, and Cu-tolerant fungal microorganisms. In addition, *Coprinopsis atramentaria* macrofungus concentrated Pb^{2+} up to 94.7% and Cd^{2+} up to 76% (Kajla *et al* 2021).

12.6.3 Nuclear wastewater

Most radioactive waste comes from the nuclear fuel cycle, which is used to make electricity and for research, military, and industrial uses, or in accidents. Although heavy metals (HMs) and radionuclides are more difficult to eradicate from the soil, they can be neutralized or changed into less dangerous metabolites (by adsorption, chemical modification, or bioavailability). Through catabolic mechanism of microbes, radionuclides and HMs can be biotransformed, biosorbed, and biomineralized. *Trichoderma aureoviride*, *T. virens*, *T. harzianum*, *Penicillium* sp., and *A. niger* are likely the best soil bio-accumulators for Cd, Cr, and Pb^{2+} (Thakare *et al* 2021). *S. cerevisiae* has the adsorption capacity to remove Zn, Cd, and Cu at high sodium conditions. Unmodified *S. cerevisiae* cells have been used to continuously remove Pb^{2+} and Cu^{2+} ions from water solutions, where cells bound Pb^{2+} and Cu^{2+} ions at 29.9 and 72.5 mg g^{-1}, respectively (Amirnia *et al* 2015). Portland cement was used to solidify nuclear waste that inoculated with the mycelium of the *Pleurotus pulmonarius* mushroom. This solidified waste acts as the first line of defense against the release of radio contaminants (Eskander *et al* 2012). Additionally, *Rhodotorula taiwanensis MD1149* showed biofilm formation under extreme gamma radiation and at low pH. Finally, *Rhizopus arrhizus* and *Mucor miehei* can accumulate uranium in their mycelia (Vandana *et al* 2021).

12.6.4 Pulp and paper industry

Lignin and its variants, chlorinated organic chemicals, halogenated hydrocarbons such as dichloroethylene and hydroxy-PCBs, and polybrominated diphenyl ethers are highly prevalent in industrial areas of pulp and paper manufacture. In addition, nonchlorinated compounds such as sterols, resin acids, and tannins have been found in the sediments of water bodies near paper manufacturers (Hubbe *et al* 2016). Several fungi have been studied for a long time as ways to clean up wastewater from paper and pulp mills. The enzyme system of WRF is made up of a group of nonspecific extracellular enzymes that break down lignin and chlorolignins, and also oxidize many aromatic and halogenated compounds, such as lindane, dichloroethylene (DCE) and trichloroethylene (TCE), and PCBs (Gupta and Gupta 2019).

During wood decay, basidiomycetes WRF such as *T. versicolor* and *Heterobasidion annosum* degrade lignin, hemicellulose, and cellulose. Various fungi break down lignin, they include soil fungi, soft rot fungi, pseudo-soft rot fungi, lignin-degrading fungi, and white and brown rot fungi. They decompose modified lignin, derivatives, absorbable organic halides (AOX), and color from pulp mill effluents (Hubbe *et al* 2016). In the treatment of pulp and paper mill effluent, a fungal consortium composed of *Nigrospora* sp. and *Curvularia lunata* reduced biological oxygen demand (BOD) (85.6%), chemical oxygen demand (COD) (80%), color (82.3%), and lignin content (76.1%). Additionally, the high degrading capacity of mixed fungal cultures (*Aspergillus flavus*, *A. niger*, and *P. digitatum*) to COD removal and decolourization of the cardboard recycling business has been investigated (Hosseini *et al* 2021).

12.6.5 Textile industry

Dyeing and finishing produce a lot of waste. Pigments and colors are made from both biological and inorganic materials. Textile effluents include suspended and dissolved particles, BOD, COD, compounds containing hazardous elements including chromium (Cr), arsenic (As), copper (Cu), and zinc (Zn), as well as dyes that are dangerous to the environment and human health (Karthik and Gopalakrishnan 2014). Various contaminants are produced throughout every step of the wool, fabric, woven, carpet, and clothing production process. The main environmental issue facing the textile sector is the large wastewater discharge of toxic recalcitrant pollutants (table 12.1). Water use per ton of textiles is 5285 m^3. Water is typically utilized to apply chemicals to clothes and rinse. In addition, recycled wastewater is used throughout production. Actually, 38% of bleaching water was used for heating, dyeing, printing, and other purposes. Textile wastewater varies by fabric and manufacturing technique. The majority of procedures, including de-sizing, scouring, dyeing, printing, and finishing, use harmful substances including salt, metals, and surfactants. Additionally, insecticide residues, BOD, COD, sulfide, urea, detergents, oils, knitting lubricants, spin finishes, and discarded solvents are among the relevant contaminants.

Chemical and physical procedures cannot destroy environmental contaminant dyes. Complicated chemical structures of dyes prevent their decolourization in water and soil. Physicochemical treatment procedures are laborious, expensive, and theoretically difficult. Fungi release highly oxidative, nonspecific ligninolytic enzymes such as LiP, MnP, and laccase, which react to dyes from diverse chemical groups. Enzymatic activity that absorbs, adsorbs, and accumulates refractory chemicals from effluents helps fungi biodegrade or decolourize textile dyes. Textile dyes are azo, diazo, nitro, phthalein, nitrated, indigo, or anthraquinones based on their chemical structure. Because they are mostly comprised of benzidine and aromatic chemicals, synthetic dyes used in textiles are poisonous and carcinogenic. Azo dyes, which makeup about 70 per cent of the overall textile dyes, produce human-mutagenic amines by reductive cleavage. These effluents also include large amounts of carcinogenic, mutagenic heavy metals and dyes containing chlorine-benzidine-cadmium should not be used. Textile effluents should be treated for hazardous dye removal before being released into the environment. (Santhilatha *et al* 2022).

12.6.5.1 Degradation of dyes by yeasts

Azo reductases enable yeast in removing azo groups (–N N–). Laccase and lignin peroxidase in yeast convert the products into aliphatic amines. *C. tropicalis* and *D. polymorphus* are yeasts that degrade azo dyes, in which the reduction of azo bonds produces amines. Methyl Red was degraded by *S. cerevisiae* MTCC 463 via unusual enzymatic activities aminopyrine N-demethylase, NADH–DCIP reductase, and azoreductase. *S. cerevisiae* absorbed reactive dyes in a medium containing molasses and dyes like Remazol Black B and Remazol Red RB (Santhilatha *et al* 2022).

Tan *et al* (2019) investigated the degradation of azo dyes using the halotolerant yeast *C. tropicalis SYF-1*. The results showed that SYF-1 yeast was capable of decolorizing all six distinct azo dyes, with Acid Red B (ARB) dye being the most successfully eliminated (Tan *et al* 2019). Guo *et al* (2019) used the isolated yeast strain *G. geotrichum* to break down several azo dyes. Under ideal circumstances, 92% of the azo dye Acid Scarlet GR was decolourized after 10 h (Guo *et al* 2019). Based on enzymatic research, NADH-dichlorophenol indophenol reductase greatly influenced the azo dye degradation process. Laccase and lignin peroxidase also performed key roles (Deng *et al* 2020).

12.6.5.2 Degradation of dyes by fungi

Many fungi can break down complex organic molecules into simpler ones, completing the mineralization process. Malachite green decolourization was shown with the use of *A. flavus* and *Alternaria solani*, and both were capable of degrading triphenylmethane dye and malachite green within 6 days. A typical fungus called *P. ostreatus* generates the laccase enzyme, which efficiently removes the color from blue HFRL dye.

A. flavus and *Aspergillus wentii* decoloured acid blue and Yellow MGR dyes in textile effluent after 24 h of aerobic incubation. The immobilized strains decolourized 94.41% of acid blue dye and 95.88% of Yellow MGR dye, respectively. Another common textile dye, Azo Dye Red 3 BN was degraded by *P. chrysogenum*, *A. niger*, and *Cladosporium* sp. when inoculated in potato dextrose medium with 0.01% Azo Dye Red 3 BN. Under ideal conditions, these fungi showed promising decolourization, reaching up to 99.56% (Santhilatha *et al* 2022). During 36 h of incubation, *A. niger MN990895* decolourized the model dye CI Direct Blue 201. (DB 201). Whereas living biomass removed all absorbed 8.4±1.2% of the dye. According to enzymatic tests, decolourization was more affected by external crude enzymes (72.7±3.3%) than intracellular ones and Lacccase was the main enzyme in DB 201 textile dye decolourization (Ekanayake and Manage 2022). Recent research indicates that immobilized *Geotrichum candidum* is more effective as a natural cleaner of dyes in textile industries. whereby decolourization was related to laccase activity (Rajhans *et al* 2021). Extracellular enzymes from several strains of WRF ex *P. pulmonaris*, *P. ostreatus*, and *Coriolus versicolor* were used to decolourize different commercial dyes (Mishra and Srivastava 2022).

12.6.6 Pharmaceutical industries

Pharmaceutical businesses use large volumes of water as a raw material, ingredient, solvent, active component, intermediary, and analytical reagent. Pharmaceutical wastewater is complicated due to its high organic content, microbial toxicity, salt, and difficulty to biodegrade (table 12.1). As a result of the diversity of pharmaceutical products and manufacturing methods, pharmaceutical effluent comes in many forms. Biopharmaceutical wastewater contains antimicrobial wastewater with significant diversity, low C/N, high SS and sulfate contents, a complex structure, and high biological toxicity. Moreover, a lack of nutrients and an abundance of salts make the chemical composition of pharmaceuticals unusual.

According to their lack of specificity, enzymes produced by fungi help manage the various chemical structures of pharmacological classes. Like many fungal species, mushrooms are hyperaccumulators that bioaccumulate xenobiotics from their surroundings. Fungi are more tolerant of environmental changes than other bioremediation species because they can adapt to adverse environmental conditions. Myco-remediation is economical, environmentally friendly, and successful in reducing wastewater toxicity. beta-blockers, psychoactive drugs, anti-inflammatory drugs, antibiotics, and hormones can all be detoxified by macromycetes, often known as mushrooms or polypores (Ortúzar *et al* 2022).

12.6.7 Mines and quarries

Mining is the process of taking valuable minerals and other earth materials from the ground. In industrial mining, water is used to reduce temperatures. Most of the wastewater is reused in the production process, so it does not negatively impact the environment. However, in acidic mining, the wastewater has a lot of harmful content, such as solids that are suspended in the water, dust water, and several metallic ions. The metal recovery and mine rehabilitation capacities of fungi have received attention in recent years. Metal is recovered by fungi via biosorption, intracellular absorption, and transformation/speciation. Nonliving fungal biomass has a high metal ion attraction because metabolism does not create protons. Biosorption is influenced by fungal species, biomass origin, processing, and external factors (for example, metal kind, ionic form, and functional site). Biosorption mechanisms include adsorption, crystallization, chelation, precipitation, ion exchange, entrapment in inter- and intrafibrillar capillaries, polysaccharide materials, and diffusion across fungal cell walls and membranes.(Kumar *et al* 2018).

Phytostabilization and phytoextraction are the principal metal cleaning methods in mining sites. However, microbes, the first living things, start to restore mines. Unicellular, they regulate geogenic changes in the soil and generate ideal conditions for multicellular life such as algae, plants, shrubs, trees, animals, etc. The naturally occurring fungi were discovered to increase plant metal absorption by being involved in nutrient cycling and regulating toxic metal uptake. Bioremediation employs plant growth-promoting (PGP) bacteria and fungi, but the large-scale application of rhizospheric and endophytic PGP bacteria is difficult, especially in mine rehabilitation. Well-studied mycorrhizal fungi help mine site restoration.

Arbuscular mycorrhizal fungi (AMF) have a mutualistic relationship with more than 80% of the plants, which can help to improve plant growth and development. In addition, AMF may assist in improving water stress tolerance and in adapting to heavy metal-contaminated environments. Fungal growth is most common in the top layer of soil, mine drainage and leachate, overburden and mine waste, the root zone of native plants, and the endophytes or epiphytes of the phyllosphere.

Glomus sp. is most common in contaminated or mined areas. *Glomus* (*G. mosseae*, *Glomus fasciculatum*, *Glomus aggregatum*, and *Glomus sclerocystis* sp.) followed by *Acaulospora* spp. were common AMF species in their rhizosphere. Members of the Chenopodiaceae and Brassicaceae are usually not mycorrhizal, but they can make mycorrhizae when they are in a polluted environment. AMF symbiosis also enhances root absorption area by expanding root surface area, hence enhancing nutrient uptake, and resistance to drought, high salinity, diseases, and harmful metals. Both AMF with endophytic fungus as consortium encouraged the development of *Verbascum lychnitis* on Pb-Zn mine waste substrate, whereas a single inoculum did not. AMF were effective because they were able to collect nutrients from the host plant and rhizosphere.

Ectomycorrhizal Fungi (EMs) work with trees of the Pinaceae, Dipterocarpaceae, Fagaceae, and Caesalpinoideae families to help rehabilitate metal-contaminated mines. Without penetrating the root cell lumen, fungal hyphae form the hyphal mantle surrounding the root tips, permeate the cell wall, and multiply between epidermal and cortical cells. Hertig net is used to exchange water, nutrients, and other elements. In tainted soil, fungal hyphae provide nutrients to plant roots (Kumar *et al* 2018).

12.6.8 Mycogenic nanoparticle-mediated heavy metal remediation

Mycogenic nanoparticles are synthesized by fungi. They have potential applications in many fields, including heavy metal removal. The mechanism behind the synthesis of these metallic nanoparticles is hypothesized to be due to the presence of ionic or electrostatic interactions in cytosol. Fungi have unique properties, such as reducing metal ions to nanoparticles, utilizing metal and metalloid compounds, tolerance to heavy metals and the ability to accumulate metals making them useful for myco-remediation. *Penicillium* sp. and *Beauveria caledonensis* produce oxalic and citric acids to reduce the toxicity of metals such as Al, Co, Cu, Cd, Pb, and Zn (Šebesta *et al* 2022). Yeast *S. cerevisiae* employs a defense mechanism against the toxic effects of Mn via its trafficking to vacuoles (Šebesta *et al* 2022). These fungi can be used to synthesize metal-containing nanoparticles that can remove heavy metals from contaminated water or soil.

Research has shown that various fungi such as *Aspergillus*, *Fusarium*, and *Trichoderma* have been used to produce silver and cadmium nanoparticles with high yield and stability (El-Baz *et al* 2016). Additionally, fungi have also been found to biosynthesize other metal nanoparticles, such as gold, copper, and lead (Loshchinina *et al* 2023).

These metal nanoparticles produced by fungi have multiple applications, including environmental remediation, biosensing, and biomonitoring. Fungi can produce other metal nanoparticles by absorbing toxic heavy metals. For example, when cleaning the environment, these nanoparticles can effectively remove heavy metals from contaminated soil and water. In biosensors, they can be used to detect heavy metal ions in environmental sensors, food safety, and medical diagnostics. In biomonitoring, these nanoparticles can be used as a tool to detect and monitor heavy metal contamination in environmental samples.

12.7 Development of new remediation techniques: advances in myco-remediation technology

These advances in myco-remediation technology have led to the development of more effective and sustainable methods for the remediation of contaminated environments. Besides increasing efficiency, the application of genetic engineering has led to a reduction in the time duration required to achieve remediation, overcoming the so-called 'Achilles heel' of bioremediation (Malik *et al* 2022).

Multi-omics studies can show the genes, pathways, metabolic pathways, biotransformation processes, enzymes and metabolites involved in fungal bioremediation. This knowledge can enhance bioremediation and improve fungal bioremediation. To understand the biology and function of a process, multi-omics studies combine genetics, transcriptomics, proteomics, and metabolomics data. The integration of data provided from different multi-omics studies is a crucial step in generating a comprehensive understanding of the bioremediation process. To achieve this, it is necessary to merge the data into a biodegradation network. Multi-omics studies can provide a comprehensive knowledge of the complex biological processes involved in myco-remediation and improve the process by choosing the best fungus species, finding the most effective enzyme systems, and optimizing breakdown conditions (Jaiswal *et al* 2019, Kumar *et al* 2021)

A recent study showed the construction of a microbial consortium consisting of four fungi (*A. flavus*, *Aspergillus nomius*, *Rhizomucor variabilis*, and *Trichoderma asperellum*) and five bacterial strains (*Klebsiella pneumoniae*, *Bacillus cereus*, *Pseudomonas aeruginosa*, *Klebsiella* sp., and *Stenotrophomonas maltophilia*) based on the evaluation of their metabolic capabilities. The metagenomic study revealed that they were successful in reorganizing soil microbial communities in favor of strains that degrade PAHs. It has also been discovered that using genetically modified fungi species in a microbial community accelerates the breakdown of contaminants. A microbial consortium containing two engineered strains of *A. niger* produced high degradation levels of low- and high-molecular-weight PAHs (Park and Choi 2020).

12.8 Conclusion

Myco-remediation is a promising technique for cleaning up industrial effluent, which employs fungi to eliminate contaminants from polluted soil and water. Among the pollutants that fungi can decay and disintegrate are heavy metals, chemicals, and dyes. Myco-remediation can be used to replace or in addition to conventional cleansing techniques. Additionally, it works well in reducing the variety of pollutants in industrial wastewater. Myco-remediation is a low cost, low tech, and ecologically friendly solution that can be applied in several industries with little to no adverse effects on the ecosystem. Additional studies are required to maximize the use of fungi in the cleaning of industrial effluent because myco-remediation is still an active trend. Furthermore, advanced methods of fungi growth and application may be necessary for the adoption of myco-remediation techniques in large-scale commercial contexts. Myco-remediation can ultimately be a vital tool for cleansing industrial wastewater, providing a realistic and efficient method to decrease pollution.

References

Ahmed E M 2015 Hydrogel: preparation, characterization, and applications: a review *J. Adv. Res.* **6** 105–21

Akhtar N and Mannan M A 2020 Mycoremediation: expunging environmental pollutants *Biotechnol. Rep.* **26** e00452

Amirnia S, Ray M B and Margaritis A 2015 Heavy metals removal from aqueous solutions using *Saccharomyces cerevisiae* in a novel continuous bioreactor–biosorption system *Chem. Eng. J.* **264** 863–72

Benghazi L, Record E, Suárez A, Gomez-Vidal J A, Martínez J and de la Rubia T 2014 Production of the Phanerochaete flavido-alba laccase in *Aspergillus niger* for synthetic dyes decolorization and biotransformation *World J. Microbiol. Biotechnol.* **30** 201–11

Bezuneh T T 2016 The role of microorganisms in distillery wastewater treatment: a review *J. Bioremediat. Biodegrad.* **2016** 1–6

Boujelben R, Ellouze M, Tóran M J, Blánquez P and Sayadi S 2022 Mycoremediation of *Tunisian tannery* wastewater under non-sterile conditions using *Trametes versicolor*: live and dead biomasses *Biomass Conv. Bioref.* **14** 299–312

Chaurasia P K, Nagraj , Sharma N, Kumari S, Yadav M, Singh S, Mani A, Yadava S and Bharati S L 2023 Fungal assisted bio-treatment of environmental pollutants with comprehensive emphasis on noxious heavy metals: recent updates *Biotechnol. Bioeng.* **120** 57–81

Colla V, Matino I, Branca T A, Fornai B, Romaniello L and Rosito F 2017 Efficient use of water resources in the steel industry *Water* **9** 874

Daccò C, Nicola L, Temporiti M E E, Mannucci B, Corana F, Carpani G and Tosi S 2020 Trichoderma: evaluation of its degrading abilities for the bioremediation of hydrocarbon complex mixtures *Appl. Sci.* **10** 3152

Danouche M, El Aroussi H, Bahafid W and El Ghachtouli N 2021 An overview of the biosorption mechanism for the bioremediation of synthetic dyes using yeast cells *Environ. Technol. Rev.* **10** 58–76

Das N and Charumathi D 2012 Remediation of synthetic dyes from wastewater using yeast—an overview *Indian J. Biotechnol.* **11** 369–80

Deng D, Lamssali M, Aryal N, Ofori-Boadu A, Jha M K and Samuel R E 2020 Textiles wastewater treatment technology: a review *Water Environ. Res.* **92** 1805–10

Ekanayake M S and Manage P 2022 Mycoremediation potential of synthetic textile dyes by *aspergillus niger* via biosorption and enzymatic degradation *Environ. Nat. Resour. J.* **20** 234–45

El-Baz A F, Sorour N M and Shetaia Y M 2016 Trichosporon jirovecii–mediated synthesis of cadmium sulfide nanoparticles *J. Basic Microbiol.* **56** 520–30

Elhussiny N I, Khattab A E-N A, El-Refai H A, Mohamed S S, Shetaia Y M and Amin H A 2020 Assessment of waste frying oil transesterification capacities of local isolated Aspergilli species and mutants *Mycoscience* **61** 136–44

Eskander S B, Abd El-Aziz S M, El-Sayaad H and Saleh H M 2012 Cementation of bioproducts generated from biodegradation of radioactive cellulosic-based waste simulates by mushroom *Int. Scholar. Res. Notice.* **2012** e329676

Gajendiran A and Abraham J 2017 Biomineralisation of fipronil and its major metabolite, fipronil sulfone, by *Aspergillus glaucus* strain AJAG1 with enzymes studies and bioformulation *3 Biotech.* **7** 212

Gopal R K, Joshi G and Kumar R 2022 Retrospective and prospective bioremediation technologies for industrial effluent treatment ed M Vasanthy, V Sivasankar and T G Sunitha *Organic Pollutants: Toxicity and Solutions, Emerging Contaminants and Associated Treatment Technologies* (Cham: Springer International Publishing) pp 343–72

Guo G, Tian F, Zhao Y, Tang M, Liu W, Liu C, Xue S, Kong W, Sun Y and Wang S 2019 Aerobic decolorization and detoxification of Acid Scarlet GR by a newly isolated salt-tolerant yeast strain Galactomyces geotrichum GG *Int. Biodeterior. Biodegrad.* **145** 104818

Gupta A and Gupta R 2019 Treatment and recycling of wastewater from pulp and paper mill ed R L Singh and R P Singh *Advances in Biological Treatment of Industrial Waste Water and Their Recycling for a Sustainable Future, Applied Environmental Science and Engineering for a Sustainable Future* (Singapore: Springer) pp 13–49

Hegde G M, Aditya S, Wangdi D and Chetri B K 2022 mycoremediation: a natural solution for unnatural problems ed V R Rajpal, I Singh and S S Navi *Fungal Diversity, Ecology and Control Management, Fungal Biology* (Singapore: Springer Nature) pp 363–86

Hosseini Z, Ghaneian M T, Ghafourzade M and Jafari N A 2021 Bioremediation of cardboard recycling industry effluents using mixed fungal culture *Pigment Resin Technol.* **51** 118–25

Hubbe M A, Metts J R, Hermosilla D, Blanco M A, Yerushalmi L, Haghighat F, Lindholm-Lehto P, Khodaparast Z, Kamali M and Elliott A 2016 Wastewater treatment and reclamation: a review of pulp and paper industry practices and opportunities *BioRes* **11** 7953–8091

Jaiswal S, Singh D K and Shukla P 2019 Gene editing and systems biology tools for pesticide bioremediation: a review *Front. Microbiol.* **10** 87

Kajla S, Nagi G K and Kumari R 2021 Microorganisms employed in the removal of contaminants from wastewater of iron and steel industries *Rend. Fis. Acc. Lincei* **32** 257–72

Kalyani D, Tiwari M K, Li J, Kim S C, Kalia V C, Kang Y C and Lee J-K 2015 A highly efficient recombinant laccase from the yeast yarrowia lipolytica and its application in the hydrolysis of biomass *PLoS One* **10** e0120156

Karas P A, Perruchon C, Exarhou K, Ehaliotis C and Karpouzas D G 2011 Potential for bioremediation of agro-industrial effluents with high loads of pesticides by selected fungi *Biodegradation* **22** 215–28

Karthik T and Gopalakrishnan D 2014 Environmental analysis of textile value chain: an overview ed S S Muthu *Roadmap to Sustainable Textiles and Clothing: Environmental and Social Aspects of Textiles and Clothing Supply Chain, Textile Science and Clothing Technology* (Singapore: Springer) pp 153–88

Kulshreshtha S 2019 Removal of pollutants using spent mushrooms substrates *Environ. Chem. Lett.* **17** 833–47

Kumar A, Tripti , Prasad M N V, Maiti S K and Favas P J C 2018 Mycoremediation for mine site rehabilitation *Bio-Geotechnologies for Mine Site Rehabilitation* (Amsterdam: Elsevier) pp 233–60

Kumar A *et al* 2021 Myco-remediation: a mechanistic understanding of contaminants alleviation from natural environment and future prospect *Chemosphere* **284** 131325

Kumar V and Dwivedi S K 2021 Mycoremediation of heavy metals: processes, mechanisms, and affecting factors *Environ. Sci. Pollut. Res.* **28** 10375–412

Liu H *et al* 2020 Biodegradation of Sulfonated lignite (SL) by fungi from waste drilling mud *IOP Conf. Ser.: Earth Environ. Sci.* **601** 012038

Loshchinina E A, Vetchinkina E P and Kupryashina M A 2023 Diversity of biogenic nano-particles obtained by the fungi-mediated synthesis: a review *Biomimetics* **8** 1

Mahmoud Y A-G and Darwesh O M 2022 Protocol for assessing mycoremediation of acidic radioactive wastes ed D Udayanga, P Bhatt, D Manamgoda and J M Saez *Mycoremediation Protocols, Springer Protocols Handbooks* (New York: Springer US) pp 109–21

Malik G, Arora R, Chaturvedi R and Paul M S 2022 Implementation of genetic engineering and novel omics approaches to enhance bioremediation: a focused review *Bull. Environ. Contam. Toxicol.* **108** 443–50

Mishra M and Srivastava D 2022 Mycoremediation: an emerging technology for mitigating environmental contaminants ed U B Singh, J P Rai and A K Sharma *Re-Visiting the Rhizosphere Eco-System for Agricultural Sustainability, Rhizosphere Biology* (Singapore: Springer Nature) pp 225–44

Mitra J, Mukherjee P K, Kale S P and Murthy N B 2001 Bioremediation of DDT in soil by genetically improved strains of soil fungus *Fusarium solani Biodegradation* **12** 235–45

Ortúzar M, Esterhuizen M, Olicón-Hernández D R, González-López J and Aranda E 2022 Pharmaceutical pollution in aquatic environments: a concise review of environmental impacts and bioremediation systems *Front. Microbiol.* **13** 869332

Pant D, Giri A and Dhiman V 2018 Bioremediation techniques for e-waste management ed S J Varjani, E Gnansounou, B Gurunathan, D Pant and Z A Zakaria *Waste Bioremediation, Energy, Environment, and Sustainability* (Singapore: Springer) pp 105–25

Park H and Choi I-G 2020 Genomic and transcriptomic perspectives on mycoremediation of polycyclic aromatic hydrocarbons *Appl. Microbiol. Biotechnol.* **104** 6919–28

Qadir G 2019 Yeast a magical microorganism in the wastewater treatment *J. Pharmacogn. Phytochem.* **8** 1498–500

Rajhans G, Sen S k, Barik A and Raut S 2021 De-colourization of textile effluent using immobilized Geotrichum candidum: an insight into mycoremediation *Lett. Appl. Microbiol.* **72** 445–57

Rhodes C J 2014 Mycoremediation (bioremediation with fungi)—growing mushrooms to clean the earth *Chem. Speciation Bioavailability* **26** 196–8

Saglam N, Yesilada O, Saglam S, Apohan E, Sam M, Ilk S, Emul E and Gurel E 2018 Bioremediation applications with fungi ed R Prasad *Mycoremediation and Environmental Sustainability: Volume 2, Fungal Biology* (Cham: Springer International Publishing) pp 1–37

Santhilatha P, Haritha B and Suseela L 2022 Role of fungi in the removal of synthetic dyes from textile industry effluents ed A Khadir and S S Muthu *Biological Approaches in Dye-Containing Wastewater Volume 2 Sustainable Textiles: Production, Processing, Manufacturing & Chemistry* (Singapore: Springer) pp 157–66

Šebesta M, Vojtková H, Cyprichová V, Ingle A P, Urík M and Kolenčík M 2022 Mycosynthesis of metal-containing nanoparticles—fungal metal resistance and mechanisms of synthesis *IJMS* **23** 14084

Seddiek H A, Shetaia Y M, Mahamound K F, El-Aassy I E and Hussien 2021 Bioleaching of Egyptian fly ash using cladosporium cladosporioides *Ann. Biol.* **37** 18–22

Sharma J, Goutam J, Dhuriya Y K and Sharma D 2021 Bioremediation of industrial pollutants ed D G Panpatte and Y K Jhala *Microbial Rejuvenation of Polluted Environment: Volume 2, Microorganisms for Sustainability* (Singapore: Springer) pp 1–31

Shukla M, Shukla R, Jha S, Singh R and Dikshit A 2022 Myco-remediation: a sustainable biodegradation of environmental pollutants ed T Aftab *Sustainable Management of Environmental Contaminants: Eco-Friendly Remediation Approaches, Environmental Contamination Remediation and Management* (Cham: Springer International Publishing) pp 425–49

Stamets P 2005 *Mycelium Running: How Mushrooms Can Help Save the World* (Berkeley, CA: Ten Speed Press)

Stiebeling D A and Labes A 2022 Mycoremediation of sewage sludge and manure with marine fungi for the removal of organic pollutants *Front. Marine Sci.* **9** 946220

Tan L, Xu B, Hao J, Wang J, Shao Y and Mu G 2019 Biodegradation and detoxification of azo dyes by a newly isolated halotolerant yeast *Candida tropicalis* SYF-1 *Environ. Eng. Sci.* **36** 999–1010

Tekere M 2019 Microbial bioremediation and different bioreactors designs applied *Biotechnology and Bioengineering* (Rijeka: IntechOpen)

Thakare M, Sarma H, Datar S, Roy A, Pawar P, Gupta K, Pandit S and Prasad R 2021 Understanding the holistic approach to plant-microbe remediation technologies for removing heavy metals and radionuclides from soil *Curr. Res. Biotechnol.* **3** 84–98

Vandana U K, Gulzar A B M, Laskar I H, Meitei L R and Mazumder P B 2021 Role of microbes in bioremediation of radioactive waste ed D G Panpatte and Y K Jhala *Microbial Rejuvenation of Polluted Environment: Volume 1, Microorganisms for Sustainability* (Singapore: Springer) pp 329–52

Vishwakarma P 2019 Role of macrofungi in bioremediation of pollutants ed P K Arora *Microbial Metabolism of Xenobiotic Compounds, Microorganisms for Sustainability* (Singapore: Springer) pp 285–304

Wang Y, Qiu L and Hu M 2018 Application of yeast in the wastewater treatment *E3S Web Conf* **53** 04025

Zhang C, Ren H-X, Zhong C-Q and Wu D 2020 Biosorption of Cr(VI) by immobilized waste biomass from polyglutamic acid production *Sci Rep.* **10** 3705

IOP Publishing

Trends in Biological Processes in Industrial Wastewater Treatment

Maulin P Shah

Chapter 13

Modern procedures for industrial effluent analysis based on gas chromatography

Farooque Ahmed Janjhi, Hameed Ul Haq and Grzegorz Boczkaj

The urbanization and industrialization of the last few decades have led to a proliferation of pollutants in water bodies. As a substantial direct and ongoing intake of pollutants, industrial effluents pose a significant threat to aquatic ecosystems. The concentration of contaminants in industrial effluents can often be both at high and low levels ($ng\ l^{-1}$ to $mg\ l^{-1}$).

Special attention should be given to volatile organic compounds (VOCs) as well as toxic organic pollutants. In both cases, even minimal volatility of pollutants allows us to control their presence by using the gas chromatography (GC) technique. This technique allows us to control hundreds of compounds during one analysis. Coupling GC with selective detectors or with mass spectrometry (MS) provides an ultra-selective and sensitive method of analysis. In some cases, a dedicated sample preparation procedure is essential to eliminate matrix components, ensure effective enrichment of analytes, as well as to gain needed selectivity.

This chapter presents a systematic presentation of sample preparation methods (including solvent-free techniques), solutions for effective GC separation, description of selective detectors, and examples of developed procedures towards analysis of several groups of organic compounds—sulfur, nitrogen, and oxygen-containing VOCs; carboxylic acids; aromatic and polycyclic aromatic hydrocarbons (PAHs)—in wastewater by procedures based on GC.

13.1 Introduction

Industrial effluents are the waste or byproducts that are generated during industrial processes, including liquids, gases, and solid waste. These effluents often contain a wide range of pollutants, including chemicals, heavy metals, and organic compounds, that can have a significant impact on the environment and human health.

Industrial effluents are typically discharged into the environment through various pathways, including air emissions, surface water, and groundwater [1]. Population expansion has raised domestic consumption, which has accelerated industrialization and, ultimately, the production of industrial waste. The discharge of industrial effluents into the environment can lead to contamination of soil, water, and air, and pose a risk to human health and wildlife. Industrial effluent from refinery industries contains VOC-based compounds [2–6]. These contaminates may be flammable, poisonous, reactive, or carcinogenic [7]. To mitigate the environmental impact of industrial effluents, it is important to monitor and control their discharge into the environment [8].

Sample preparation is the critical first step in the GC determination of an analyte. Depending on the complexity of the sample, there may be several processes involved in sample preparation, and the concentration level of the analyte in the sample needs to be analyzed by the GC instrument. However, sample preparation is often a challenging process that contributes to the complexity of the analyte analysis. For example, in the case of organics and volatile organics, sample preparation procedures may include extraction, cleanup, derivatization, and transfer to vapor phase [9]. GC involves injection of the sample via a dedicated injector (which is responsible for vaporization of the sample if it is injected in liquid state and transferred to the GC column) connected with the head of a chromatographic column, with elution occurring due to the flow of an inert gas (mobile phase) such as helium, argon, nitrogen, carbon dioxide, or hydrogen. The mobile phase in GC does not interact with the analyte molecules, it only transports them through the column [10]. There are two main types of GC: gas–solid chromatography (GSC) and gas–liquid chromatography (GLC), with the mechanisms of analyte retention in the column differing between them. In GLC, analytes are partitioned between a gaseous mobile phase and a liquid stationary phase, whereas in GSC retention occurs through physical adsorption onto a solid stationary phase. GLC is more commonly used in various scientific fields when compared to GSC [9].

The primary drawback of using GC to analyze water samples is that it cannot directly analyze them because water samples typically contain various non-volatile compounds and have a high-water content. Compounds such as dissolved inorganic and organic compounds, salts, and minerals can potentially interfere with the GC analysis. Introducing water as a solvent into a GC system is generally discouraged due to its tendency to damage the coatings of GC columns and reduce detector sensitivity. However, these drawbacks can be avoided by separating water from the analytes before they reach the GC column, which can be accomplished through the following techniques [11]:

Low volatility: GC is suitable for analyzing compounds that are volatile, meaning they have a low boiling point and high vapor pressure. However, detecting certain compounds in water samples using GC can be challenging because they may have low volatility.

Matrix effects: Water samples can contain numerous substances such as dissolved solids, salts, and other non-volatile compounds. These substances can impede the analysis process and result in matrix effects that may cause erroneous outcomes.

Sensitivity: GC may lack the sensitivity required to identify small amounts of certain substances in water samples, especially non-volatile or polar compounds. In such cases, more sensitive techniques such as liquid chromatography may be necessary.

Sample preparation: Prior to GC analysis of water samples, it is recommended to extract and concentrate them, which can be a tedious but very effortful procedure. Furthermore, it may involve the utilization of organic solvents that can pose additional analytical difficulties.

Incomplete extraction: Conventional extraction methods may not effectively extract certain compounds present in water samples, resulting in incomplete extraction and reduced recovery rates.

Sample complexity: Water samples can be complex, and may contain a wide range of compounds at varying concentrations. Sample preparation can help to reduce sample complexity and isolate the target compounds, thereby improving the accuracy and sensitivity of the analysis.

Reproducibility: Sample preparation can help to improve the reproducibility of the analysis by redu/cing variability in the sample matrix, and ensuring consistent extraction and concentration of the target compounds.

13.1.1 Types of industrial effluents

Industrial effluents can be classified based on the type of industry that generates them. Some common types of industrial effluents include:

(i) Chemical industry effluent: This type of effluent is generated by the chemical industry and contains a wide range of chemicals, including acids, bases, and solvents, that can be harmful to the environment and human health [12].

(ii) Petrochemical industry effluent: This type of effluent is generated by the petrochemical industry and contains a mixture of organic and inorganic compounds, including hydrocarbons, sulfur compounds, and heavy metals [13].

(iii) Food and beverage industry effluent: This type of effluent is generated by the food and beverage industry and contains organic matter, nutrients, and microorganisms that can contribute to water pollution and eutrophication [14].

(iv) Textile industry effluent: This type of effluent is generated by the textile industry and contains a mixture of dyes, salts, and organic matter that can be harmful to the environment and human health [15].

(v) Mining industry effluent: This type of effluent is generated by the mining industry and contains heavy metals and other contaminants that can pose a risk to human health and the environment [16].

(vi) Power generation industry effluent: This type of effluent is generated by the power generation industry and contains a mixture of organic and inorganic compounds, including heavy metals, sulfur compounds, and organic matter, that can be harmful to the environment and human health [17].

These are just some examples of the types of industrial effluents that are generated by various industries. The composition of industrial effluents can vary widely depending on the specific industry and processes involved.

13.2 Sample preparation methods for GC

13.2.1 Introduction to sample preparation techniques

The technique for preparing samples is extremely important in analytical chemistry, particularly when dealing with complex matrices found in biological and environmental samples. The difficulty with sample preparation in these types of analyses lies in the ability to recover analytes in a form that is suitable for high sensitivity and specificity determinations [18]. Sample preparation refers to the process of treating a sample to make it suitable for analysis or experimentation. This can involve a range of techniques, such as homogenization, extraction, purification, size reduction, and preservation. The goal of sample preparation is to obtain a representative and uniform sample that accurately reflects the properties of the original sample. Some common techniques used in sample preparation include centrifugation, filtration, freeze-drying, evaporation, and chemical treatments. The choice of technique depends on the type of sample and the desired outcome of the analysis or experiment [18].

13.2.2 Sample preparation techniques for GC

GC is a widely used analytical technique for the separation and quantification of volatile and semi-volatile organic compounds (SVOCs) in complex mixtures. The success of GC analysis depends heavily on the quality of the sample preparation because the sample must be properly prepared to ensure accurate and reliable results.

Sample preparation for GC involves a series of steps to extract, purify, and concentrate the analytes of interest from the sample matrix. These steps can include extraction, derivatization, concentration, and purification, depending on the nature of the sample and the desired analytes [19].

Sample preparation is a critical aspect of GC analysis because it can influence the accuracy, precision, and sensitivity of the results. Proper sample preparation can also help to eliminate interference from the sample matrix and improve the stability and reproducibility of the analysis. Sample preparation is an essential step in GC analysis, and plays a critical role in ensuring the accuracy and reliability of the results. It is important to choose an appropriate sample preparation method that is compatible with the sample matrix and the desired analytes, and to carefully follow the sample preparation procedures to ensure high-quality results [19]. Sample preparation techniques for GC include:

   (i) Solid-phase extraction (SPE): This is a technique designed for rapid, selective sample preparation and purification prior to chromatographic analysis. In SPE, one or more analytes from a liquid sample are isolated by extracting, partitioning, and/or adsorbing onto a solid stationary phase

A method to selectively extract specific components from a sample using a solid-phase sorbent. A well-designed SPE protocol can provide several advantages, such as enabling the use of sample matrices that are more compatible with the desired chromatographic method, enhancing sensitivity by concentrating trace amounts of analytes, eliminating interferences that could result in high background or inaccurate peaks, safeguarding the analytical column from contaminants, and allowing for automation of the extraction process [20].

(ii) Liquid–liquid extraction (LLE): This is a method for separating a sample into two immiscible liquid phases, with the target compounds partitioning into one of the phases. LLE is a sample preparation technique that is used in GC to isolate and concentrate target analytes from a complex sample matrix. LLE is particularly useful for samples that are not amenable to direct injection into the GC, such as those containing high levels of interfering compounds or those with low analyte concentrations [21].

(iii) Quick, easy, cheap, effective, rugged, and safe (QuEChERS): This is a multi-step extraction technique that combines liquid–liquid extraction with a dispersive solid-phase extraction (dSPE) step. The sample is first extracted with a polar solvent, and the extract is then mixed with a dSPE sorbent to remove interfering compounds. The analytes are then eluted from the sorbent using a nonpolar solvent and concentrated prior to GC analysis [22].

(iv) Molecularly imprinted polymer (MIP) based extraction: MIPs are highly selective materials that are used in various extraction techniques, including SPE, solid-phase microextraction (SPME), and stir bar sorptive extraction. MIPs are prepared by polymerizing monomers in the presence of a template molecule, which creates cavities in the polymer that are complementary to the shape, size, and functional groups of the template molecule. Once the template molecule is removed, the resulting MIP can be used as a highly selective sorbent for the target molecule. MIP-based extraction techniques are particularly useful for the selective extraction of low-concentration target analytes from complex matrices, such as environmental, food, and biological samples. In GC, MIP-based extraction techniques can be used as a sample preparation step to isolate and concentrate target analytes prior to GC analysis [23].

(v) Soxhlet extraction: In this technique, the sample is placed in a thimble, and then extracted using a continuous cycle of solvent extraction and distillation. The solvent vapor condenses on the cooler thimble and drips back into the sample, allowing for continuous extraction until the analytes are fully extracted. Soxhlet extraction has been the most widely used extraction technique for several decades. Some of the advantages of this method are its simplicity, low cost, specified consumption of the solvent, and repetition of extraction cycles [24].

(vi) Steam distillation: This is a sample preparation technique that can be used in conjunction with GC to extract and analyze VOCs from a wide range of sample matrices, including plant materials, essential oils, and food products. In the steam distillation process, the sample is placed in a distillation apparatus and steam is passed through the sample, causing the VOCs to vaporize. The vaporized compounds are then condensed and collected in a separate vessel. The collected compounds can be further purified or concentrated using a suitable technique, such as SPE or evaporation. The purified analytes are then analyzed by GC using a suitable stationary phase and detector [25]. The benefits of steam distillation for GC analysis include high extraction efficiency for volatile compounds, minimal use of organic solvents, and the ability to extract a wide range of analytes, including polar and nonpolar compounds. However, steam distillation can also be time-consuming and requires specialized equipment and expertise [26].

(vii) Supercritical fluid extraction (SFE): This is a sample preparation technique that can be used in conjunction with GC to extract and purify analytes from solid, liquid, or gaseous matrices. SFE uses a supercritical fluid, typically carbon dioxide (CO_2), as the extraction solvent, which allows for the efficient extraction of analytes with minimal use of organic solvents [27]. In the SFE process, the sample is loaded into an extraction vessel and pressurized with supercritical CO_2 to extract the analytes of interest. The analytes are then collected in a separate vessel, typically by reducing the pressure and allowing the CO_2 to evaporate. The collected analytes can then be further purified or concentrated using a suitable technique, such as SPE or evaporation. The purified analytes are then analyzed by GC using a suitable stationary phase and detector [27]. The benefits of SFE for GC analysis include reduced use of organic solvents, high extraction efficiency, and the ability to extract a wide range of analytes, including both polar and nonpolar compounds. However, the SFE process can also be time-consuming and requires specialized equipment and expertise [28].

(viii) Headspace extraction (HSE): This is a method for extracting volatile compounds from a sample by heating the sample and collecting the gases that are evolved into the headspace. Headspace extraction is a sample preparation technique used in GC for the analysis of VOCs and SVOCs in solid, liquid, and gaseous samples. This technique involves the equilibration of the analytes between the headspace (gas phase) of a sample and a sealed vial, followed by extraction and concentration of the headspace using a nonpolar solvent for GC analysis [29].

The headspace extraction technique typically involves the following steps:

1. Place a small amount of the sample (typically 1–5 ml) into a sealed vial and heat it to a temperature that is sufficient to promote the release of volatile compounds into the headspace (e.g., 60°C–100°C for liquids and 50°C–120°C for solids).

2. Allow the vial to equilibrate for a specific amount of time (e.g., 30–60 min) to allow the VOCs and SVOCs to partition into the headspace.

3. Use a syringe to withdraw a small amount of the headspace (typically 0.5–5 ml) and inject it into the GC inlet for analytes separation using a selected stationary phase, and a suitable detection, such as a flame ionization detector (FID) or mass spectrometer (MS).

(ix) FAMEs formation: This is a method for converting carboxylic acids and alcohols into their corresponding esters by reaction with a strong base such as sodium hydroxide. This technique is particularly useful for the analysis of fatty acid methyl esters (FAMEs), which are commonly used as biomarkers in environmental, food, and pharmaceutical applications [30]. Once esters are formed, they can be extracted using a nonpolar solvent, such as hexane or heptane or dichloromethane, and concentrated using a rotary evaporator or other concentration technique. The resulting extract can then be analyzed by GC using a proper stationary phase and detected by FID or MS [21].

(x) Derivatization: This is a method for chemically modifying a sample to enhance its volatility, stability, or detectability by GC. Derivatization is a sample preparation technique commonly used in GC to improve the separation, detection, and quantification of polar and non-volatile analytes. This technique involves the chemical modification of analytes prior to GC analysis to make them more volatile, less polar, or more amenable to detection by the GC detector. Derivatization can be used for a wide range of analytes, including amino acids, carbohydrates, steroids, and fatty acids, among others. Some of the commonly used derivatization reagents for GC analysis include silylating agents, such as trimethylsilyl (TMS) reagents, and alkylating agents, such as diazomethane and pentafluorobenzyl bromide (PFPA) derivatives [31].

The choice of sample preparation technique for GC depends on several factors, including the nature of the sample, the target analytes, the sensitivity and selectivity of the GC method, and the required detection limit and accuracy of the analysis.

13.2.3 Sample preparation based on green solvents

Sample preparation based on green solvents refers to the use of environmentally friendly solvents in the process of extracting and purifying target compounds from a sample prior to analysis by techniques such as GC. The use of green solvents is driven by the need to reduce the environmental impact of sample preparation and to minimize the release of toxic and hazardous substances into the environment.

13.2.3.1 Ionic liquids

Ionic liquids (ILs) are salts that are liquid at room temperature and can serve as solvents for various applications, including sample preparation. ILs, as green solvents, are widely used due to their appealing properties such as negligible vapor pressure, large liquid range, high thermal stability, high ionic conductivity, large electrochemical window, and ability to solvate compounds of widely varying polarity [32]. ILs are a class of green solvents that can be used for sample preparation in GC to extract and purify analytes from a wide range of sample matrices, including environmental samples, pharmaceuticals, and food products.

13.2.3.2 Deep eutectic solvents

Deep eutectic solvents (DESs) are a relatively new class of solvents that are gaining attention as alternative solvents for sample preparation in the analysis of industrial effluents. DESs are composed of two or more components, typically a salt and a low molecular weight organic compound, that form a non-volatile and low-melting mixture with unique solubilizing properties [33–35]. The use of DESs in sample preparation for the analysis of industrial effluents has several advantages, including greenness, enhanced extraction efficiency, improved sample stability, and ease of use [35, 36].

While DESs have several advantages in sample preparation for the analysis of industrial effluents, it is important to note that not all DESs are suitable for all samples and applications, and that the choice of DES will depend on the specific sample matrix and analytes of interest. DESs are an emerging class of solvents for sample preparation in the analysis of industrial effluents, offering several advantages over traditional solvents, including improved extraction efficiency, stability, and ease of use. The use of DESs in sample preparation is a promising area of research and further studies are needed to fully understand their potential in this application.

13.2.4 Advances in sample preparation for analysis of industrial effluents: future perspective

In recent years, there have been several advances in sample preparation techniques for the analysis of industrial effluents, aimed at improving the efficiency, accuracy, and sustainability of the analysis. Some of these advances include:

Automated sample preparation: Automated sample preparation systems have been developed that allow for rapid, consistent, and reproducible preparation of samples for analysis. These systems can help to improve the efficiency and accuracy of the analysis, and reduce the risk of human error.

Microfluidic sample preparation: Microfluidic systems have been developed that allow for small-scale, on-chip sample preparation, reducing the amount of sample needed and minimizing waste.

Green solvents: The use of green solvents in sample preparation has gained popularity in recent years due to the need to minimize the environmental impact of sample preparation and reduce exposure to hazardous chemicals.

Hybrid sample preparation: Hybrid sample preparation techniques that combine different sample preparation methods, such as SPE and LLE, have been developed to improve the efficiency and selectivity of the analysis.

These advances in sample preparation have led to improved efficiency, accuracy, and sustainability in the analysis of industrial effluents, helping to better monitor and control the release of pollutants into the environment.

The sample preparation for industrial effluents using GC is an active area of research. In addition, there are several trends and directions that are likely to shape the future of this field.

Multi-analyte methods: The development of multi-analyte methods for the simultaneous analysis of multiple contaminants in industrial effluents is an active area of research. These methods will help to improve the accuracy, sensitivity, and efficiency of the analysis, and to reduce the need for multiple individual analyses.

On-site analysis: There is a growing need for on-site analysis of industrial effluents, to reduce the time and costs associated with sample transportation and to minimize sample degradation. This is likely to drive the development of compact, portable GC systems for sample preparation and analysis.

In summary, the future of sample preparation for industrial effluents using GC is likely to be shaped by the development of green solvents, automation, multi-analyte methods, and on-site analysis. These trends will help to improve the efficiency, accuracy, and sustainability of the analysis, and will play a key role in addressing the challenges associated with the analysis of industrial effluents.

13.3 Types of GC detectors dedicated to the analysis of specific groups of pollutants in water and wastewater

13.3.1 FID (universal detection)

The FID works by passing a sample of the gas through a flame (combustion of hydrogen in air(rich)). The detector is heated by an electric heating element and the gas sample is ionized (carbo-ions are formed) by the flame. The ionized sample is then passed through an electrometer, which measures the generated electrical current. This current is then correlated to the number of ions present in the sample, which is then used to calculate the concentration of the gas.

The FID is able to detect low levels of VOCs in the atmosphere due to its high sensitivity. Furthermore, the FID is a highly sensitive instrument and is used in many different applications, including air pollution [37], metabolite fingerprinting [38], dioxane determination [39, 40], and aromatic hydrocarbon (BTEX) analysis [41, 42]. Overall, the FID is a powerful and sensitive instrument that can be used in many different applications. It is able to detect low levels of VOCs in the environment, detect fires, and measure the composition of gases in industrial settings.

13.3.2 Selective detectors for sulfur containing VOCs

Flame photometric detectors (FPDs) utilize chemiluminescence phenomenon, which involves detecting the amount of light emitted by substances when they are introduced into a flame and undergo chemical reaction and excitation. A hydrogen(rich)-air flame is used in FPDs to excite atoms (* denotes the excited state), which then fluoresce or chemiluminesce. The wavelength of emitted light provides qualitative information, while the intensity of the emitted light yields quantitative information [43].

FPDs favor halogens and specific elements, such as S, P, Sn, B, As, Ge, Se, and Cr. The sensitivity of FPD is on the scale of ng ml^{-1} for P and S and other elements [44]. Spectral and chemical interference can affect the detection of FPD. Spectral interference can occur due to either the flame's background or the inability of the detector to distinguish between the emission lines of two different species. When flames reach high temperatures of at least 3000°C, metal atoms get ionized, leading to chemical interference. These ionized metal atoms emit distinct spectra, different from those of excited atoms [45]. The way that an FPD reacts is influenced by the chemical compound it interacts with, and its response is significantly reduced by substances such as hydrocarbons and thick fluids such as sucrose [46]. A multiple flame detector, designed by Clark and Thurbide, features interconnected fluid channels milled into a planar stainless-steel plate. This design reduces background noise by 50% compared to the quartz tube mFPD [47]. Sevčík and Tranchida provided a comprehensive explanation of the remaining detection methods [48].

The FPD is able to detect low levels of VOCs in the atmosphere due to its sensitivity. It can also detect fires because it can detect hydrocarbons released by burning materials. In industrial settings, the FPD is used to measure the composition of gases released during a process, allowing for optimization and improved safety measures. The potential application FPD includes the analysis of organophosphorus pesticides in samples of water [49], vegetables [50], human urine [51], and fish [52].

The dual flame FPD, pulsed flame photometric detector (PFPD), and sulfur chemilumescence detector (SCD) are more advanced alternatives to the FPD. The dual flame FPD uses two separate flames to improve signal-to-noise ratio and provide more accurate readings. The PFPD uses a pulsed ignition in the detector and time delayed (in miliseconds) signal aquisiton adjusted to specific time of chem-iluminescence after combustion (e.g., it is 6–24 ms for S), providing greater sensitivity and a lower detection limit. The SCD is the most advanced alternative.

13.3.3 Selective detectors for nitrogen containing VOCs

One of the most commonly used methods for the determination of nitrogen-organic compounds in water and wastewater samples is based on the use of GC in combination with a nitrogen-phosphorus selective detector (NPD), also known as a thermionic detector (TID). The first detector of this type was developed in 1964 by Karmen and Giuffrid for the determination of phosphoric and chlorinated hydro-carbons [53]. It was not until three years later that its usefulness in identifying nitrogen compounds was also found [54]. The former detector, i.e., alkali flame ionization detector, in which the signal was obtained by using an ionization source in the form of alkali metal salts and a burner powered by hydrogen and air, was characterized by poor stability and the need for frequent replacement of the ion source. Many solutions were developed to eliminate these disadvantages [55], but the best results were obtained in 1974 by Kolb and Bisschof, who replaced alkali metal salts with non-volatile rubidium silicate deposited on a glass bed, while the burner was replaced with platinum wire and a much smaller volumetric flow was used. Thanks to these developments, higher sensitivity, selectivity, and improved stability were obtained [56]. The principle of operation of modern nitrogen-phosphorus detectors is to create an active boundary layer or plasma in the detector at a temperature of 600°C–800°C, at which the compounds are decomposed into electronegative products (NO_2, CN, and PO_2), and then negative ions (CN^-, PO^-, PO_2^-, and PO_3^-) at elevated temperature with the use of alkaline catalysts are placed in a ceramic source [57].

Most analytical methods based on GC-NPD are suitable for the determination of nitrosamines in wastewater streams where the concentration of these compounds is very high. However, despite the use of extraction techniques, i.e., solid-phase microextraction (SPME), continuous liquid–liquid extraction (CLLE), and SPE, the technique is not sensitive enough to meet the requirements of drinking water analytics. Current standards require the use of methods with a limit of detection in the range of 1–10 ng l^{-1} [58].

Another selective detector used in GC to detect nitrogen-organic compounds is the nitrogen chemiluminescence detector (CLND). Its principle of operation, in accordance with reactions (13.1) and (13.2), consists in burning the sample at high temperature (>1000°C) and generating nitric oxide (NO) that then reacts with ozone to generate excited nitrogen dioxide (NO_2^*), which contributes to the production of a photon of light in the range from 600 to 900 nm, i.e., the phenomenon of chemiluminescence [59]. In the case of nitrosamines, this phenomenon is conditioned by the formation of nitrosyl radicals under the influence of high temperature reaction (13.3) and (13.4) [60]. The use of the SPME extraction system in combination with the GC-CLND provides the most reliable results in the determination of N-nitrosamines in wastewater samples when compared to methods using MS and NPD detectors.

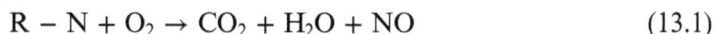

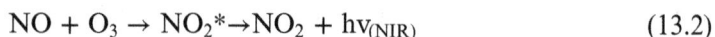

$$R - N + O_2 \rightarrow CO_2 + H_2O + NO \tag{13.1}$$

$$NO + O_3 \rightarrow NO_2^* \rightarrow NO_2 + h\nu_{(NIR)} \tag{13.2}$$

$$R - N = NO + O_2 \rightarrow CO_2 + H_2O + 2NO \tag{13.3}$$

$$R - N = NO \rightarrow R - N + NO^* \tag{13.4}$$

13.3.4 Gas chromatography coupled with mass spectrometry

Gas chromatography–mass spectrometry (GC–MS) is an analytical technique that combines the features of GC and mass spectrometry to identify different substances within a sample. GC–MS is a powerful analytical technique and is used in a variety of fields, such as environmental chemistry, forensic science, and food science. GC–MS is used to identify and quantify volatile and semi-volatile compounds in complex mixtures. It is also used to determine the structure of unknown compounds, confirm purity of compounds, and monitor environmental pollutants. The technique is also used in clinical laboratories to confirm the presence of drugs in biological samples. Additionally, GC–MS is used for the authentication of food products and trace analysis of foods for contaminants.

The principle of mass spectrometer operation is based on the measurement of the ratio of mass to electric charge (m/z). MS is operated under high vacuum. Compounds are first subjected to ion source for ionization. General types of ionization modes in GC–MS are electron impact ionization (EI) or chemical ionization. The formed ions are then separated in an analyzer. Several types of analyzer are available, including quadrupole, ion-trap, and time of flight. Finally, the ions are detected in dedicated system, a photomultiplier is typically used. Depending on operating mode, all of the produced ions can be acquired (full scan mode, identification purpose) or only a specific ion is detected (single ion monitoring mode, increased sensitivity, quantitative analysis purpose) [61]. SCAN mode is a type of data acquisition that is used to acquire a full mass spectrum of separated compounds, which is used to obtain basic information about sample components. In this mode, the user can select a range of m/z values to scan. The compounds can be identified on the basis of the obtained spectrum and a comparison made by the software (or manually by the user) with mass spectra stored in a library.

A GC–MS detector with EI is used in conjunction with a suitable extraction technique to increase sensitivity and eliminate the water matrix., e.g. SPE [61], SPME [62], microwave-assisted dispersive liquid–liquid microextraction (DUSA-DLLME) [63], SDME [64], liquid phase membrane microextraction (HF -LPME) [66]. However, the use of SPE with conventional C-18 sorbents does not bring satisfactory results because they show only strong hydrophobic interactions. Much higher selectivity is achieved by using a silica gel sorbent modified with phenthothiazine because it provides both hydrophobic and charge transfer interactions [65]. Good results are also obtained with the use of ultrasound-assisted DLLME [63]. Comparable limit of detection (LOD) values are obtained using hollow fiber liquid phase microextraction (HF-LPME) and single droplet microextraction. However, SDME is more challenging and often does not provide satisfactory reproducibility [66].

Due to the lack of characteristic ions, aliphatic amines must be derivatized prior to GC–MS analysis for high sensitivity applications. Typical derivatizing reagents for selective detection of amines are: benzosulfonyl chloride [67], iodine [68], and pentafluorobenzaldehyde (PFBAY) [69]. These reagents make it possible to obtain derivatives within a minute. After using the appropriate isolation/enrichment technique, the obtained derivatives can be determined in at $\mu g\ l^{-1}$ level.

In the case of the determination of aromatic amines, an interesting solution is to use a combination of fiber-assisted microextraction with the formation of an emulsion (fiber-assisted emulsification microextraction, FAEME) combined with GC–MS. This method not only eliminates the need to use an additional solvent but also enables the determination of compounds at even lower concentration levels [69]. In addition, carboxylic acid-based compounds [70, 87], aromatic hydrocarbons (HCs) and PAHs [41, 42], and oxygen-containing VOCs such as alcohol, ketones, esters can be analyzed by GC–MS [21, 71, 72]. Derivatization is required to increase the volatility of oxygen-containing compounds in order to facilitate their separation capabilities.

13.4 Examples of GC methods

13.4.1 Determination of aromatic hydrocarbons and PAHs in water and wastewater by GC-based procedures

Aromatic hydrocarbons and PAHs are a significant group of organic compounds that result from natural processes such as carbonization, which involve incomplete combustion of organic materials. PAHs, which are present in the atmosphere, soil, and water, have considerable toxicity and can cause carcinogenic, mutagenic, and endocrine-disrupting effects. The US Environmental Protection Agency has identified PAHs as priority pollutants [73]. Therefore, the detection of PAHs in the environment is crucial and indispensable for ensuring human well-being.

A very popular technique for aromatic HC (BTEX) and PAHs is to use DLLME aromatic hydrocarbons [41, 42, 74, 75]. This method involves the use of a ternary solvent system where a combination of extracting and dispersive solvents is quickly added to an aqueous sample containing the desired analytes. This causes rapid dispersion of the extractant into very small droplets. The solution becomes 'cloudy', allowing for the efficient and rapid extraction of the analytes. The two phases are then separated (often assisted by centrifugation) and the compounds of interest are extracted into the organic phase. The organic phase can be additionally reconcentrated by partial evaporation of solvent. The obtained extract is injected into the GC–MS instrument. DLLME is a relatively simple and cost-effective method for extracting these compounds and is used in a variety of environmental analysis applications, including PAHs [76].

This method offers several benefits, including its simplicity, speed, ability to achieve high enrichment, and minimal usage of extraction solvent. Additionally, DLLME is compatible with many different analytical instruments and can be easily integrated with other sample preparation methods. Leong *et al* introduced a new method that combines DLLME with the solidification of floating organic drops (DLLME-SFO) [77]. Solvents with lower densities than water are used, and the extractant that floats on the top is solidified for easy collection and analysis. This technique offers a novel approach to microextraction and has potential for various applications.

13.4.2 Determination of carboxylic acids in water and wastewater by GC-based procedures

The task of reducing the impact of environmental pollutants, specifically those that are organic in nature, is difficult. There is a growing concern regarding wastewater containing carboxylic acids and other organic components [78]. Municipal wastewater often contains carboxylic acids, including volatile fatty acids, benzoic acid, and hydroxybenzoic acid [79]. Carboxylic acids have a strong odor, high toxicity, and can have harmful effects on aquatic environments. Therefore, various studies have been conducted to develop methods for their removal and detection at low concentrations [70, 87]. These efforts aim to improve the technology for managing carboxylic acids.

Carboxylic acid analysis in aqueous samples currently relies on chromatographic techniques, with GC being the primary method used. However, due to the unique properties of carboxylic acids, such as high polarity, high boiling point, and low volatility, only a small portion of these acids can undergo direct GC analysis. Specifically, only volatile fatty acids ranging from C1 to C12 can be analyzed directly, while the majority of carboxylic acids cannot [80]. To prepare analytes for analysis, it is often required to undergo derivatization techniques such as esterification, alkylation, or silylation. The commonly used approach is the use of alkyl silane derivatizing agents, which create unstable derivatives that require a reaction time of up to 24 h [81]. Alternative derivatization methods often involve highly toxic reagents [82] or reagents that have a low yield of derivatization [83]. Only a few of the accessible processes are ecologically friendly and free of the aforementioned flaws.

An example involves the utilization of alkyl chloroformates to facilitate esterification reactions [84]. Another popular method involves the formation of ion pairs through derivatization, using safe and non-toxic quaternary ammonium salts such as tetramethylammonium chloride (TMA-Cl), tetramethylammonium acetate (TMAAc), tetrabutylammonium hydrogen sulfate (TBA-HSO4), tetrabutylammonium chloride (TBA-Cl), tetrabutylammonium bromide (TBA-Br), or tetra butyl ammonium iodide (TBA-I) [85, 88]. The second method is advantageous because it is simple to modify by adding a derivatizing agent and a buffer solution. The resulting ions pairs are then transformed into esters through the use of a hot GC injection port. An effective extraction technique is needed to detect low amounts of organic acids in wastewater samples. This technique should follow the principles of green chemistry by being straightforward, quick, automated, and requiring minimal organic solvents. Dispersive liquid–liquid microextraction (DLLME) can meet these requirements [86, 87].

13.5 Conclusions

Modern analytical GC is a powerful method for analyzing industrial effluents. This method is widely used in environmental monitoring and assessment because it provides precise and accurate measurements of the chemical composition of complex mixtures. Sample preparation is a critical step in the analysis of complex samples by techniques such as GC. A wide range of sample preparation techniques are available, each with its own strengths and limitations. The choice of sample preparation technique will depend on the type of sample, the target compounds, and the desired level of accuracy and efficiency. Extraction techniques, such as SPE and LLE, are used to concentrate and isolate target compounds from the sample matrix. Derivatization techniques, such as silylation and acetylation, are used to modify target compounds to enhance their detectability by GC. Concentration techniques based on partial evaporation of extractant are used to reduce the volume of the sample prior to analysis.

Advances in sample preparation, such as the use of green solvents, automated sample preparation systems, microfluidic systems, and hybrid sample preparation techniques, have helped to improve the efficiency, accuracy, and sustainability of the analysis. In summary, sample preparation plays a crucial role in the success of GC analysis and careful consideration of the appropriate sample preparation technique is essential for obtaining reliable and meaningful results.

Many advances in GC detectors have occurred throughout the years, resulting in enhanced sensitivity, selectivity, and accuracy. Increased analytical capabilities and potential innovations for GC detectors have benefited several fields of science. Furthermore, advances in technology have made GC more accessible, faster, and more efficient, enabling scientists to analyze larger numbers of samples and generate more accurate results.

In conclusion, GC is an indispensable technique for the examination of industrial effluent water, and it will continue to play a crucial role in environmental monitoring and management throughout the coming decades.

References

[1] Matei E, Predescu A M, Şăulean A A, Râpă M, Sohaciu M G, Coman G, Berbecaru A-C, Predescu C, Vâju D and Vlad G 2022 Ferrous industrial wastes—valuable resources for water and wastewater decontamination *Int. J. Environ. Res. Public Health* **19** 13951

[2] Fernandes A, Gągol M, Makoś P, Khan J A and Boczkaj G 2019 Integrated photocatalytic advanced oxidation system (TiO_2/UV/O_3/H_2O_2) for degradation of volatile organic compounds *Sep. Purif. Technol.* **224** 1–14

[3] Fernandes A, Makoś P, Wang Z and Boczkaj G 2020 Synergistic effect of TiO_2 photocatalytic advanced oxidation processes in the treatment of refinery effluents *Chem. Eng. J.* **391** 123488

[4] Boczkaj G, Fernandes A and Makoś P 2017 Study of different advanced oxidation processes for wastewater treatment from petroleum bitumen production at basic pH *Ind. Eng. Chem. Res.* **56** 8806–14

[5] Fernandes A, Makoś P and Boczkaj G 2018 Treatment of bitumen post oxidative effluents by sulfate radicals based advanced oxidation processes (S-AOPs) under alkaline pH conditions *J. Clean. Prod.* **195** 374–84

[6] Fernandes A, Makoś P, Khan J A and Boczkaj G 2019 Pilot scale degradation study of 16 selected volatile organic compounds by hydroxyl and sulfate radical based advanced oxidation processes *J. Clean. Prod.* **208** 54–64

[7] Boczkaj G, Przyjazny A and Kamiński M 2014 New procedures for control of industrial effluents treatment processes *Ind. Eng. Chem. Res.* **53** 1503–14

[8] Moloantoa K M, Khetsha Z P, Van Heerden E, Castillo J C and Cason E D 2022 Nitrate water contamination from industrial activities and complete denitrification as a remediation option *Water (Basel)* **14** 799

[9] Poole C F 2021 Sample preparation for gas chromatography *Gas Chromatography* (IntechOpen) pp 615–53

[10] Hroboňová K, Jablonský M, Králik M and Vizárová K 2022 Advanced sampling, sample preparation and combination of methods applicable in analysis of compounds in aged and deacidified papers. A minireview *J. Cult. Herit.* **60** 95–107

[11] Namieśnik J, Górecki T, Biziuk M and Torres L 1990 Isolation and preconcentration of volatile organic compounds from water *Anal. Chim. Acta* **237** 1–60

[12] Alawa B, Galodiya M N and Chakma S 2022 Source reduction, recycling, disposal, and treatment *Hazardous Waste Management* (Amsterdam: Elsevier) pp 67–88

[13] Boczkaj G, Makoś P, Fernandes A and Przyjazny A 2017 New procedure for the examination of the degradation of volatile organonitrogen compounds during the treatment of industrial effluents *J. Sep. Sci.* **40** 1301–9

[14] Phelan A A, Meissner K, Humphrey J and Ross H 2022 Plastic pollution and packaging: Corporate commitments and actions from the food and beverage sector *J. Clean. Prod.* **331** 129827

[15] Adane T, Adugna A T and Alemayehu E 2021 Textile industry effluent treatment techniques *J. Chem.* **2021** 1–14

[16] Etteieb S, Magdouli S, Zolfaghari M and Brar S K 2020 Monitoring and analysis of selenium as an emerging contaminant in mining industry: a critical review *Sci. Total Environ.* **698** 134339

[17] Loni R, Najafi G, Bellos E, Rajaee F, Said Z and Mazlan M 2021 A review of industrial waste heat recovery system for power generation with organic rankine cycle: recent challenges and future outlook *J. Clean. Prod.* **287** 125070

[18] Chen Y, Guo Z, Wang X and Qiu C J J C A 2008 *Sample preparation Light Scattering from Polymer Solutions and Nanoparticle Dispersions* (Springer Laboratory vol 1184) (Berlin: Springer) pp 191–219

[19] McNair H M, Miller J M and Snow N H 2019 *Basic Gas Chromatography* (New York: Wiley)

[20] Antunes M, Sequeira M, de Caires Pereira M, Caldeira M J, Santos S, Franco J, Barroso M and Gaspar H J J A T 2021 Determination of selected cathinones in blood by solid-phase extraction and GC–MS *J. Anal. Toxicol.* **45** 233–42

[21] Boczkaj G, Makoś P and Przyjazny A 2016 Application of dispersive liquid–liquid microextraction and gas chromatography with mass spectrometry for the determination of oxygenated volatile organic compounds in effluents from the production of petroleum bitumen *J. Sep. Sci.* **39** 2604–15

[22] Di X, Wang X, Liu Y and Guo X J J C B 2019 Microwave assisted extraction in combination with solid phase purification and switchable hydrophilicity solvent-based homogeneous liquid-liquid microextraction for the determination of sulfonamides in chicken meat *J. Chromatogr. B Analyt. Technol. Biomed. Life Sci.* **1118** 109–15

[23] Wang C, Ding C, Wu Q and Xiong X J F A M 2019 Molecularly imprinted polymers with dual template and bifunctional monomers for selective and simultaneous solid-phase extraction and gas chromatographic determination of four plant growth regulators in plant-derived tissues and foods *Food Anal. Methods* **12** 1160–9

[24] Zhou T, Xiao X and Li G 2012 Microwave accelerated selective Soxhlet extraction for the determination of organophosphorus and carbamate pesticides in ginseng with gas chromatography/mass spectrometry *Anal. Chem.* **84** 5816–22

[25] Pires V P, Almeida R N, Wagner V M, Lucas A M, Vargas R M F and Cassel E J J E O R 2019 Extraction process of the Achyrocline satureioides (Lam) DC. essential oil by steam distillation: modeling, aromatic potential and fractionation *J. Essent. Oil Res.* **31** 286–96

[26] Orav A, Kailas T and Liiv M J C 1996 Analysis of terpenoic composition of conifer needle oils by steam distillation/extraction, gas chromatography and gas chromatography–mass spectrometry *Chromatographia* **43** 215–9

[27] Smith R M 1988 Supercritical fluid chromatography *The Application of Green Solvents in Separation Processes* (Amsterdam: Elsevier) pp 483 516

[28] Tejedor-Calvo E, García-Barreda S, Sánchez S, Morales D, Soler-Rivas C, Ruiz-Rodriguez A, Sanz M Á, Garcia A P, Morte A and Marco P J L W T 2021 Supercritical CO2 extraction method of aromatic compounds from truffles *LWT* **150** 111954

[29] Snow N H and Bullock G P J J C A 2010 Novel techniques for enhancing sensitivity in static headspace extraction-gas chromatography *J. Chromatogr.* A **1217** 2726–35

[30] Moret S, Scolaro M, Barp L, Purcaro G and Conte L S 2016 Microwave assisted saponification (MAS) followed by on-line liquid chromatography (LC)–gas chromatography (GC) for high-throughput and high-sensitivity determination of mineral oil in different cereal-based foodstuffs *Food Chem.* **196** 50–7

[31] Sajid M, Płotka-Wasylka J J T T and A C 2018 Green' nature of the process of derivatization in analytical sample preparation *TrAC—Trends Anal. Chem.* **102** 16–31

[32] Plechkova N V and Seddon K R 2007 Ionic liquids: 'designer' solvents for green chemistry *Methods and Reagents for Green Chemistry: An Introduction* (Hoboken, NJ: Wiley) pp 103–30

[33] Ullah S, Haq H U, Salman M, Jan F, Safi F, Arain M B, Khan M S, Castro-Muñoz R and Boczkaj G 2022 Ultrasound-assisted dispersive liquid-liquid microextraction using deep eutectic solvents (DESs) for neutral red dye spectrophotometric determination *Molecules* **27** 6112

[34] Ul Haq H, Bibi R, Balal Arain M, Safi F, Ullah S, Castro-Muñoz R and Boczkaj G 2022 Deep eutectic solvent (DES) with silver nanoparticles (Ag-NPs) based assay for analysis of lead (II) in edible oils *Food Chem.* **379** 132085

[35] Haq H U, Balal M, Castro-Muñoz R, Hussain Z, Safi F, Ullah S and Boczkaj G 2021 Deep eutectic solvents based assay for extraction and determination of zinc in fish and eel samples using FAAS *J. Mol. Liq.* **333** 115930

[36] Faraz N, Haq H U, Balal Arain M, Castro-Muñoz R, Boczkaj G and Khan A 2021 Deep eutectic solvent based method for analysis of Niclosamide in pharmaceutical and wastewater samples—a green analytical chemistry approach *J. Mol. Liq.* **335** 116142

[37] Vargas-Muñoz M A, Cerdà V, Cadavid-Rodríguez L S and Palacio E 2021 Automated method for volatile fatty acids determination in anaerobic processes using in-syringe magnetic stirring assisted dispersive liquid-liquid microextraction and gas chromatography with flame ionization detector *J. Chromatogr.* A **1643** 462034

[38] Jumhawan U, Putri S P, Yusianto , Bamba T and Fukusaki E 2015 Application of gas chromatography/flame ionization detector-based metabolite fingerprinting for authentication of Asian palm civet coffee (Kopi Luwak) *J. Biosci. Bioeng.* **120** 555–61

[39] Sonawane S, Fedorov K, Rayaroth M P and Boczkaj G 2022 Degradation of 1,4-dioxane by sono-activated persulfates for water and wastewater treatment applications *Water Resour. Ind.* **28** 100183

[40] Fedorov K, Rayaroth M P, Shah N S and Boczkaj G 2023 Activated sodium percarbonate-ozone (SPC/O$_3$) hybrid hydrodynamic cavitation system for advanced oxidation processes (AOPs) of 1,4-dioxane in water *Chem. Eng. J.* **456** 141027

[41] Fedorov K, Plata-Gryl M, Khan J A and Boczkaj G 2020 Ultrasound-assisted heterogeneous activation of persulfate and peroxymonosulfate by asphaltenes for the degradation of BTEX in water *J. Hazard. Mater.* **397** 122804

[42] Fedorov K, Sun X and Boczkaj G 2021 Combination of hydrodynamic cavitation and SR-AOPs for simultaneous degradation of BTEX in water *Chem. Eng. J.* **417** 128081

[43] Sevcik J G K 2011 *Detectors in Gas Chromatography* (Amsterdam: Elsevier)

[44] Brody S S and Chaney J E 1966 Flame photometric detector: the application of a specific detector for phosphorus and for sulfur compounds—sensitive to subnanogram quantities *J. Chromatogr. Sci.* **4** 42–6

[45] Banerjee P and Prasad B 2020 Determination of concentration of total sodium and potassium in surface and ground water using a flame photometer *Appl. Water Sci.* **10** 113

[46] Caton R D and Bremmer R W 1954 Some interferences in flame photometry *Anal. Chem.* **26** 805–13

[47] Clark A G and Thurbide K B 2015 An improved multiple flame photometric detector for gas chromatography *J. Chromatogr. A* **1421** 154–61

[48] Snow N H 2020 *Basic Multidimensional Gas Chromatography* (New York: Academic)

[49] Boczkaj G, Makoś P, Fernandes A and Przyjazny A 2016 New procedure for the control of the treatment of industrial effluents to remove volatile organosulfur compounds *J. Sep. Sci.* **39** 3946–56

[50] Sapahin H A, Makahleh A and Saad B 2019 Determination of organophosphorus pesticide residues in vegetables using solid phase micro-extraction coupled with gas chromatography–flame photometric detector *Arab. J. Chem.* **12** 1934–44

[51] Prapamontol T, Sutan K, Laoyang S, Hongsibsong S, Lee G, Yano Y, Hunter R E, Ryan P B, Barr D B and Panuwet P 2014 Cross validation of gas chromatography–flame photometric detection and gas chromatography–mass spectrometry methods for measuring dialkylphosphate metabolites of organophosphate pesticides in human urine *Int. J. Hyg. Environ. Health* **217** 554–66

[52] Gao Z, Deng Y, Yuan W, He H, Yang S and Sun C 2014 Determination of organo-phosphorus flame retardants in fish by pressurized liquid extraction using aqueous solutions and solid-phase microextraction coupled with gas chromatography-flame photometric detector *J. Chromatogr. A* **1366** 31–7

[53] Maier-Bode H and Riedmann M 1975 Gas chromatographic determination of nitrogen-containing pesticides using the nitrogen flame ionization detector (N-FID) *Residue Reviews* ed F A Gunther and J D Gunther (New York: Springer) pp 113–81

[54] Aue W A, Gehrke C W, Tindle R C, Stalling D L and Ruyle C D 1967 Application of the alkali-flame detector to nitrogen containing compounds *J. Chromatogr. Sci.* **5** 381–2

[55] Conte E D and Barry E F 1993 Alkali flame ionization detector for gas chromatography using an alkali salt aerosol as the enhancement source *J. Chromatogr. A* **644** 349–55

[56] Kolb B and Bischoff J 1974 A new design of a thermionic nitrogen and phosphorus detector for GC *J. Chromatogr. Sci.* **12** 625–9

[57] Burgett C A, Smith D H and Bente H B 1977 The nitrogen-phosphorus detector and its applications in gas chromatography *J. Chromatogr. A* **134** 57–64

[58] Anon 1982 *Method 607, Nitrosamines. Code of Federal Regulations: Protection of the Environment, Part 136, Title 40, US GPO* (Washington, DC)

[59] Greaves J C and Garvin D 1959 Chemically induced molecular excitation: excitation spectrum of the nitric oxide-ozone system *J. Chem. Phys.* **30** 348–9

[60] Fine D H, Lieb D and Rufeh F 1975 Principle of operation of the thermal energy analyzer for the trace analysis of volatile and non-volatile N-nitroso compounds *J. Chromatogr. A* **107** 351–7

[61] Jönsson S, Gustavsson L and van Bavel B 2007 Analysis of nitroaromatic compounds in complex samples using solid-phase microextraction and isotope dilution quantification gas chromatography–electron-capture negative ionisation mass spectrometry *J. Chromatogr. A* **1164** 65–73

[62] Berg M, Bolotin J and Hofstetter T B 2007 Compound-specific nitrogen and carbon isotope analysis of nitroaromatic compounds in aqueous samples using solid-phase microextraction coupled to GC/IRMS *Anal. Chem.* **79** 2386–93

[63] Cortada C, Vidal L and Canals A 2011 Determination of nitroaromatic explosives in water samples by direct ultrasound-assisted dispersive liquid–liquid microextraction followed by gas chromatography–mass spectrometry *Talanta* **85** 2546–52

[64] Ebrahimzadeh H, Yamini Y, Kamarei F and Khalili-Zanjani M 2007 Application of headspace solvent microextraction to the analysis of mononitrotoluenes in waste water samples *Talanta* **72** 193–8

[65] Peng X T, Zhao X and Feng Y Q 2011 Preparation of phenothiazine bonded silica gel as sorbents of solid phase extraction and their application for determination of nitrobenzene compounds in environmental water by gas chromatography–mass spectrometry *J. Chromatogr.* A **1218** 9314–20

[66] Psillakis E, Mantzavinos D and Kalogerakis N 2004 Development of a hollow fibre liquid phase microextraction method to monitor the sonochemical degradation of explosives in water *Anal. Chim. Acta* **501** 3–10

[67] Zhang H, Ren S, Yu J and Yang M 2012 Occurrence of selected aliphatic amines in source water of major cities in China *J. Environ. Sci.* **24** 1885–90

[68] Rubio L, Sanllorente S, Sarabia L A and Ortiz M C 2014 Optimization of a headspace solid-phase microextraction and gas chromatography/mass spectrometry procedure for the determination of aromatic amines in water and in polyamide spoons *Chemometr. Intell. Lab. Syst.* **133** 121–35

[69] Feng W, Jiang R, Chen B and Ouyang G 2014 Fiber-assisted emulsification microextraction coupled with gas chromatography–mass spectrometry for the determination of aromatic amines in aqueous samples *J. Chromatogr.* A **1361** 16–22

[70] Makoś P, Fernandes A, Przyjazny A and Boczkaj G 2018 Sample preparation procedure using extraction and derivatization of carboxylic acids from aqueous samples by means of deep eutectic solvents for gas chromatographic-mass spectrometric analysis *J. Chromatogr.* A **1555** 10–9

[71] Makoś P, Przyjazny A and Boczkaj G 2019 Methods of assaying volatile oxygenated organic compounds in effluent samples by gas chromatography—a review *J. Chromatogr.* A **1592** 143–60

[72] Boczkaj G, Makoś P and Przyjazny A 2016 Application of dynamic headspace and gas chromatography coupled to mass spectrometry (DHS-GC-MS) for the determination of oxygenated volatile organic compounds in refinery effluents *Anal. Methods* **8** 3570–7

[73] Wong P K and Wang J 2001 The accumulation of polycyclic aromatic hydrocarbons in lubricating oil over time—a comparison of supercritical fluid and liquid–liquid extraction methods *Environ. Pollut.* **112** 407–15

[74] Makoś P, Fernandes A and Boczkaj G 2018 Method for the simultaneous determination of monoaromatic and polycyclic aromatic hydrocarbons in industrial effluents using dispersive liquid–liquid microextraction with gas chromatography–mass spectrometry *J. Sep. Sci.* **41** 2360–7

[75] Makoś P, Przyjazny A and Boczkaj G 2018 Hydrophobic deep eutectic solvents as 'green' extraction media for polycyclic aromatic hydrocarbons in aqueous samples *J. Chromatogr.* A **1570** 28–37

[76] Barro R, Regueiro J, Llompart M and Garcia-Jares C 2009 Analysis of industrial contaminants in indoor air: Part 1. Volatile organic compounds, carbonyl compounds, polycyclic aromatic hydrocarbons and polychlorinated biphenyls *J. Chromatogr.* A **1216** 540–66

[77] Fattahi N, Assadi Y, Hosseini M R M and Jahromi E Z 2007 Determination of chlorophenols in water samples using simultaneous dispersive liquid-liquid microextraction and derivatization followed by gas chromatography-electron-capture detection *J. Chromatogr.* A **1157** 23–9

[78] Karunanithi S, Kapoor A and Delfino P 2019 Separation of carboxylic acids from aqueous solutions using hollow fiber membrane contactors *J. Membr. Sci. Res.* **5** 233–9

[79] Ábalos M, Bayona J M and Pawliszyn J 2000 Development of a headspace solid-phase microextraction procedure for the determination of free volatile fatty acids in waste waters *J. Chromatogr.* A **873** 107–15

[80] Ullah M A, Kim K-H, Szulejko J E and Cho J 2014 The gas chromatographic determination of volatile fatty acids in wastewater samples: evaluation of experimental biases in direct injection method against thermal desorption method *Anal. Chim. Acta* **820** 159–67

[81] Latorre A, Rigol A, Lacorte S and Barceló D 2003 Comparison of gas chromatography–mass spectrometry and liquid chromatography–mass spectrometry for the determination of fatty and resin acids in paper mill process waters. *J. Chromatogr.* A **991** 205–15

[82] Ngan F and Ikesaki T 1991 Determination of nine acidic herbicides in water and soil by gas chromatograpy using an electron-capture detector *J. Chromatogr.* A **537** 385–95

[83] Ferreira A M C, Laespada M E F, Pavón J L P and Cordero B M 2013 In situ aqueous derivatization as sample preparation technique for gas chromatographic determinations *J. Chromatogr.* A **1296** 70–83

[84] Shah M P 2020 *Microbial Bioremediation and Biodegradation* (Berlin: Springer)

[85] Shah M P 2021 *Removal of Refractory Pollutants from Wastewater Treatment Plants* (Boca Raton, FL: CRC Press)

[86] Husek P and Simek P 2006 Alkyl chloroformates in sample derivatization strategies for GC analysis. Review on a decade use of the reagents as esterifying agents *Curr. Pharm. Anal.* **2** 23–43

[87] Makoś P, Fernandes A and Boczkaj G 2017 Method for the determination of carboxylic acids in industrial effluents using dispersive liquid-liquid microextraction with injection port derivatization gas chromatography–mass spectrometry *J. Chromatogr.* A **1517** 26–34

[88] Rezaee M, Assadi Y, Milani Hosseini M R, Aghaee E, Ahmadi F and Berijani S 2006 Determination of organic compounds in water using dispersive liquid-liquid microextraction *J. Chromatogr.* A **1116** 1–9

IOP Publishing

Trends in Biological Processes in Industrial Wastewater Treatment

Maulin P Shah

Chapter 14

Novel anammox-based biological nitrogen removal process for high-strength industry wastewater treatment

Zhetai Hu and Shihu Hu

Industrial development enables a human life that is full of variety but it also generates a large amount of unmanageable industrial wastewater. In the case of wastewater containing a high concentration of nitrogen, the conventional treatment process is often costly due to the high aeration demand and inefficient because of the inhibitory effect of ammonium, nitrite, or nitrate on the microbes in the engineering system. The anammox-based process, which can efficiently remove nitrogen with low aeration requirements and less organic carbon, has been successfully demonstrated in treating many types of industrial wastewater. Based on the literature that has been published so far, this review summarizes and discusses the performance of the anammox-based process in treating various industrial wastewaters. The main challenges in the application of the anammox-based process in treating industrial wastewater are then identified. Finally, the future prospects of applying anammox -based approaches in industrial wastewater treatment are proposed.

14.1 Introduction

Anammox bacteria were first discovered in the early-1990s in a denitrifying pilot plant in the Netherlands. Anammox microorganisms can autotrophically convert ammonium and nitrite to nitrogen gas ($NH_4^+ + 1.32NO_2^- \rightarrow 1.02N_2 + 0.26NO_3^- + 2.03H_2O$) (Mulder *et al* 1995). Many anammox-based nitrogen removal processes have since been proposed and demonstrated (Cao and Zhou 2019, Liu *et al* 2019). Anammox-based processes are known to be energy- and carbon-efficient for nitrogen removal. For example, the partial nitrition and anammox (PN/A) process, can save the demand of oxygen and organic carbon by 60% and 100%, respectively, compared to the conventional nitrification-denitrification (N/DN) process.

Industrialization and urbanization have resulted in a substantial increase in the diversity and amount of industrial wastewater being generated. Some industrial wastewater contains high ammonium and low biodegradable carbon, such as liquor from anaerobic digestion (AD) in wastewater treatment plants (WWTPs) (referred as sidestream wastewater below), mature landfill leachate, and pharmaceutical wastewater (Ren *et al* 2022b). Removing nitrogen from these industrial wastewaters using conventional N/DN methods is not only uneconomical but also environmentally unsustainable due to the large amounts of external carbon sources required. To address this issue, anammox-based technologies have been proposed as an alternative solution (Li *et al* 2018).

Extensive research has been dedicated to evaluating the feasibility of implementing anammox-based approaches for nitrogen removal in high-strength industrial wastewater (Li *et al* 2018). Ammonium is typically the dominating nitrogen source in industrial wastewater and, as such, the use of the PN/A process has emerged as an optimal solution for its removal. The key for achieving PN/A process is to selectively suppress nitrite oxidizing bacteria (NOB), which is relatively easy to achieve in high-strength wastewater condition by introducing NOB inhibitor, including low DO, free ammonia (FA), and free nitrite acid (FNA). Indeed, PN/A is the most widely used approach for eliminating nitrogen from industrial wastewater at present (Li *et al* 2018). Hundreds of full-scale PN/A processes have been installed and operated to efficiently remove nitrogen from industrial wastewater (Lackner *et al* 2014, Ren *et al* 2022a). For industrial wastewater containing high nitrate, another anammox--based technology, partial denitrification and anammox (PD/A), has been proposed and successfully demonstrated (Cao and Zhou 2019). In addition, as a polish process, PD/A process can eliminate the nitrate from the PN/A process, thereby further improving the nitrogen removal efficiency (NRE) of the anammox-based process (Ren *et al* 2022a).

Although the PN/A and PD/A process have been demonstrated to be promising treatment options, previous works have shown that the toxic substances that are commonly present in industrial wastewater, including refractory organics and antibiotics, can significantly reduce the activity of anammox bacteria (Li *et al* 2018, Fu *et al* 2021). This led to a decrease in the efficiency of anammox-based nitrogen removal process and an increase in the cost of detoxifying industrial wastewater, which needs to be addressed through process design and special operational strategies.

This review aims to summarize the critical findings of applying anammox-based processes to treat four different types of industrial wastewater, i.e., sidestream wastewater, landfill leachate, pharmaceutical wastewater, and swine wastewater. Furthermore, the key challenges limiting the development of anammox-based process in industrial wastewater treatment are revealed. Finally, several prospects of future applications of anammox-based process for industrial wastewater treatment are proposed.

14.2 Removing nitrogen from industrial wastewater using an anammox-based process

14.2.1 Sidestream wastewater

AD is widely used to manage sludge in large-scale WWTPs. This process can reduce the amount of sludge, recover the organic carbon as bioenergy, and stabilize the sludge. The digested sludge will be dewatered, which generates dewatered sludge to be further processed (e.g. land application) and high-ammonia-content sidestream wastewater. Typically, the ammonia and chemical oxygen demand (COD) concentrations are 500–1500 mg N l^{-1} and 400–1000 mg COD l^{-1} in sidestream wastewater, respectively (Lackner *et al* 2014, Zhang *et al* 2016, Qiu *et al* 2021). This means the sidestream wastewater is so deficient in organic carbon that the ammonia cannot be removed via the conventional N/DN process.

The PN/A process is a promising technology to manage the sidestream wastewater because it does not rely on organic carbon to remove nitrogen. It took about 3.5 years from 2002 to setup the first full-scale PN/A process in Rotterdam, Netherlands (Abma *et al* 2007). It was a two-stage PN/A process treating sidestream wastewater, using granular anammox. The volumetric loading rate of this anammox reactor reached above 10 kg N m^{-3} d^{-1}, with a NRE of 90%–95%. After that, the full-scale installations of PN/A process increased rapidly. By 2014, more than 100 full-scale installations of the PN/A process were operated worldwide, with 75% of them setup for sidestream wastewater treatment (Lackner *et al* 2014).

Table 14.1 summarises the performance of some full-scale PN/A processes treating sidestream wastewater. The PN/A process has two configurations: one-stage, where PN and anammox reactions occur in a single reactor, and two-stage, where PN and anammox reactions are separated in two reactors. For full-scale applications, both one-stage and two-stage configurations can effectively remove the ammonia in the sidestream wastewater, with a NRE of >82% and nitrogen loading rate (NLR) of >0.21 kg N m^{-3} d^{-1}. However, the NOB suppression mechanism is different between the one-stage and two-stage processes. The one-stage process mainly relies on the low DO control and the introduction of anammox bacteria as a substance (i.e., nitrite) competitor to selectively suppress NOB (table 14.1) (Joss *et al* 2009, Wang *et al* 2022b). There are two common operating modes in the PN process of the two-stage configuration, i.e., continuous stirred tank reactor (CSTR) and sequencing batch reactor (SBR). The PN process in the CSTR configuration is maintained using an operating condition of low sludge retention time (SRT) and high temperature (30°C–35°C), in which NOB can be selectively washed out from the system because of its lower growth rate compared to that of ammonia oxidizing bacteria (AOB). The NOBs in SBR are effectively inhibited by the alternant effects of FA, formed at a condition of high pH and high-ammonia concentration after feeding, and FNA, formed at a condition of low pH and high nitrite concentration at the end of aeration phase. Both FA and FNA can selectively suppress NOB, which has been widely demonstrated in the literature (Wang *et al* 2014, Wang *et al* 2017, Duan *et al* 2019, Hu *et al* 2023).

Table 14.1. Summary of full-scale anammox-based processes for sidestream wastewater treatment.

Influent NH_4^+ concentration (mg N l^{-1})	Process	Nitrogen loading rate (kg N m^{-3} d^{-1})	Nitrogen removal efficiency (%)	NOB suppression strategy	References
650 ± 50	One-stage PN/A (Floc sludge; SBR)	~0.45	~94.6%	DO < 1 mg O_2 l^{-1} NH_4^+ > 10 mg N l^{-1}	Joss et al (2009)
890 ± 100	One-stage PN/A (Floc sludge; SBR)	~0.36	~82.7%		
760 ± 75	One-stage PN/A (Floc sludge; SBR)	~0.35	~90.8%		
1200	Two-stage PN/A (Granular anammox; SBR)	up to 10	>90%	SHARON system: low SRT and high temperature	Abma et al (2007), Van der Star et al (2007)
255–705	One-stage PN/A (Floc sludge and biofilms; IFAS)	0.48	85%	DO: 0.3–0.5 mg O_2 l^{-1}	Zhang et al (2015)
1407	One-stage PN/A (Floc sludge and biofilms; IFAS)	0.21	>85%	DO: 0.1–0.3 mg O_2 l^{-1}	Han et al (2020)
983 ± 233	Two-stage PNA (Granular PN + MBBR anammox)	~1.86	~87%	SHARON system: low SRT	Jung et al (2021)

Note: IFAS: integrated fixed-film activated sludge; MBBR: moving bed biofilm reactor; SHARON: single reactor system for high activity ammonium removal over nitrite.

Although both one-stage and two-stage PN/A processes showed excellent performance in removing ammonium from sidestream wastewater, the one-stage configuration, accounting for >80% of installed full-scale PN/A process, is more popular than the two-stage process, for several reasons. First, the PN process of the two-stage configuration often contains a high concentration of nitrite, which makes it challenging to suppress NOB because its substance is unlimited. Second, the accumulation of nitrite significantly increases the emission of nitrous oxide (N_2O), which is about 300 times as potent as carbon dioxide in terms of greenhouse effect. Finally, since high nitrite concentration is toxic to the anammox bacteria, unstable control of the PN process may increase the nitrite concentration in the following anammox process, thereby reducing its efficiency. In comparison, the one-stage process has relatively low *in situ* nitrite concentration because the nitrite produced by AOB is immediately consumed by the anammox bacteria.

14.2.2 Landfill leachate

Landfill leachate can be divided into young, medium, and mature leachate according to the age of the landfill (Ren *et al* 2022a). As the age of the landfill grows, the concentration of the organic matter in the leachate decreases, while the nitrogen (mainly ammonium) concentration is nearly constant (Ma *et al* 2022). The ratio of COD to ammonia nitrogen (C/N) in young, medium, and mature leachate is above 10, 5–10, and 3–5, respectively (Ma *et al* 2022). Although traditional N/DN processes can remove nitrogen from landfill leachate using the organic carbon present, the anammox-based process offers several advantages as an energy- and carbon-efficient nitrogen removal technology. Therefore, implementing anammox--based processes for nitrogen removal in landfill leachate holds promise for achieving superior results and multiple benefits. After upgrading from traditional N/DN to anammox-based technology, the specific energy consumption of a reported full-scale landfill leachate plant decreased from 1.6 to 0.2 kWh m^{-3}, and the additional organic carbon requirement and sludge production reduced by 91% and 96%, respectively (Azari *et al* 2017).

Table 14.2 summarizes the performance of several anammox-based technologies for landfill leachate treatment. The ammonium concentration in landfill leachate is about 1000–2000 mg N l^{-1}, which can be efficiently removed by anammox-based technology, with a NRE of higher than 80% in laboratory and pilot-scale studies. There are several full-scale applications worldwide that show excellent nitrogen removal performance with NLR of 0.4–0.7 kg N m^{-3} d^{-1} and NRE of 75%–94%. The anammox process configuration of the one-stage and two-stage processes, and the anammox bacteria morphology of floc, granule, and biofilm are all widely applied to treat leachate (Ren *et al* 2022a).

The most popular anammox-based technology for treating landfill leachate is the PN/A process, normally using the low DO control, intermittent aeration, FNA, and FA to suppress NOB. In addition, because organic carbon is typically abundant in leachate, it is common to combine the PN/A process with other biological nitrogen conversion processes, such as partial and full denitrification. By coupling with the denitrification, the NRE of the anammox-based bioprocess reached nearly 99%, with about 15% of the nitrogen removed via the denitrification process (Zhang *et al* 2017, 2019a). The addition of easily biodegradable organic carbon triggered the PD/A process, which further reduced the nitrate concentration in the effluent of PN/A process from 111.9 to 4.0 mg N l^{-1}, therefore achieving a total NRE of 98.8% (Wang *et al* 2020).

Landfill leachate contains a large amount of dissolved and particle organic matter, potentially impacting anammox's activity due to its biorefractory characteristic. A pretreatment unit is normally introduced to reduce the concentration of organic carbon or recover the organic carbon as bioenergy (i.e., methane) (Wu *et al* 2019, Li *et al* 2020b, Wang *et al* 2022a). In this way, the impact of organic carbon on

Table 14.2. Summary of the performance of anammox-based processes for mature landfill leachate treatment.

Influent NH$_4^+$ concentration (mg N l^{-1})	Process	Nitrogen loading rate (kg N m^{-3} d^{-1})	Nitrogen removal efficiency (%)	Temperature	Scale	NOB suppression strategy	References
1950 ± 250	One-stage SBR (SPNAD Floc Sludge)	~0.13	99 ± 0.1	28–31	Lab scale	Low DO: <0.5 and Intermittent aeration	Zhang et al (2017)
1000 ± 250	One-stage SBR (SPNAD Floc Sludge)	~0.23	98.7	25–31	Lab scale	Low DO: <0.5and Intermittent aeration	Zhang et al (2019a)
2250 ± 200	D-PN/A(Granular Sludge)	0.45	96.7	34	Lab scale	Low DO: 0.2–0.4	Li et al (2021)
1370.8 ± 188.7	D-PN-A(Floc Sludge)	0.49 ± 0.02	92.1 ± 2.1	25 ± 5	Lab scale	Low DO: 0.8–1.0	Choi et al (2022)
1500	Two-stage PN/A (Granular Sludge)	10	~95.2	25–30	Pilot scale	—	Phan et al (2017)
1500 ± 400	Two-stage PNA (Floc Sludge)	0.4 ± 0.5	86 ± 9	20–35	Full scale	FA	Magri et al (2021)
758	Two-stage PNA (Granular sludge)	0.71 ± 0.1	94 ± 2.7	31–34	Full scale	Low DO: 0.1–0.4	Azari et al (2017)
634 ± 143	One-stage PNA (IFAS)	0.39–0.62	75	28–35	Full scale	Low DO: 0.3	Kaewyai et al (2022)

Note: SPNAD: simultaneous partial nitritation, anaerobic ammonium oxidation and denitrification; D: denitrification; IFAS: integrated fixed-film activated sludge.

the downstream anammox process can be relieved. For instance, in a two-stage PN/A process, the addition of a pre-denitrification unit increased the NRR from 0.31 to 0.45 kg N m^{-3} d^{-1} and NRE from 76.3% to 96.7% (Li et al 2020b).

14.2.3 Pharmaceutical wastewater

The high diversity of the pharmaceutical industry leads to the production of many different kinds of pharmaceutical wastewater with a wide range of ammonium concentrations, from a few hundred to several tens of thousands mg nitrogen per liter (Mai et al 2020, Zhang et al 2022, Chen et al 2023b). The PN/A process is the most widely used anammox-based technology for handling pretreated ammonium-rich pharmaceutical wastewater. In addition, the feasibility of using the PD/A process to treat pharmaceutical wastewater containing high concentrations of nitrate has also been experimentally demonstrated (Zhang et al 2022).

The summary of system performance in table 14.3 indicates that nitrogen in pharmaceutical wastewaters can be efficiently eliminated using anammox-based technology. The NLR can reach 9.4 kg N m^{-3} d^{-1} and the NRE can be as high as 90% (Tang et al 2011). Due to its high toxicity, pharmaceutical wastewater normally needs to be pretreated by AD or conventional N/DN, or diluted with other low-strength wastewater, before being fed to the anammox-based process (Tang et al 2011, Mai et al 2020, Zuo et al 2020, Zhang et al 2022, Chen et al 2023a). For example, the potential anammox inhibitors, organic carbon and oxytetracycline (OTC), were reduced by about 91.4% and 27.9%, respectively, after AD treatment (Chen et al 2021a).

Anammox biomass in granule structure has been recommended by the literature to alleviate the inhibition of pharmaceutical wastewater to anammox bacteria (Chen et al 2023a). Chen et al (2023a) observed that the stable granular structure can retain high biomass, and anammox bacteria can produce more humic acid-like substance to generate a fine and coarse granule in response to the inhibitory conditions.

14.2.4 Swine wastewater

The waste from large-scale pig farms is normally collected for bioenergy or valuable chemical recovery in AD (Ishimoto et al 2020). After waste collection, the waste storage room needs to be flushed, generating a considerable amount of flushing wastewater, also called manure-free piggery wastewater (Meng et al 2015). Therefore, two streams of swine wastewater are generated: AD liquor and manure-free piggery wastewater. The ammonium concentration in AD liquor is above 2000 mg N l^{-1}, while the COD within AD liquor is normally refractory, and the ratio of COD to nitrogen of manure-free piggery wastewater is often below 3.6 (Deng et al 2019). This means that nitrogen in swine wastewater is difficult to remove through the traditional N/DN process without an external carbon source, which is limited by the low ratio of biodegradable COD to nitrogen. Thus, anammox-based technology is a promising approach to remove nitrogen in swine wastewater.

Table 14.4 shows the nitrogen removal performance of anammox-based processes in treating different kinds of swine wastewater. Ammonium is the primary nitrogen

Table 14.3. Summary of the performance of anammox-based processes for pharmaceutical wastewater treatment.

Type of wastewater	Influent N concentration (mg N l^{-1})	Process	Nitrogen loading rate (kg N m^{-3} d^{-1})	Nitrogen removal efficiency (%)	Pretreatment	NOB suppression strategy	References
Diluted Vitamin B2	NH_4^+: 329 NO_2^-: 9.2 NO_3^-: 32	PN/A	0.35	75	AD	—	Mai et al (2020)
Colistin sulfate and kitasamycin	NH_4^+: 123–257 NO_2^-: 133–264 NO_3^-: 0–32	Anammox	9.4	~90	N/DN and PN	—	Tang et al (2011)
Chlortetracycline wastewater	NH_4^+: 524 ± 106	PN/A	1.0	75.2	—	DO: 0.4–0.8	Zuo et al (2020)
Diluted bismuth nitrate and bismuth potassium citrate	NH_4^+: 77.9 ± 2.6 NO_3^-: 104.1 ± 4.4	PD/A	~0.2	81.2 ± 6.3	—	—	Zhang et al (2022)
Oxytetracycline wastewater	NH_4^+: 798 ± 17	PN/A	~0.7	80 ± 12.78	AD	—	Chen et al (2023a)

Table 14.4. Summary of the performance of anammox-based processes for swine wastewater treatment.

Influent N concentration (mg N l⁻¹)	Process	Nitrogen loading rate (kg N m⁻³ d⁻¹)	Nitrogen removal efficiency (%)	Type of swine wastewater	Scale	NOB suppression strategy	References
NH_4^+: 299.7	One-stage anammox process	~1.1	~85.5	manure-free piggery wastewater	Lab scale	Low DO < 1	Meng et al (2015)
NH_4^+: 276.8	One-stage PN/A	~0.2	91.3 ± 1.2	manure-free piggery wastewater	Lab scale	DO control: 2.5	Deng et al (2019)
NH_4^+: 670 ± 16 NO_2^-: 700 ± 240 NO_3^-: 1650 ± 580	Two-stage PN/A	~1.5	~80	AD liquor pretreated by aeration	Lab scale	—	Molinuevo et al (2009)
NH_4^+: 2000 ± 30	One-stage PN/A	3.27 ± 0.13	83	AD liquor	Lab scale	Intermittent aeration and Low DO control: 0.5 ± 0.5 mg l⁻¹	Chini et al (2020)
NH_4^+: 373 ± 63	AAO with anammox	0.03	83.5 ± 6.7	AD liquor	Full scale	Low DO control: 0.3–0.5	Chen et al (2021b)
NH_4^+: 457 ± 128	One-stage PN/A	0.1	~70	—	Full scale	0.06–2.0	Ishimoto et al (2020)

source in swine wastewater, with a concentration of about several thousand milligrams per litre in the AD liquor and several hundred in the manure-free piggery wastewater. As shown in table 14.4, nitrogen in both AD liquor and manure-free piggery wastewater can be efficiently eliminated via the anammox--based approach, showing a NRE of higher than 80% in laboratory scale experiments and above 70% in full-scale applications.

PN/A is the most popular anammox-based nitrogen removal approach for swine wastewater treatment. The NLR reached 3.27 ± 0.13 kg N m^{-3} d^{-1} in a laboratory one-stage PN/A process (Chini *et al* 2020). For full-scale application, a NLR of about 0.1 kg N m^{-3} d^{-1} was achieved in a one-stage PN/A process (Ishimoto *et al* 2020). A denitrification process is normally introduced to remove the nitrate produced by anammox and NOB, thereby further increase the NRE (Chen *et al* 2021b). In a full-scale swine AD liquor treatment plant, with upgraded intermittent aeration and low DO control, the anammox process gradually dominates the nitrogen removal process, which increases the overall NRE from $65.5\% \pm 6.0\%$ to $83.5\% \pm 6.7\%$ (Chen *et al* 2021b). As a result, this simple upgrade reduces the energy consumption of treating swine wastewater from 1.93 to 0.9 kWh m^{-3}, or 4.18 to 2.57 kWh kg^{-1} N.

14.3 The main challenges of applying an anammox-based process in treating industrial wastewater

As mentioned earlier, the anammox-based process has been widely used in treating many different kinds of industrial wastewater. The success of many full-scale applications demonstrates that upgrading the nitrogen removal approach from the conventional N/DN process to the anammox-based process can save a considerable amount of cost. However, there are several challenges that limit the range of applications of the anammox-based process.

14.3.1 NOB suppression

PN/A is the most popular anammox-based process for removing nitrogen in ammonium-rich industrial wastewater. Notably, stable NOB suppression is critical for satisfactory nitrogen performance of the PN/A process. Normally, the one-stage PN/A process relies on low DO control and intermittent aeration to control NOB. For the two-stage PN/A process, apart from aeration control, an additional NOB inhibitor, namely, FA or FNA, is frequently applied for NOB suppression. However, some studies observed that NOB was able to adapt to the low DO, FA, and FNA conditions, making the established PN/A system unstable (Liu and Wang 2013, Ma *et al* 2017, Li *et al* 2020a). *Nitrobacter*, *Nitrospira*, *Candidatus* (Ca.) Nitrotoga, and *Nitrolancea* are four common NOBs in the wastewater treatment process, with different levels of tolerance to the different inhibitors. Under the specific inhibitory condition, the shift of the dominating NOB species enables adaptation (Zheng *et al* 2020). Therefore, multiple NOB control strategies are recommended to simultaneously suppress different NOB species to achieve long term stable performance.

14.3.2 Substrate inhibition

Ammonium and nitrite, which are two substrates for anammox, also show inhibitory effects on anammox when their concentrations are too high (Fernández *et al* 2012). Indeed, FA rather than ammonium is regarded as the real inhibitor for anammox bacteria. FA concentration at 38 and 100 mg N l^{-1} reduced the anammox activity by 50% and 80%, respectively (Fernández *et al* 2012). FA of 38 mg N l^{-1} can be achieved at a condition of ammonium of 500 mg N l^{-1}, pH of 8.0, and temperature of 30°C. This condition can occur in a PN/A system treating high-strength industrial wastewater, especially in the one-stage configuration. In full-scale application, appropriate dilution is often needed to reduce the ammonium concentration in the feeding, thereby avoiding the inhibitory effect of FA on anammox--based process.

Both nitrite and FNA can significantly decrease the anammox activity. It was reported that nitrite concentration of 400 mg N l^{-1} and FNA concentration of 11 μg N l^{-1} resulted in a reduction of anammox activity of 50% (Fernández *et al* 2012, Lotti *et al* 2012). The nitrite concentration in raw industrial wastewater is normally negligible. However, the nitrite concentration in anammox-based process can potentially be high in the case of a higher nitrite production rate than the nitrite consumption rate, leading to the anammox suppression.

14.3.3 Refractory organics

Most industrial wastewater contains plenty of refractory organics, even after AD treatment. Each type of industrial wastewater has organic matters that are specific to its characteristics. The results of many studies show that the refractory organics significantly reduce the NRE of anammox-based process (Zhang *et al* 2018, Cao *et al* 2021).

To increase the biodegradability of the sludge, the thermal hydrolysis process (THP), which is normally operated at a temperature and pressure condition of 130°C–190°C and 4.8–12.6 bar, respectively, is widely applied in present WWTPs (Cao *et al* 2021, Ngo *et al* 2021, Yan *et al* 2022). However, the digester liquor from THP-AD contains a mass of refractory organics, which significantly impacts the activity of AOB and anammox, resulting in the reduction of efficiency of PN/A process (Zhang *et al* 2018, Cao *et al* 2021). Zhang *et al* (2018) reported a reduction of nitrogen removal rate of PN/A process from 0.55 to 0.37 kg N m^{-3} d^{-1} when the proportion of THP-AD liquor in the feed increased from 60% to 100%. The abundance of AOB and anammox bacteria also reduced with the increasing THP-AD liquor proportion. The soluble compounds produced in THP-AD process induced the main suppression for the anammox bacteria (Zhang *et al* 2018). The THP-AD liquor often contains some solids, which can induce the sludge washout of PN/A process (Cao *et al* 2021). The AOB was reported to be strongly sensitive to the large colloids in the THP-AD liquor (Zhang *et al* 2018).

Antibiotics, which are plentiful in pharmaceutical and swine wastewaters, have been found to dramatically suppress the activity of anammox bacteria (Fu *et al* 2021). For example, OTC, a widely used antibiotic, almost destroyed the anammox performance within three weeks of exposure to 2 mg l^{-1}, with a reduction of specific anammox activity of 81.3% (Shi *et al* 2017). Similarly, another commonly used antibiotic, sulfamethoxazole (SMZ), led to a decrease in specific anammox activity by 68.6% ± 10.7% at a concentration of 1 mg l^{-1} (Zhang *et al* 2019b). The concentration of antibiotics in the pharmaceutical wastewater can be as high as 200 mg l^{-1} (Chen *et al* 2021a), which is a level that is known to induce significant inhibition of anammox activity. Thus, directly feeding antibiotic-rich wastewater to the anammox process can result in a significant inhibition of anammox activity.

14.4 Future perspectives

14.4.1 The impacts of wastewater from upgraded AD in WWTPs on PN/A process

WWTPs are exploring various promising technologies, including THP, FNA, and FA pretreatment, and co-digestion, to increase the capacity of their existing AD process. These upgrade strategies have been shown to effectively improve AD performance. However, studies have also demonstrated that the wastewater from the THP-enhanced AD reduced the efficiency of downstream PN/A process (Cao *et al* 2021). Moreover, the impacts of these upgrading technologies on wastewater compositions and the subsequent anammox-based nitrogen removal process are still unclear. A comprehensive understanding of the effects of these technologies on the downstream nitrogen removal process could facilitate their full-scale application.

14.4.2 Additional nitrite suppliers to support anammox reaction

For the substances of the anammox reaction, ammonium is typically present in abundance, while nitrite is often negligible or present at a low level in the feeding. The PN/A process is the most widely applied anammox-based process in treating industrial wastewater, which relies on AOB to provide nitrite. However, the effluent of the PN/A process normally contains superfluous nitrate because about 11% of the nitrogen (ammonium + nitrite) fed for anammox reaction is converted to nitrate. Moreover, some types of industrial wastewater contain abundant amounts of nitrate, which is difficult to be handled by the PN/A process. Introducing denitrification is a commonly used technology to reduce nitrate in effluent. However, it needs a considerable amount of external organic carbon, which increases the operating costs and environmental impacts. PD/A is a promising approach, which can be coupled with PN/A to reduce the nitrate concentration in effluent. By dosing a small amount of organic carbon, nitrate will be converted to nitrite, and then converted to nitrogen gas by anammox. The effectiveness of the PD/A process has only been evaluated in several types of industrial wastewater.

AD is normally applied to recover energy from industrial wastewater that is rich in organics. However, the emission of methane from the effluent AD will intensify the greenhouse effect. A novel microbial process of nitrate-dependent anaerobic methane oxidation (n-DAMO archaea) process ($CH_4 + 4NO_3^- \rightarrow CO_2 + 4NO_2^- + 2H_2O$), which is able to simultaneously remove methane and reduce nitrate to nitrite, can potentially be applied to further improve the NRE of the anammox-based process (Liu *et al* 2019). The efficiency of the process of incorporating PN, anammox, and n-DAMO has been successfully demonstrated in treating synthetic high-strength wastewater (Liu *et al* 2019).

Overall, simultaneously achieving multiple nitrite suppliers for the anammox reaction can stabilize the nitrogen removal process and increase its efficiency. However, it remains unclear how to efficiently connect these nitrite suppliers into a comprehensive system in treating real industrial wastewater. Therefore, it is of significant practical significance to evaluate the effects of coupling these processes in the treatment of various types of industrial wastewater.

14.4.3 Promising approaches to detoxify industrial wastewater

Industrial wastewater contains several elements that are often toxic to anammox bacteria, including refractory organics and antibiotics. This toxicity can be reduced to some extent by dilution with low-strength wastewater or pretreatment using AD. However, the need for hundreds of times dilution significantly increases capital and operating costs, and AD is an inefficient method for detoxification. Therefore, there is a need to develop promising detoxification technologies, e.g., advance oxidation process, to improve the efficiency of anammox in eliminating nitrogen from industrial wastewater.

14.5 Conclusion

This chapter has reviewed the key findings and main challenges of using anammox--based processes to remove the nitrogen from different types of industrial wastewater. The PN/A process is the most widely implemented anammox-based approach in current full-scale applications due to the fact that ammonium is the primary nitrogen source in the majority of industrial wastewater. The anammox-based PD/A process is applied to remove nitrogen from industrial wastewater that contains high levels of nitrate. PD/A is also used to further remove the nitrogen (mainly nitrate) in the effluent of the PN/A process. However, the operation of the anammox-based process is challenged by NOB adaptation and the toxic materials in the industrial wastewater. To address these problems, future studies of coupling multiple nitrite suppliers and developing promising detoxify processes were proposed.

References

Abma W, Schultz C, Mulder J, Van der Star W, Strous M, Tokutomi T and Van Loosdrecht M 2007 Full-scale granular sludge Anammox process *Water Sci. Technol.* **55** 27–33

Azari M, Walter U, Rekers V, Gu J-D and Denecke M 2017 More than a decade of experience of landfill leachate treatment with a full-scale anammox plant combining activated sludge and activated carbon biofilm *Chemosphere* **174** 117–26

Cao S, Yan W, Yu L, Zhang L, Lay W and Zhou Y 2021 Challenges of THP-AD centrate treatment using partial nitritation-anammox (PN/A)—inhibition, biomass washout, low alkalinity, recalcitrant and more *Water Res.* **203** 117555

Cao S and Zhou Y 2019 New direction in biological nitrogen removal from industrial nitrate wastewater via anammox *Appl. Microbiol. Biotechnol.* **103** 7459–66

Chen H, Li X, Liu G, Zhu J, Ma X, Piao C, You S and Wang K 2023a Decoding the carbon and nitrogen metabolism mechanism in anammox system treating pharmaceutical wastewater with varying COD/N ratios through metagenomic analysis *Chem. Eng. J.* **457** 141316

Chen H, Liu G, Wang K, Piao C, Ma X and Li X-K 2021a Characteristics of microbial community in EGSB system treating with oxytetracycline production wastewater *J. Environ. Manage.* **295** 113055

Chen H, Liu G, Zhu J, Ma X, Piao C, Li X and Wang K 2023b Investigation of the mechanism of anammox granules alleviating the inhibition of organic matter in pharmaceutical wastewater *J. Clean. Prod.* **398** 136129

Chen Y, Zheng R, Sui Q, Ritigala T, Wei Y, Cheng X, Ren J, Yu D, Chen M and Wang T 2021b Coupling anammox with denitrification in a full-scale combined biological nitrogen removal process for swine wastewater treatment *Bioresour. Technol.* **329** 124906

Chini A, Hollas C E, Bolsan A C, Venturin B, Bonassa G, Cantão M E, Ibelli A M G, Antes F G and Kunz A 2020 Process performance and anammox community diversity in a deammonification reactor under progressive nitrogen loading rates for swine wastewater treatment *Bioresour. Technol.* **311** 123521

Choi D, Shin H and Jung J 2022 Control parameters in three-stage deammonification process for a mature leachate treatment based on a traditional Modified Ludzack-Ettinger process *J. Water Process. Eng.* **48** 102863

Deng K, Tang L, Li J, Meng J and Li J 2019 Practicing anammox in a novel hybrid anaerobic-aerobic baffled reactor for treating high-strength ammonium piggery wastewater with low COD/TN ratio *Bioresour. Technol.* **294** 122193

Duan H, Ye L, Lu X and Yuan Z 2019 Overcoming nitrite oxidizing bacteria adaptation through alternating sludge treatment with free nitrous acid and free ammonia *Environ. Sci. Technol.* **53** 1937–46

Fernández I, Dosta J, Fajardo C, Campos J, Mosquera-Corral A and Méndez R 2012 Short-and long-term effects of ammonium and nitrite on the Anammox process *J. Environ. Manage.* **95** S170–4

Fu J, Zhang Q, Huang B, Fan N and Jin R 2021 A review on anammox process for the treatment of antibiotic-containing wastewater: linking effects with corresponding mechanisms *Front. Environ. Sci. Eng.* **15** 1–15

Han X, Zhang S, Yang S, Zhang L and Peng Y 2020 Full-scale partial nitritation/anammox (PN/A) process for treating sludge dewatering liquor from anaerobic digestion after thermal hydrolysis *Bioresour. Technol.* **297** 122380

Hu Z, Liu T, Wang Z, Meng J and Zheng M 2023 Toward energy neutrality: novel wastewater treatment incorporating acidophilic ammonia oxidation *Environ. Sci. Technol.* **57** 4522–32

Ishimoto C, Sugiyama T, Matsumoto T, Uenishi H, Fukumoto Y and Waki M 2020 Full-scale simultaneous partial nitrification, anammox, and denitrification process for treating swine wastewater *Water Sci. Technol.* **81** 456–65

Joss A, Salzgeber D, Eugster J, König R, Rottermann K, Burger S, Fabijan P, Leumann S, Mohn J and Siegrist H 2009 Full-scale nitrogen removal from digester liquid with partial nitration and anammox in one SBR *Environ. Sci. Technol.* **43** 5301–6

Jung M, Oh T, Rhu D, Liberzon J, Kang S J, Daigger G T and Kim S 2021 A high-rate and stable nitrogen removal from reject water in a full-scale two-stage AMX® system *Water Sci. Technol.* **83** 652–63

Kaewyai J, Noophan P L, Lin J-G, Munakata-Marr J and Figueroa L A 2022 A comparison of nitrogen removal efficiencies and microbial communities between anammox and de-ammonification processes in lab-scale ASBR, and full-scale MBBR and IFAS plants *Int. Biodeterior. Biodegrad.* **169** 105376

Lackner S, Gilbert E M, Vlaeminck S E, Joss A, Horn H and van Loosdrecht M C 2014 Full-scale partial nitritation/anammox experiences—an application survey *Water Res.* **55** 292–303

Li J, Li J, Gao R, Wang M, Yang L, Wang X, Zhang L and Peng Y 2018 A critical review of one-stage anammox processes for treating industrial wastewater: optimization strategies based on key functional microorganisms *Bioresour. Technol.* **265** 498–505

Li S, Duan H, Zhang Y, Huang X, Yuan Z, Liu Y and Zheng M 2020a Adaptation of nitrifying community in activated sludge to free ammonia inhibition and inactivation *Sci. Total Environ.* **728** 138713

Li X, Lu M-y, Huang Y, Yuan Y and Yuan Y 2021 Influence of seasonal temperature change on autotrophic nitrogen removal for mature landfill leachate treatment with high-ammonia by partial nitrification-Anammox process *J. Environ. Sci.* **102** 291–300

Li X, Lu M-y, Qiu Q-c, Huang Y, Li B-l, Yuan Y and Yuan Y 2020b The effect of different denitrification and partial nitrification-Anammox coupling forms on nitrogen removal from mature landfill leachate at the pilot-scale *Bioresour. Technol.* **297** 122430

Liu G and Wang J 2013 Long-term low DO enriches and shifts nitrifier community in activated sludge *Environ. Sci. Technol.* **47** 5109–17

Liu T, Hu S, Yuan Z and Guo J 2019 High-level nitrogen removal by simultaneous partial nitritation, anammox and nitrite/nitrate-dependent anaerobic methane oxidation *Water Res.* **166** 115057

Lotti T, Van Der Star W, Kleerebezem R, Lubello C and Van Loosdrecht M 2012 The effect of nitrite inhibition on the anammox process *Water Res.* **46** 2559–69

Ma B, Yang L, Wang Q, Yuan Z, Wang Y and Peng Y 2017 Inactivation and adaptation of ammonia-oxidizing bacteria and nitrite-oxidizing bacteria when exposed to free nitrous acid *Bioresour. Technol.* **245** 1266–70

Ma S, Zhou C, Pan J, Yang G, Sun C, Liu Y, Chen X and Zhao Z 2022 Leachate from municipal solid waste landfills in a global perspective: characteristics, influential factors and environmental risks *J. Clean. Prod.* **333** 130234

Magrí A, Ruscalleda M, Vilà A, Akaboci T R, Balaguer M D, Llenas J M and Colprim J 2021 Scaling-up and long-term operation of a full-scale two-stage partial nitritation-Anammox system treating landfill leachate *Processes* **9** 800

Mai W, Hu T, Li C, Wu R, Chen J, Shao Y, Liang J and Wei Y 2020 Effective nitrogen removal of wastewater from vitamin B2 production by a potential anammox process *J. Water Process. Eng.* **37** 101515

Meng J, Li J, Li J, Antwi P, Deng K, Wang C and Buelna G 2015 Nitrogen removal from low COD/TN ratio manure-free piggery wastewater within an upflow microaerobic sludge reactor *Bioresour. Technol.* **198** 884–90

Molinuevo B, García M C, Karakashev D and Angelidaki I 2009 Anammox for ammonia removal from pig manure effluents: effect of organic matter content on process performance *Bioresour. Technol.* **100** 2171–5

Mulder A, Van de Graaf A A, Robertson L and Kuenen J 1995 Anaerobic ammonium oxidation discovered in a denitrifying fluidized bed reactor *FEMS Microbiol. Ecol.* **16** 177–83

Ngo P L, Udugama I A, Gernaey K V, Young B R and Baroutian S 2021 Mechanisms, status, and challenges of thermal hydrolysis and advanced thermal hydrolysis processes in sewage sludge treatment *Chemosphere* **281** 130890

Phan T N, Van Truong T T, Ha N B, Nguyen P D, Bui X T, Dang B T, Park J, Guo W and Ngo H H 2017 High rate nitrogen removal by ANAMMOX internal circulation reactor (IC) for old landfill leachate treatment *Bioresour. Technol.* **234** 281–8

Qiu S, Li Z, Hu Y, Shi L, Liu R, Shi L, Chen L, Zhan X and Technology 2021 What's the best way to achieve successful mainstream partial nitritation-anammox application? *Crit. Rev. Environ. Sci. Technol.* **51** 1045–77

Ren S, Zhang L, Zhang Q, Zhang F, Jiang H, Li X, Wang S and Peng Y 2022a Anammox-mediated municipal solid waste leachate treatment: a critical review *Bioresour. Technol.* **361** 127715

Ren Z-Q, Wang H, Zhang L-G, Du X-N, Huang B-C and Jin R-C 2022b A review of anammox-based nitrogen removal technology: From microbial diversity to engineering applications *Bioresour. Technol.* **363** 127896

Shi Z-J, Hu H-Y, Shen Y-Y, Xu J-J, Shi M-L and Jin R-C 2017 Long-term effects of oxytetracycline (OTC) on the granule-based anammox: process performance and occurrence of antibiotic resistance genes *Biochem. Eng. J.* **127** 110–8

Tang C-J, Zheng P, Chen T-T, Zhang J-Q, Mahmood Q, Ding S, Chen X-G, Chen J-W and Wu D-T 2011 Enhanced nitrogen removal from pharmaceutical wastewater using SBA-ANAMMOX process *Water Res.* **45** 201–10

Van der Star W R, Abma W R, Blommers D, Mulder J-W, Tokutomi T, Strous M, Picioreanu C and van Loosdrecht M C 2007 Startup of reactors for anoxic ammonium oxidation: experiences from the first full-scale anammox reactor in Rotterdam *Water Res.* **41** 4149–63

Wang H, Wang J, Zhou M, Wang W, Liu C and Wang Y 2022a A versatile control strategy based on organic carbon flow analysis for effective treatment of incineration leachate using an anammox-based process *Water Res.* **215** 118261

Wang Q, Duan H, Wei W, Ni B-J, Laloo A and Yuan Z 2017 Achieving stable mainstream nitrogen removal via the nitrite pathway by sludge treatment using free ammonia *Environ. Sci. Technol.* **51** 9800–7

Wang Q, Ye L, Jiang G, Hu S and Yuan Z 2014 Side-stream sludge treatment using free nitrous acid selectively eliminates nitrite oxidizing bacteria and achieves the nitrite pathway *Water Res.* **55** 245–55

Wang Z, Zhang L, Zhang F, Jiang H, Ren S, Wang W and Peng Y 2020 A continuous-flow combined process based on partial nitrification-Anammox and partial denitrification-Anammox (PN/A + PD/A) for enhanced nitrogen removal from mature landfill leachate *Bioresour. Technol.* **297** 122483

Wang Z, Zheng M, Duan H, Yuan Z and Hu S 2022b A 20-year journey of partial nitritation and anammox (PN/A): from sidestream toward mainstream *Environ. Sci. Technol.* **56** 7522–31

Wu L, Li Z, Huang S, Shen M, Yan Z, Li J and Peng Y 2019 Low energy treatment of landfill leachate using simultaneous partial nitrification and partial denitrification with anaerobic ammonia oxidation *Environ. Int.* **127** 452–61

Yan W, Xu H, Lu D and Zhou Y 2022 Effects of sludge thermal hydrolysis pretreatment on anaerobic digestion and downstream processes: mechanism, challenges and solutions *Bioresour. Technol.* **344** 126248

Zhang F, Peng Y, Miao L, Wang Z, Wang S and Li B 2017 A novel simultaneous partial nitrification Anammox and denitrification (SNAD) with intermittent aeration for cost-effective nitrogen removal from mature landfill leachate *Chem. Eng. J.* **313** 619–28

Zhang F, Peng Y, Wang S, Wang Z and Jiang H 2019a Efficient step-feed partial nitrification, simultaneous Anammox and denitrification (SPNAD) equipped with real-time control parameters treating raw mature landfill leachate *J. Hazard. Mater.* **364** 163–72

Zhang J, Peng Y, Li X and Du R 2022 Feasibility of partial-denitrification/anammox for pharmaceutical wastewater treatment in a hybrid biofilm reactor *Water Res.* **208** 117856

Zhang L, Zhang S, Peng Y, Han X and Gan Y 2015 Nitrogen removal performance and microbial distribution in pilot-and full-scale integrated fixed-biofilm activated sludge reactors based on nitritation-anammox process *Bioresour. Technol.* **196** 448–53

Zhang Q-Q, Bai Y-H, Wu J, Zhu W-Q, Tian G-M, Zheng P, Xu X-Y and Jin R-C 2019b Microbial community evolution and fate of antibiotic resistance genes in anammox process under oxytetracycline and sulfamethoxazole stresses *Bioresour. Technol.* **293** 122096

Zhang Q, De Clippeleir H, Su C, Al-Omari A, Wett B, Vlaeminck S E and Murthy S 2016 Deammonification for digester supernatant pretreated with thermal hydrolysis: overcoming inhibition through process optimization *Appl. Microbiol. Biotechnol.* **100** 5595–606

Zhang Q, Vlaeminck S E, DeBarbadillo C, Su C, Al-Omari A, Wett B, Pümpel T, Shaw A, Chandran K and Murthy S 2018 Supernatant organics from anaerobic digestion after thermal hydrolysis cause direct and/or diffusional activity loss for nitration and anammox *Water Res.* **143** 270–81

Zheng M, Li S, Ni G, Xia J, Hu S, Yuan Z, Liu Y and Huang X 2020 Critical factors facilitating candidatus nitrotoga to be prevalent nitrite-oxidizing bacteria in activated sludge *Environ. Sci. Technol.* **54** 15414–23

Zuo L, Yao H, Li H, Fan L and Jia F 2020 Nitrogen removal efficiency for pharmaceutical wastewater with a single-stage anaerobic ammonium oxidation process *Int. J. Env. Res. Public Health* **17** 7972

Chapter 15

Bioprocesses in industrial wastewater treatment: trends and prospects

Prasann Kumar and Joginder Singh

Industrial wastewater treatment is a critical aspect of environmental sustainability and public health. Significant advances and trends have been made in applying biological processes to treat industrial wastewater in recent years. This chapter provides a concise overview of the emerging trends in biological treatment methods, focusing on their application, performance, and potential for enhancing wastewater treatment efficiency. This chapter will first give an introduction to industrial wastewater treatment, which briefly outlines the challenges associated with industrial wastewater and the need for effective treatment methods to minimize environmental impact. It will then describe the biological treatment processes used in industrial wastewater treatment, including aerobic and anaerobic treatment, activated sludge systems, membrane bioreactors (MBRs), and biofilm reactors. Each process's principles, advantages, and limitations are briefly touched upon. Next, it will discuss the emerging trends in biological treatment, such as the implementation of high-rate bioreactors, membrane-aerated biofilm reactors, and granular sludge systems. It emphasizes the integration of innovative process configurations and the utilization of microbial consortia to enhance pollutant removal. It then outlines the performance and efficiency of biological treatment processes, including factors influencing treatment efficiency, process optimization, and monitoring techniques. It highlights the importance of resource recovery and the potential for generating energy from wastewater treatment. Finally, this chapter briefly mentions the prospects of biological processes in industrial wastewater treatment, including incorporating emerging technologies like artificial intelligence and nanotechnology. It acknowledges the challenges of scaling up these processes and the ongoing need for research and development. Overall, this chapter provides a succinct overview of the recent trends in biological processes for industrial wastewater treatment.

It highlights this field's advancements, challenges, and prospects, contributing to the existing knowledge and fostering further research in sustainable wastewater treatment practices.

15.1 Introduction

Industrial wastewater treatment is pivotal in ensuring environmental sustainability and safeguarding public health (Huang *et al* 2023a, Ma *et al* 2024). The treatment of industrial wastewater involves the removal of pollutants and contaminants generated from various industrial processes before their discharge into water bodies. Over the years, significant advancements have been made in the biological processes for industrial wastewater treatment (Liu *et al* 2023). These biological treatment methods have proven effective, cost-efficient, and environmentally friendly alternatives to conventional treatment approaches, as follows:

1. Importance of industrial wastewater treatment: Industrial wastewater contains a diverse range of pollutants, including organic compounds, heavy metals, nutrients, and toxic substances. Industrial wastewater can pose severe environmental and public health risks if left untreated or inadequately treated. It can contaminate water sources, disrupt ecosystems, harm aquatic life, and potentially find its way into the food chain (Han *et al* 2022). Therefore, efficient industrial wastewater treatment is paramount to mitigate these risks and ensure sustainable industrial practices.

2. Traditional approaches to industrial wastewater treatment: Conventional industrial wastewater treatment methods typically include physical and chemical processes, such as sedimentation, coagulation, flocculation, and chemical oxidation (Mondal *et al* 2023). While these methods can effectively remove certain pollutants, they may be limited in their ability to treat complex and recalcitrant compounds found in industrial wastewater. Additionally, these approaches often require extensive infrastructure, consume significant energy, and generate substantial amounts of chemical sludge as a byproduct.

3. Advantages of biological processes: Biological treatment processes have gained considerable attention as a viable alternative to traditional treatment methods. These processes harness the power of microorganisms, such as bacteria, fungi, and algae, to degrade and transform pollutants in industrial wastewater (Li *et al* 2021, Fayyaz Shahandashty *et al* 2023). Biological processes offer several advantages, including high pollutant removal efficiency, lower energy requirements, reduced chemical consumption, and the potential for resource recovery by producing biogas or biofertilizers. Furthermore, they can be adapted to treat a wide range of contaminants and exhibit greater resilience to variations in wastewater composition.

4. Emerging trends in biological processes: Recent years have witnessed notable advancements and trends in biological processes for industrial wastewater treatment (Zhang *et al* 2021, Jia *et al* 2023, Zhang *et al* 2023a). These trends encompass technological and operational aspects,

aiming to enhance treatment efficiency, sustainability, and cost-effectiveness. Some of the key trends include:

a. High-rate bioreactors: The implementation of high-rate bioreactors, such as upflow anaerobic sludge blanket (UASB) reactors and expanded granular sludge bed reactors, allows for enhanced organic matter and nutrient removal in a compact footprint.

b. Membrane-based technologies: MBRs and membrane-aerated biofilm reactors (MABRs) have gained prominence due to their ability to achieve high-quality effluent, minimize sludge production, and provide excellent process control.

c. Biofilm reactors: Biofilm reactors, including moving bed biofilm reactors (MBBRs) and integrated fixed-film activated sludge (IFAS) systems, promote the growth of microbial communities on biofilm carriers, leading to improved pollutant removal and increased treatment capacity.

d. Microbial consortia and engineered microorganisms: The use of engineered microbial consortia and genetically modified microorganisms is emerging as a promising approach to enhance the degradation of recalcitrant compounds and address specific wastewater treatment challenges.

While biological processes offer immense potential for industrial wastewater treatment, several challenges must be addressed. These challenges include maintaining stable microbial communities, effectively managing inhibitory substances, optimizing process parameters, and scaling up these technologies for industrial-scale applications. Additionally, integrating emerging technologies, such as artificial intelligence, nanotechnology, and advanced monitoring systems, holds promise for further improving treatment efficiency and process control.

The advances and trends in biological processes for industrial wastewater treatment signify a shift towards sustainable and efficient treatment approaches. These trends focus on enhancing pollutant removal, reducing energy consumption, minimizing chemical usage, and promoting resource recovery (Yang *et al* 2022a, Gao *et al* 2023, Irshad *et al* 2023, Martín-González *et al* 2023). By embracing these emerging trends, industries can effectively address the challenges associated with industrial wastewater treatment, contributing to a cleaner environment and sustainable water resources for future generations (table 15.1).

Electrochemical treatment systems combined with biological processes are an innovative and practical approach to wastewater remediation. These systems integrate electrochemical techniques, such as electrocoagulation, electrooxidation, and electroflotation, with biological processes, such as bioremediation and bioelectrochemical systems, to treat and purify wastewater (Egbuikwem *et al* 2020, Faggiano *et al* 2023, Gobelius *et al* 2023, Wu *et al* 2023b). Comprehensive information about electrochemical treatment systems combined with biological processes for wastewater remediation follows:

Table 15.1. Physical, chemical, and biological techniques for the remediation of phenol from wastewater, along with their explanations, advantages, disadvantages, formulations, microbes, their role, and classification.

Technique category	Technique name	Explanation	Advantages	Disadvantages	Formulation	Microbes and their role	Classification
Physical	Activated carbon adsorption	Adsorption of phenol onto activated carbon surfaces	High adsorption capacity	Limited regeneration capability	Activated carbon	N/A	Adsorption
	Adsorption onto biochar	Adsorption of phenol onto biochar surfaces	Cost-effective and sustainable	Adsorbent saturation over time	Biochar	N/A	Adsorption
	Adsorption onto zeolite	Adsorption of phenol onto zeolite surfaces	High adsorption capacity	Limited regeneration capability	Zeolite	N/A	Adsorption
	Membrane separation techniques	Separation of phenol using membranes with different pore sizes	Selective removal of phenol	Membrane fouling and maintenance requirements	Membranes	N/A	Separation
	Coagulation-flocculation	Aggregation of phenol particles using chemical coagulants	Effective removal of phenol particles	Generation of sludge during coagulation process	Coagulants, such as aluminum sulfate	N/A	Flocculation
	Sparging	Aeration technique to remove volatile phenol compounds	Efficient removal of volatile phenols	Limited effectiveness for non-volatile phenols	Air or oxygen	N/A	Aeration
	Thermal desorption	Heating of phenol-contaminated media to volatilize phenol	Complete removal of phenol by volatilization	Energy-intensive process	Heat	N/A	Volatilization
	Freeze–thaw extraction	Extraction of phenol using repeated freezing and thawing cycles	Enhanced extraction efficiency	Time-consuming process	Temperature cycling	N/A	Extraction

Solvent extraction	Extraction of phenol using organic solvents	Efficient extraction of phenol from water	Generation of hazardous waste from solvent recovery	Organic solvents	N/A	Extraction
Volatile organic compound (VOC) recovery	Removal of phenol and other VOCs using various techniques such as condensation	Simultaneous removal of VOCs and phenol	High energy requirements for VOC recovery	Condensation or adsorption techniques	N/A	Recovery
Supercritical fluid extraction	Extraction of phenol using supercritical fluids	High extraction efficiency	High capital and operational costs	Supercritical fluids	N/A	Extraction
Wet air oxidation	Oxidation of phenol in the presence of water and oxygen	Complete oxidation of phenol to carbon dioxide and water	High energy consumption	Oxygen and water	N/A	Oxidation
Sonolysis	Degradation of phenol using ultrasound waves	Enhanced degradation efficiency	Equipment and operational costs	Ultrasound waves	N/A	Oxidation
Thermal desorption	Heating of phenol-contaminated media to volatilize phenol	Complete removal of phenol by volatilization	Energy-intensive process	Heat	N/A	Volatilization
Microwave-assisted treatment	Phenol degradation using microwave irradiation	Rapid and efficient degradation	Limited penetration depth in media	Microwave irradiation	N/A	Oxidation
Membrane distillation	Separation of phenol from water by membrane distillation	High separation efficiency	Limited application for low phenol concentrations	Membranes	N/A	Separation

(Continued)

Table 15.1. (*Continued*)

Technique category	Technique name	Explanation	Advantages	Disadvantages	Formulation	Microbes and their role	Classification
	Electrochemical membrane processes	Removal of phenol using electrochemical techniques combined with membranes	Selective removal of phenol	High energy consumption	Membranes and electrochemical cells	N/A	Separation
	Sonochemical treatment	Degradation of phenol using ultrasonic waves	Enhanced degradation efficiency	Limited effectiveness for high phenol concentrations	Ultrasound waves	N/A	Oxidation
Chemical	Advanced oxidation processes (AOPs)	Degradation of phenol using powerful oxidizing agents like ozone, hydrogen peroxide, or UV radiation.	Complete degradation of phenol	High operational costs and handling of hazardous chemicals	Ozone, hydrogen peroxide, UV radiation	N/A	Oxidation
	Fenton process	Oxidation of phenol using iron catalyst and hydrogen peroxide	Effective oxidation of phenol	Generation of sludge containing iron residuals	Iron catalyst and hydrogen peroxide	N/A	Oxidation
	Photocatalysis	Degradation of phenol using photocatalysts and light energy	Efficient degradation under light irradiation	High cost and limited photocatalyst stability	Photocatalysts and light source	N/A	Oxidation
	Peroxide enhanced processes	Phenol degradation using chemical oxidation with hydrogen peroxide	Enhanced oxidation efficiency	High cost of hydrogen peroxide	Hydrogen peroxide	N/A	Oxidation

	Sustainable coagulants	Use of environmentally friendly coagulants for phenol removal	Effective removal of phenol	Limited availability and higher cost of sustainable coagulants	Sustainable coagulants	N/A	Flocculation
	Magnetic nanoparticle adsorption	Adsorption of phenol onto magnetic nanoparticles	Enhanced adsorption capacity	Recovery and recycling of magnetic nanoparticles	Magnetic nanoparticles	N/A	Adsorption
	Advanced oxidation-biological hybrid systems	Combination of oxidation and biological processes for phenol degradation	Synergistic removal of phenol	Complex system design and operation	Oxidizing agents and microbial cultures	N/A	Hybrid
	Carbon-based materials for adsorption	Adsorption of phenol onto carbon-based materials	High adsorption capacity	Limited regeneration capability	Carbon-based materials, such as activated carbon	N/A	Adsorption
Biological	Biodegradation	Microbial degradation of phenol into less toxic compounds	Complete degradation of phenol	Slower degradation rate compared to other techniques	Phenol-degrading microorganisms	Microbes: bacteria, fungi, or algae	Biodegradation
	Bioaugmentation	Introduction of specialized phenol-degrading microorganisms to enhance degradation	Accelerated degradation of phenol	Need for continuous addition of specialized microorganisms	Phenol-degrading microorganisms	Microbes: bacteria, fungi, or algae	Biodegradation
	Biostimulation	Enhancement of phenol-degrading microbial activity through nutrient addition	Accelerated degradation of phenol	Nutrient requirements and potential eutrophication risk	Nutrient sources and microbial inoculum	Microbes: bacteria, fungi, or algae	Biodegradation

(Continued)

Table 15.1. (*Continued*)

Technique category	Technique name	Explanation	Advantages	Disadvantages	Formulation	Microbes and their role	Classification
	Constructed wetlands	Use of wetland plants and microorganisms to remove phenol	Enhanced removal through plant uptake and microbial degradation	Long-term sustainability and habitat creation	Wetland plants and microbial communities	Microbes: bacteria, fungi, or algae	Biodegradation
	Phytoremediation	Use of plants to take up and degrade phenol from wastewater	Sustainable and esthetically pleasing treatment method	Limited to certain plant species and growth conditions	Phenol-accumulating plants	Microbes: rhizosphere microorganisms	Biodegradation
	Rhizoremediation	Use of plant roots and associated microorganisms for phenol degradation	Enhanced degradation in the root zone	Limited effectiveness in low-nutrient environments	Plants with phenol-degrading microbes	Microbes: rhizosphere microorganisms	Biodegradation
	Mycoremediation	Use of fungi for phenol degradation in wastewater	Broad substrate specificity	Longer treatment duration	Phenol-degrading fungi	Microbes: fungi	Biodegradation
	Anaerobic digestion	Biodegradation of phenol in the absence of oxygen	Energy recovery through biogas production	Slow degradation rate and potential inhibition	Anaerobic digesters	Microbes: anaerobic bacteria	Biodegradation
	Composting	Biological degradation of phenol in a controlled aerobic environment	Sustainable and organic treatment method	Limited applicability for high-strength phenol wastewater	Composting facilities or piles	Microbes: bacteria, fungi	Biodegradation
	Biofiltration	Passage of phenol-laden wastewater through biologically active media for degradation	Efficient removal of phenol	Media clogging and maintenance requirements	Biological filter media	Microbes: bacteria, fungi	Biodegradation

Trickling filters	Phenol removal through the growth of biofilms on solid media	Effective removal through biofilm activity	Media clogging and maintenance requirements	Media filled with support material	Microbes: bacteria, fungi	Biodegradation
Microbial fuel cells	Use of microorganisms to degrade phenol and generate electricity	Simultaneous treatment and energy production	Low power generation efficiency	Electrodes and microbial cultures	Microbes: bacteria, archaea	Biodegradation
Aerobic granular sludge process	Biodegradation of phenol using dense granular microbial aggregates	High treatment efficiency and biomass retention	Granule stability and maintenance requirements	Granular sludge	Microbes: bacteria, fungi	Biodegradation
Membrane bioreactors (MBRs)	Combination of biological treatment and membrane filtration for phenol removal	Effective solids-liquid separation	Membrane fouling and maintenance requirements	Biological reactor with membrane filtration	Microbes: bacteria, fungi	Biodegradation
Biofilm reactors	Use of biofilms on solid surfaces for phenol degradation	Enhanced degradation due to biofilm activity	Biofilm detachment and maintenance requirements	Biofilm carriers or packed bed reactors	Microbes: bacteria, fungi	Biodegradation
Anaerobic fixed-film reactors	Biodegradation of phenol using microorganisms attached to fixed media without oxygen	Efficient phenol removal in anaerobic conditions	Media clogging and maintenance requirements	Fixed-film media or packed bed reactors	Microbes: anaerobic bacteria	Biodegradation
Constructed microbial ecosystems	Use of engineered microbial communities for phenol degradation	Robust treatment performance	High system design and operation complexity	Engineered microbial consortia	Microbes: engineered microbial communities	Biodegradation

(Continued)

Table 15.1. (*Continued*)

Technique category	Technique name	Explanation	Advantages	Disadvantages	Formulation	Microbes and their role	Classification
	Bioelectrochemical systems	Use of microorganisms and electrodes for phenol degradation	Simultaneous treatment and electricity production	High capital and operational costs	Electrodes and microbial cultures	Microbes: bacteria, archaea	Biodegradation
	Sequential batch reactors (SBRs)	Biological treatment of phenol in a series of batch stages	Flexible operation and process control	Longer treatment time and lower throughpu	Sequential batch reactors	Microbes: bacteria, fungi	Biodegradation
	Biological nutrient removal	Simultaneous removal of phenol and nutrients by microorganisms	Integrated treatment for phenol and nutrient removal	Additional process complexity and control	Biological reactors with nutrient addition	Microbes: bacteria, fungi	Biodegradation
	Membrane-aerated biofilm reactors (MABRs)	Use of biofilms in membrane-aerated systems for phenol degradation	Efficient removal in oxygen-rich conditions	Membrane fouling and maintenance requirements	Biofilm carriers and membrane aeration	Microbes: bacteria, fungi	Biodegradation
	Upflow anaerobic sludge blanket (UASB) reactors	Anaerobic degradation of phenol using a blanket of sludge particles	High treatment efficiency and low sludge production	Sensitivity to fluctuations in phenol concentration	Sludge blanket and anaerobic reactor	Microbes: anaerobic bacteria	Biodegradation
	Microbial desalination cells	Simultaneous phenol removal and desalination using microbial activity	Integrated treatment for phenol and desalination	Limited desalination efficiency	Saltwater and microbial cultures	Microbes: bacteria, archaea	Biodegradation
	Anammox process	Anaerobic ammonium oxidation process for phenol degradation	Nitrogen removal and phenol degradation	Slower degradation rate compared to aerobic processes	Anammox reactors	Microbes: anaerobic bacteria	Biodegradation

Method	Description	Benefit	Limitation	Combination of techniques	Microbes	Category
Hybrid biological-physicochemical systems	Combination of biological and physicochemical processes for phenol removal	Synergistic treatment performance	Complex system design and operation	Combination of techniques	Microbes: bacteria, fungi	Hybrid
Advanced microbial systems	Use of engineered microbial systems for efficient phenol degradation	Enhanced treatment performance	Complex system design and operation	Engineered microbial systems	Microbes: engineered microbial systems	Biodegradation
Bioleaching	Use of microorganisms to extract phenol from solid waste or contaminated soil	Enhanced extraction of phenol	Limited to solid waste or contaminated soil	Microorganisms and solid waste or soil	Microbes: bacteria, fungi	Extraction
Bioabsorption	Absorption of phenol by living or dead microbial biomass	Enhanced phenol removal	Limited to low concentrations of phenol	Microbial biomass	Microbes: bacteria, fungi	Adsorption
Biosorption	Binding of phenol to microbial cell surfaces or extracellular polymeric substances	Efficient phenol removal	Limited capacity for higher phenol concentrations	Microbial biomass or extracellular substances	Microbes: bacteria, fungi	Adsorption
Enzymatic degradation	Phenol degradation using enzymes produced by microorganisms	Specific and efficient phenol degradation	Enzyme production and cost	Enzymes	Microbes: bacteria, fungi	Oxidation
Immobilized cell systems	Entrapment of microorganisms in a matrix for phenol degradation	Enhanced microbial activity and stability	Limited diffusion of phenol to immobilized cells	Matrix material with immobilized cells	Microbes: bacteria, fungi	Biodegradation

(Continued)

Table 15.1. (*Continued*)

Technique category	Technique name	Explanation	Advantages	Disadvantages	Formulation	Microbes and their role	Classification
	Microbial consortium development	Engineering of diverse microbial communities for phenol degradation	Enhanced treatment efficiency and resilience	Complexity in designing and maintaining microbial consortia	Engineered microbial consortia	Microbes: engineered microbial communities	Biodegradation
	Rhodococcus strains for phenol degradation	Use of Rhodococcus strains for efficient phenol degradation	Broad substrate specificity	Limited to specific Rhodococcus strains	Rhodococcus bacteria	Microbes: bacteria	Biodegradation
	Mixed culture systems	Use of mixed microbial cultures for phenol degradation	Enhanced treatment efficiency and degradation capability	Difficulty in controlling and maintaining mixed cultures	Mixed microbial cultures	Microbes: mixed microbial cultures	Biodegradation
	Genetically modified organisms	Engineering of microorganisms with enhanced phenol degradation capabilities	Enhanced degradation efficiency and specificity	Regulatory concerns and public acceptance	Engineered microorganisms	Microbes: engineered microorganisms	Biodegradation
	Microbial consortia	Use of diverse microbial consortia for phenol degradation	Enhanced treatment efficiency and resilience	Complexity in designing and maintaining microbial consortia	Microbial consortia	Microbes: bacteria, fungi	Biodegradation
	Microbial diversity enhancement	Enrichment of phenol-degrading microorganisms through selective conditions	Enhanced phenol degradation capacity	Selective conditions and long-term stability	Microbial enrichment techniques	Microbes: bacteria, fungi	Biodegradation

Microbial adaptation strategies	Optimization of microbial phenol degradation through adaptation strategies	Enhanced degradation capability and resilience	Time-consuming and complex optimization processes	Microbial adaptation strategies	Microbes: bacteria, fungi	Biodegradation
Microbial metabolic engineering	Engineering microbial metabolic pathways for efficient phenol degradation	Enhanced degradation efficiency and pathway optimization	Complexity in pathway engineering and regulatory concerns	Engineered metabolic pathways	Microbes: engineered microorganisms	Biodegradation
Microbial biodegradation enzymes	Use of purified enzymes for efficient phenol degradation	High specificity and catalytic efficiency	High cost of enzyme production and purification	Purified enzymes	Microbes: bacteria, fungi	Biodegradation
Microbial adaptation through acclimation	Acclimation of microorganisms to phenol for enhanced degradation	Improved phenol degradation capability	Time-consuming acclimation process	Microbial acclimation techniques	Microbes: bacteria, fungi	Biodegradation
Natural attenuation	Natural degradation of phenol by indigenous microbial populations	Passive treatment method	Slow degradation rate and potential long-term contamination	Indigenous microbial populations	Microbes: indigenous microbial populations	Biodegradation
Microbial decolorization	Simultaneous removal of phenol and color from wastewater	Treatment of phenol-containing colored wastewater	Limited to phenol-containing colored wastewater	Phenol-degrading microbes and dyes	Microbes: bacteria, fungi	Biodegradation
Microbial consortia engineering	Construction of synthetic microbial consortia for phenol degradation	Enhanced treatment efficiency and resilience	Complexity in designing and maintaining synthetic consortia	Synthetic microbial consortia	Microbes: engineered microbial communities	Biodegradation

(Continued)

15-13

Table 15.1. (*Continued*)

Technique category	Technique name	Explanation	Advantages	Disadvantages	Formulation	Microbes and their role	Classification
	Bioelectrochemical phenol degradation	Phenol degradation using electrochemical systems with microbial activity	Simultaneous treatment and electricity production	High capital and operational costs	Electrodes and microbial cultures	Microbes: Bacteria, archaea	Biodegradation
	Microbial desulfurization	Use of microorganisms for the removal of phenol and sulfur compounds	Integrated treatment for phenol and sulfur removal	Limited to sulfur-containing phenol compounds	Microbial cultures	Microbes: bacteria, fungi	Biodegradation
	Microbial aggregates	Use of microbial aggregates for enhanced phenol degradation	Enhanced phenol degradation capability	Aggregation and disintegration of microbial aggregates	Microbial aggregates	Microbes: bacteria, fungi	Biodegradation
	Microbial rhodanese activity	Use of microbial rhodanese enzyme for phenol degradation	Enhanced degradation and detoxification of phenol	Limited to specific microbial species	Microbial cultures or purified enzyme	Microbes: bacteria, fungi	Biodegradation
	Bioimmobilization	Immobilization of microorganisms for phenol degradation in solid matrices	Enhanced microbial activity and stability	Limited diffusion of phenol to immobilized cells	Solid matrices with immobilized cells	Microbes: bacteria, fungi	Biodegradation
	Microbial adhesion	Enhanced phenol degradation through microbial adhesion to surfaces	Increased contact between microorganisms and phenol	Risk of biofouling and microbial detachment	Surface-adhered microorganisms	Microbes: bacteria, fungi	Biodegradation

Microbial consortia enrichment	Enrichment of phenol-degrading microbial consortia for enhanced degradation	Enhanced treatment efficiency and resilience	Selective conditions and long-term stability	Microbial consortia enrichment techniques	Microbes: bacteria, fungi	Biodegradation
Microbial-plant interactions	Use of microbial-plant interactions for enhanced phenol degradation	Synergistic removal of phenol and plant-mediated degradation	Limited to certain plant-microbe combinations	Microbes and phenol-accumulating plants	Microbes: bacteria, fungi	Biodegradation
Microbial precipitation	Precipitation of phenol from wastewater using microbial activities	Efficient removal of phenol from solution	Limited to certain microbial species	Microbial cultures	Microbes: bacteria, fungi	Precipitation
Microbial entrapment	Entrapment of microorganisms in polymeric matrices for phenol degradation	Enhanced phenol degradation capability and stability	Limited mass transfer and diffusion of phenol	Polymeric matrices with entrapped cells	Microbes: bacteria, fungi	Biodegradation
Microbial exopolymeric substances	Utilization of microbial exopolymeric substances for phenol degradation	Enhanced phenol removal and biofilm formation	Risk of biofilm detachment and sludge formation	Exopolymeric substances	Microbes: bacteria, fungi	Biodegradation
Bioadsorption	Adsorption of phenol onto microbial cell surfaces or extracellular substances	Enhanced phenol removal	Limited capacity for higher phenol concentrations	Microbial biomass or extracellular substances	Microbes: bacteria, fungi	Adsorption
Immobilized enzymes	Immobilization of phenol-degrading enzymes for enhanced degradation	Improved stability and reusability of enzymes	Limited to specific enzymes and immobilization techniques	Enzymes immobilized on supports	Microbes: bacteria, fungi	Oxidation

(Continued)

Table 15.1. (*Continued*)

Technique category	Technique name	Explanation	Advantages	Disadvantages	Formulation	Microbes and their role	Classification
	Genetic engineering of microbial consortia	Engineering of microbial consortia for enhanced phenol degradation	Enhanced degradation efficiency and pathway optimization	Complexity in consortium engineering and regulatory concerns	Engineered microbial consortia	Microbes: engineered microbial communities	Biodegradation
	Microbial consortia in biofilms	Utilization of microbial consortia in biofilms for phenol degradation	Enhanced treatment efficiency and biofilm activity	Risk of biofilm detachment and sludge formation	Microbial consortia in biofilms	Microbes: bacteria, fungi	Biodegradation
	Microbial fermentation	Phenol degradation through microbial fermentation processes	Efficient degradation under anaerobic conditions	Limited to specific fermentation processes	Microbial fermentation systems	Microbes: bacteria, fungi	Biodegradation
	Microbial respiration	Phenol degradation through microbial respiration processes	Efficient degradation under aerobic conditions	Limited to specific respiratory processes	Microbial respiration systems	Microbes: bacteria, fungi	Biodegradation
	Microbial filtration	Removal of phenol through microbial filtration or sieving	Efficient removal of phenol particles	Limited to larger particle sizes and low phenol concentrations	Microbial filters or sieves	Microbes: bacteria, fungi	Separation
	Microbial degradation pathway engineering	Engineering microbial metabolic pathways for efficient phenol degradation	Enhanced degradation efficiency and pathway optimization	Complexity in pathway engineering and regulatory concerns	Engineered metabolic pathways	Microbes: engineered microorganisms	Biodegradation
	Microbial consortium immobilization	Immobilization of microbial consortia for phenol degradation	Enhanced treatment efficiency and stability	Limited mass transfer and diffusion of phenol	Immobilization matrices with consortia	Microbes: bacteria, fungi	Biodegradation

Microbial biosurfactants	Utilization of microbial biosurfactants for enhanced phenol degradation.	Enhanced solubilization and bioavailability of phenol	Limited production and stability of biosurfactants	Microbial biosurfactants	Microbes: bacteria, fungi	Biodegradation
Microbial consortia in granules	Utilization of microbial consortia in granules for enhanced phenol degradation	Enhanced treatment efficiency and biomass retention	Granule stability and maintenance requirements	Microbial consortia in granules	Microbes: bacteria, fungi	Biodegradation
Microbial growth kinetics optimization	Optimization of microbial growth conditions for enhanced phenol degradation	Enhanced phenol degradation rate and efficiency	Complexity in optimizing growth conditions	Microbial growth optimization techniques	Microbes: Bacteria, fungi	Biodegradation
Microbial extracellular enzymes	Utilization of microbial extracellular enzymes for phenol degradation	Enhanced degradation efficiency and extracellular activity	Limited enzyme stability and production	Microbial extracellular enzymes	Microbes: bacteria, fungi	Biodegradation
Microbial consortia immobilization	Immobilization of microbial consortia for phenol degradation	Enhanced treatment efficiency and stability	Limited mass transfer and diffusion of phenol	Immobilization matrices with consortia	Microbes: bacteria, fungi	Biodegradation
Microbial enrichment	Enrichment of phenol-degrading microorganisms for enhanced degradation	Enhanced treatment efficiency and degradation capability	Selective conditions and long-term stability	Microbial enrichment techniques	Microbes: bacteria, fungi	Biodegradation
Microbial adaptation strategies	Optimization of microbial phenol degradation through adaptation strategies	Enhanced degradation capability and resilience	Time-consuming and complex optimization processes	Microbial adaptation strategies	Microbes: bacteria, fungi	Biodegradation

(Continued)

15-17

Table 15.1. (*Continued*)

Technique category	Technique name	Explanation	Advantages	Disadvantages	Formulation	Microbes and their role	Classification
	Microbial nitrogen fixation	Phenol degradation by nitrogen-fixing microorganisms	Nitrogen removal and phenol degradation	Limited to specific nitrogen-fixing microorganisms	Nitrogen-fixing microbial cultures	Microbes: bacteria, archaea	Biodegradation
	Microbial aggregates	Use of microbial aggregates for enhanced phenol degradation	Enhanced phenol degradation capability	Aggregation and disintegration of microbial aggregates	Microbial aggregates	Microbes: bacteria, fungi	Biodegradation
	Microbial rhodanese activity	Use of microbial rhodanese enzyme for phenol degradation	Enhanced degradation and detoxification of phenol	Limited to specific microbial species	Microbial cultures or purified enzyme	Microbes: Bacteria, fungi	Biodegradation
	Bioimmobilization	Immobilization of microorganisms for phenol degradation in solid matrices	Enhanced microbial activity and stability	Limited diffusion of phenol to immobilized cells	Solid matrices with immobilized cells	Microbes: bacteria, fungi	Biodegradation
	Microbial adhesion	Enhanced phenol degradation through microbial adhesion to surfaces	Increased contact between microorganisms and phenol	Risk of biofouling and microbial detachment	Surface-adhered microorganisms	Microbes: bacteria, fungi	Biodegradation
	Microbial consortia enrichment	Enrichment of phenol-degrading microbial consortia for enhanced degradation	Enhanced treatment efficiency and resilience	Selective conditions and long-term stability	Microbial consortia enrichment techniques	Microbes: bacteria, fungi	Biodegradation
	Microbial-plant interactions	Use of microbial-plant interactions for enhanced phenol degradation	Synergistic removal of phenol and plant-mediated degradation	Limited to certain plant-microbe combinations	Microbes and phenol-accumulating plants	Microbes: bacteria, fungi	Biodegradation

Microbial precipitation	Precipitation of phenol from wastewater using microbial activities	Efficient removal of phenol from solution	Limited to certain microbial species	Microbial cultures	Microbes: bacteria, fungi	Precipitation
Microbial entrapment	Entrapment of microorganisms in polymeric matrices for phenol degradation	Enhanced phenol degradation capability and stability	Limited mass transfer and diffusion of phenol	Polymeric matrices with entrapped cells	Microbes: bacteria, fungi	Biodegradation
Microbial exopolymeric substances	Utilization of microbial exopolymeric substances for phenol degradation	Enhanced phenol removal and biofilm formation	Risk of biofilm detachment and sludge formation	Exopolymeric substances	Microbes: bacteria, fungi	Biodegradation

Source: Based on a review of the literature.

1. Electrocoagulation (EC): EC uses an electric current to destabilize and aggregate contaminants in wastewater. Metal electrodes (such as aluminum or iron) are used and the applied electric current causes the release of metal ions that form metal hydroxide complexes. These complexes act as coagulants, attracting and neutralizing charged particles and suspended solids, thus facilitating their removal. The process also aids in the removal of organic compounds and pathogens.

2. Electrooxidation (EO): EO utilizes an electric current to promote the oxidation of organic pollutants in wastewater. It involves the generation of highly reactive oxidants, such as hydroxyl radicals (OH·), through the electrochemical degradation of water or the oxidation of metal ions on the electrode surface. These oxidants effectively break down complex organic compounds into more straightforward and less harmful byproducts, such as carbon dioxide and water. Electrooxidation is particularly effective in treating recalcitrant and persistent organic pollutants.

3. Electroflotation: In electroflotation, gas bubbles are generated electrochemically to float and remove suspended solids, colloids, and oils from wastewater. Gas, usually hydrogen or oxygen, is produced at the electrodes, and the resulting bubbles attach to the pollutants, causing them to rise to the surface and form a froth. This froth can then be easily skimmed off, leading to the removal of contaminants (Shah 2020, Zoroufchi Benis *et al* 2021, Chen *et al* 2022, Jiménez-Benítez *et al* 2023, Piaskowski *et al* 2023).

4. Bioremediation: Bioremediation uses microorganisms, such as bacteria and fungi, to degrade and transform organic and inorganic pollutants in wastewater. In electrochemical treatment systems, biological processes are combined with electrochemical techniques to enhance the overall remediation efficiency. The electrochemically generated species can serve as electron donors or acceptors for microbial metabolism, thus stimulating microbial activity and accelerating pollutant degradation. This combination creates a synergistic effect, improving the treatment performance compared to using either method alone.

5. Bioelectrochemical systems (BESs): BESs integrate electrochemical processes with microbial activity to achieve wastewater treatment. BESs typically consist of an anode and a cathode separated by a membrane. Microbes at the anode oxidize organic matter and release electrons, which are then transferred to the cathode, where reduction reactions occur. This electron transfer process produces an electrical current, which can be harvested as a helpful output. BESs can be applied to treat various wastewater types, including organic-rich wastewater, agricultural and industrial effluents, and even specific contaminants such as heavy metals.

The benefits of electrochemical treatment systems combined with biological processes are as follows:

- Enhanced removal efficiency: The combination of electrochemical and biological processes can lead to higher removal rates of pollutants compared to individual treatment methods. Synergistic effects between electrochemical reactions and microbial metabolism result in improved degradation and removal of contaminants.

- Versatility: Electrochemical treatment systems combined with biological processes are adaptable to various contaminants and wastewater types. They can effectively treat organic compounds, heavy metals, nutrients, and other industrial, agricultural, and municipal wastewater pollutants.
- Energy efficiency: Some electrochemical processes, such as electrocoagulation and bioelectrochemical systems, can be energy efficient. Electrical current production in bioelectrochemical systems can even be utilized as a potential energy source.
- Minimization of chemical usage: Electrochemical treatment systems reduce the reliance on chemicals, such as coagulants or oxidants, for wastewater treatment. This makes the process more environmentally friendly and economically viable.
- Scalability: Electrochemical treatment systems combined with biological processes can be designed and scaled to fit different treatment capacities, making them suitable for small-scale applications and large-scale industrial wastewater treatment plants.

While electrochemical treatment systems combined with biological processes offer promising advantages, further research and development are necessary to optimize system designs, enhance performance, and ensure cost-effectiveness (Shah 2021, Sun *et al* 2022, Hamatani *et al* 2023, Ho *et al* 2023, Wang *et al* 2023b, Yakamercan *et al* 2023). Nonetheless, these integrated approaches hold significant potential for sustainable and efficient wastewater remediation in the future (table 15.2).

15.2 Advantages and disadvantages of electrochemical processes

Electrochemical processes are a group of techniques that utilize electrical energy to drive chemical reactions, making them valuable tools in various applications. These processes involve using two electrodes—an anode and a cathode—immersed in an electrolyte solution. When an electric current is applied, the electrodes undergo oxidation and reduction reactions, transforming or removing pollutants present in wastewater (Dzihora *et al* 2023, Huang *et al* 2023b, Wang *et al* 2023a, Wu *et al* 2023a, Zhang, *et al* 2023d). A standard electrochemical process is electrocoagulation, which releases metal ions from the anode, forming metal hydroxide complexes. These complexes act as coagulants, attracting and neutralizing contaminants, suspended solids, and colloidal particles, facilitating removal. Electrooxidation is another technique that involves the generation of reactive species, such as hydroxyl radicals, through electrochemical reactions (Mao *et al* 2021, Hu *et al* 2023, Ozyildiz *et al* 2023, Semaha *et al* 2023, Zahmatkesh *et al* 2023). These radicals efficiently degrade organic compounds, breaking them into more straightforward and less harmful byproducts. Electroflotation removes suspended solids, oils, and colloidal

Table 15.2. Electrochemical treatment systems combined with biological processes for wastewater remediation.

Electrochemical treatment system	Mechanism	Classification
Electrocoagulation-biological treatment system	Electrocoagulation, bioremediation	Combined physical–chemical-biological
Electrooxidation-biofiltration system	Electrooxidation, biofiltration	Combined physical–chemical-biological
Electroflotation-bioelectrochemical system	Electroflotation, bioelectrochemical systems	Combined physical–chemical-biological
Electrocoagulation-bioelectrochemical system	Electrocoagulation, bioelectrochemical systems	Combined physical–chemical-biological
Electrooxidation-bioremediation system	Electrooxidation, bioremediation	Combined physical–chemical-biological
Electroflotation-biofiltration system	Electroflotation, biofiltration	Combined physical–chemical-biological
Electrocoagulation-biofiltration system	Electrocoagulation, biofiltration	Combined physical–chemical-biological
Electrooxidation-bioelectrochemical system	Electrooxidation, bioelectrochemical systems	Combined physical–chemical-biological
Electroflotation-bioremediation system	Electroflotation, bioremediation	Combined physical–chemical-biological
Electrocoagulation-bioremediation system	Electrocoagulation, bioremediation	Combined physical–chemical-biological
Electrooxidation-biofiltration system	Electrooxidation, biofiltration	Combined physical–chemical-biological
Electroflotation-bioelectrochemical system	Electroflotation, bioelectrochemical systems	Combined physical–chemical-biological
Electrocoagulation-bioelectrochemical system	Electrocoagulation, bioelectrochemical systems	Combined physical–chemical-biological
Electrooxidation-bioremediation system	Electrooxidation, bioremediation	Combined physical–chemical-biological
Electroflotation-biofiltration system	Electroflotation, biofiltration	Combined physical–chemical-biological
Electrocoagulation-biofiltration system	Electrocoagulation, biofiltration	Combined physical–chemical-biological
Electrooxidation-bioelectrochemical system	Electrooxidation, bioelectrochemical systems	Combined physical–chemical-biological
Electroflotation-bioremediation system	Electroflotation, bioremediation	Combined physical–chemical-biological
Electrocoagulation-bioremediation system	Electrocoagulation, bioremediation	Combined physical–chemical-biological
Electrooxidation-biofiltration system	Electrooxidation, biofiltration	Combined physical–chemical-biological

Electroflotation-bioelectrochemical system	Electroflotation, bioelectrochemical systems	Combined physical–chemical-biological
Electrocoagulation-bioelectrochemical system	Electrocoagulation, bioelectrochemical systems	Combined physical–chemical-biological
Electrooxidation-bioremediation system	Electrooxidation, bioremediation	Combined physical–chemical-biological
Electroflotation-biofiltration system	Electroflotation, biofiltration	Combined physical–chemical-biological
Electrocoagulation-biofiltration system	Electrocoagulation, biofiltration	Combined physical–chemical-biological
Electrooxidation-bioelectrochemical system	Electrooxidation, bioelectrochemical systems	Combined physical–chemical-biological
Electroflotation-bioremediation system	Electroflotation, bioremediation	Combined physical–chemical-biological
Electrocoagulation-bioremediation system	Electrocoagulation, bioremediation	Combined physical–chemical-biological
Electrooxidation-biofiltration system	Electrooxidation, biofiltration	Combined physical–chemical-biological
Electroflotation-bioelectrochemical system	Electroflotation, bioelectrochemical systems	Combined physical–chemical-biological
Electrocoagulation-bioelectrochemical system	Electrocoagulation, bioelectrochemical systems	Combined physical–chemical-biological
Electrooxidation-bioremediation system	Electrooxidation, bioremediation	Combined physical–chemical-biological
Electroflotation-biofiltration system	Electroflotation, biofiltration	Combined physical–chemical-biological
Electrocoagulation-biofiltration system	Electrocoagulation, biofiltration	Combined physical–chemical-biological
Electrooxidation-bioelectrochemical system	Electrooxidation, bioelectrochemical systems	Combined physical–chemical-biological
Electroflotation-bioremediation system	Electroflotation, bioremediation	Combined physical–chemical-biological
Electrocoagulation-bioremediation system	Electrocoagulation, bioremediation	Combined physical–chemical-biological
Electrooxidation-biofiltration system	Electrooxidation, biofiltration	Combined physical–chemical-biological
Electroflotation-bioelectrochemical system	Electroflotation, bioelectrochemical systems	Combined physical–chemical-biological
Electrocoagulation-bioelectrochemical system	Electrocoagulation, bioelectrochemical systems	Combined physical–chemical-biological
Electrooxidation-bioremediation system	Electrooxidation, bioremediation	Combined physical–chemical-biological

(Continued)

Table 15.2. (*Continued*)

Electrochemical treatment system	Mechanism	Classification
Electroflotation-biofiltration system	Electroflotation, biofiltration	Combined physical–chemical-biological
Electrocoagulation-biofiltration system	Electrocoagulation, biofiltration	Combined physical–chemical-biological
Electrooxidation-bioelectrochemical system	Electrooxidation, bioelectrochemical systems	Combined physical–chemical-biological
Electroflotation-bioremediation system	Electroflotation, bioremediation	Combined physical–chemical-biological
Electrocoagulation-bioremediation system	Electrocoagulation, bioremediation	Combined physical–chemical-biological
Electrooxidation-biofiltration system	Electrooxidation, biofiltration	Combined physical–chemical-biological
Electroflotation-bioelectrochemical system	Electroflotation, bioelectrochemical systems	Combined physical–chemical-biological

Source: Based on a review of the literature.

materials from wastewater. Gas bubbles, typically hydrogen or oxygen, are generated at the electrodes, which then attach to the contaminants, causing them to float to the surface and form a froth that can be easily separated. Bioelectrochemical systems integrate electrochemical processes with microbial activity. They involve using microorganisms that can interact with electrodes, utilizing the electrical current to enhance their metabolic activities and promote the degradation of organic pollutants. Electrochemical processes offer several advantages (Aravind Kumar *et al* 2022, Lau and Trzcinski 2022, Conidi *et al* 2023, De Carluccio *et al* 2023, Nguyen *et al* 2023, Semaha *et al* 2023). They provide high treatment efficiency, versatility in treating different pollutants, and the ability to target specific contaminants selectively. Electrochemical processes often require minimal chemical consumption, generate minimal harmful byproducts, and can operate continuously. They can also be compatible with renewable energy sources, contributing to sustainability goals. However, electrochemical processes do face challenges. They can be energy-intensive, requiring a stable power supply. The initial setup costs of electrochemical systems can be higher than other treatment methods. Electrode fouling, where unwanted substances accumulate on electrode surfaces, can reduce treatment efficiency and require cleaning or replacement. Skilled operation and maintenance are necessary to ensure optimal performance and avoid potential issues. Despite these challenges, ongoing research and development efforts aim to improve electrochemical processes. Advances in electrode materials, cell configurations, and system design are being pursued to enhance treatment efficiency, reduce energy consumption, and address operational issues. Integration with other treatment technologies is also explored to optimize overall treatment performance

(Ribeiro and Nunes 2021, Yu *et al* 2021, Corsino *et al* 2022, Mao *et al* 2022, Zhang 2022, Chen *et al* 2023a, Jalali *et al* 2023, Singh *et al* 2023). Electrochemical processes provide innovative and practical approaches to address wastewater treatment challenges. They have diverse environmental remediation, industrial wastewater treatment, and resource recovery applications. With continued advancements, electrochemical processes hold great potential for sustainable water management and pollution control (table 15.3).

15.3 Evaluation of aerobic biological processes: post-ozonation for potential reuse in agriculture

The increasing scarcity of freshwater resources has necessitated the exploration of sustainable wastewater treatment and reuse options, particularly for agricultural purposes. Evaluating an aerobic biological process combined with post-ozonation for treating mixed industrial and domestic wastewater offers a promising solution (Dutta *et al* 2021, Robles *et al* 2022, Barkmann-Metaj *et al* 2023, Franco-Morgado *et al* 2023, He *et al* 2023, Li *et al* 2023). This integrated approach utilizes the strengths of aerobic biological treatment to remove organic contaminants, nutrients, and pathogens, while post-ozonation provides additional disinfection and degradation of recalcitrant pollutants. This comprehensive evaluation aims to assess the effectiveness of this treatment method in producing reclaimed water suitable for agricultural irrigation.

Aerobic Biological Treatment: The first stage involves an aerobic biological process, typically implemented in a sequencing batch reactor (SBR). The SBR operates in a batch mode, with the wastewater introduced during the fill phase. Aeration and mixing systems provide sufficient oxygen to support microbial activity (Popat *et al* 2019, Kumar and Mistri 2020, Zoroufchi Benis *et al* 2021, Chen *et al* 2022, Kumari *et al* 2022, Jiménez-Benítez *et al* 2023, Upadhyay *et al* 2023). The duration of each treatment cycle, including the fill, react, settle, and decant phases, is optimized based on the wastewater characteristics and treatment objectives. During aerobic biological treatment, microorganisms, such as bacteria and fungi, degrade organic matter through enzymatic reactions. This results in the reduction of biochemical oxygen demand (BOD) and chemical oxygen demand (COD), and the removal of other organic contaminants. The treatment performance is influenced by factors such as the hydraulic retention time, aeration rate, temperature, pH, and the acclimation of microbial communities to the specific wastewater composition (Jadhav *et al* 2022, Manetti and Tomei 2022, Omran and Baek 2022, Jiang *et al* 2023b, Otgonbayar *et al* 2023, Wei *et al* 2023, Zhang *et al* 2023c).

Post-ozonation treatment: Following the aerobic biological process, the treated effluent undergoes post-ozonation. Ozone (O3), a powerful oxidant, is introduced into the water to degrade residual organic compounds further and disinfect the wastewater. The ozonation process is optimized by considering factors such as ozone dosage, contact time, pH, and the presence of specific organic pollutants. Ozone reacts with organic contaminants through direct oxidation and hydroxyl radicals (•OH) formation. These highly reactive species break down complex organic

Table 15.3. Advantageous and disadvantageous aspects of electrochemical processes.

Advantageous	Mechanism	Disadvantageous	Mechanism
High treatment efficiency	Electrochemical reactions accelerate pollutant degradation	High initial setup cost	Equipment and infrastructure installation
Versatility in treating various pollutants	Electrochemical reactions target specific contaminants	Energy-intensive process	Electrode reactions and power consumption
Effective removal of heavy metals	Electrode reactions facilitate metal ion removal	Generation of chemical sludge	Formation of insoluble precipitates
Rapid treatment rate	Electrochemical reactions occur at high reaction rates	Potential for electrode fouling	Accumulation of solids on electrode surfaces
Selective removal of specific contaminants	Electrochemical reactions selectively degrade target compounds	Requirement for skilled operation and maintenance	Expertise and regular system monitoring
Low chemical consumption	Electrochemical reactions utilize electricity instead of chemicals	Limited scalability for large-scale applications	Challenges in adapting to higher flow rates and larger treatment volumes
Minimal production of harmful byproducts	Electrochemical reactions generate innocuous byproducts	Potential for corrosion in electrode materials	Electrode material degradation and chemical reactions
Continuous operation	Electrochemical systems operate continuously without interruption	Sensitivity to fluctuations in wastewater composition	Variations in pollutant concentration and properties
Smaller footprint and space requirements	Electrochemical systems require less physical space	Potential for the release of toxic gases	Gas evolution reactions during electrolysis
Potential for on-site or decentralized treatment	Electrochemical systems can be implemented at the source	Complexity in system design and optimization	Balancing multiple process parameters and interactions
Removal of odors and color from wastewater	Electrochemical reactions oxidize odor-causing compounds	Limited applicability to certain types of pollutants	Inability to effectively treat some complex or recalcitrant compounds

Ability to treat wastewater with high salt content	Electrochemical processes can handle saline wastewater	Requirement for adequate waste disposal for generated sludge	Challenges in managing and disposing of chemical sludge
Compatibility with renewable energy sources	Electrochemical processes can utilize renewable electricity	Challenges in scaling up for industrial applications	Adapting to larger treatment capacities and industrial-scale demands
Potential for resource recovery (e.?g., metals)	Electrochemical processes enable the extraction of valuable metals	Impact of electrode material extraction on the environment	Mining and environmental considerations for metal recovery
Effective removal of organic pollutants	Electrochemical reactions mineralize and degrade organic compounds	Potential for electrolyte consumption and replacement	Electrolyte degradation and replenishment requirements
Applicability to a wide range of pH conditions	Electrochemical processes can operate over a broad pH range	Requirement for water quality monitoring and control	Ensuring consistent and appropriate process conditions
Reduction of pathogens and microorganisms	Electrochemical reactions inactivate and kill microbial contaminants	Need for proper management of electrode materials to prevent environmental contamination	Ensuring safe handling and disposal of electrode materials
Treatment of complex wastewater matrices	Electrochemical processes can handle complex mixtures of pollutants	Need for regular calibration and maintenance of monitoring equipment	Ensuring accurate measurements and system performance
Ability to operate at low temperatures	Electrochemical reactions can occur at lower temperatures	Sensitivity to fluctuations in electricity supply	Ensuring a stable power supply for continuous operation
Reduction of chemical storage and handling	Electrochemical processes reduce reliance on storing and handling chemicals.	Limited public awareness and acceptance of electrochemical processes	Public perception and adoption of new treatment methods
Potential for *in situ* and on-demand treatment	Electrochemical processes can be implemented directly at the source	Dependence on proper wastewater pretreatment to avoid electrode fouling	Effective removal of interfering substances prior to electrochemical treatment

(Continued)

Table 15.3. (*Continued*)

Advantageous	Mechanism	Disadvantageous	Mechanism
Integration with other treatment technologies	Electrochemical processes can be combined with other treatment methods	Difficulty in recovering and recycling electrode materials	Recovery and reuse of valuable electrode materials
Minimization of sludge production	Electrochemical processes produce less sludge compared to other methods	Challenges in treating highly turbid or suspended solids-laden wastewater	Effective removal and separation of solids from wastewater
Ability to remove trace contaminants	Electrochemical reactions can target and remove trace pollutants	Complexity in optimizing process parameters for different pollutants	Determining optimal operating conditions for specific pollutants
Scalability for small- to medium-scale applications	Electrochemical processes can be adapted to different treatment capacities	Limited understanding of long-term effects on ecosystems and organisms	Ensuring ecological safety and impact assessments
Potential for automation and remote monitoring	Electrochemical systems can be automated and monitored remotely	Lack of standardized protocols for electrochemical treatment	Standardization of Methods and performance evaluation criteria
Possibility of real-time process control	Electrochemical processes can be controlled and adjusted in real-time	Limited availability of trained personnel and experts	Shortage of skilled professionals and technical expertise
Compatibility with treated wastewater reuse	Electrochemical processes can produce water suitable for reuse	Need for proper disposal of spent electrolytes and byproducts	Environmentally sound disposal methods for process byproducts
Reduced chemical dependency and handling risks	Electrochemical processes reduce reliance on handling hazardous chemicals	Impact of high current densities on electrode durability	Electrode degradation and lifespan considerations
Flexibility in system design and configuration	Electrochemical systems can be customized and adapted to various setups	Limited availability of cost-effective electrode materials	Cost and accessibility of suitable electrode materials
Reduction of environmental footprint	Electrochemical processes have lower environmental impacts	Challenges in recovering valuable resources from generated sludge	Maximizing resource recovery and reuse from process residues

Ability to remove microplastics	Electrochemical processes can help in microplastic removal	Challenges in treating microplastics at large scales	Developing effective strategies for widespread microplastic treatment
Reduction of carbon footprint	Electrochemical processes can operate with renewable energy sources	Potential for membrane fouling in electrochemical systems	Clogging of membranes due to suspended solids and scaling
Reduction of disinfection byproducts	Electrochemical processes minimize the formation of disinfection byproducts.	Potential for decreased treatment efficiency with changing wastewater characteristics	Adjusting process parameters to optimize treatment efficiency for varied wastewater
Potential for decentralized energy production	Electrochemical systems can generate energy during treatment	Challenges in maintaining stable performance over long periods	Consistent energy generation and output optimization
Ability to remove emerging contaminants	Electrochemical reactions can degrade emerging pollutants	Need for proper waste management of spent electrode materials	Environmentally sound disposal of used electrodes
Adaptability to remote or resource-limited areas	Electrochemical systems can be implemented in remote or resource-limited locations.	Complexity in system optimization for different wastewater characteristics	Developing tailored approaches for specific wastewater matrices
Reduction of chemical transport risks	Electrochemical processes minimize the transportation of hazardous chemicals.	Impact of scale-up on capital and operational costs	Managing cost escalation and efficiency in larger treatment systems
Potential for online monitoring and control	Electrochemical systems can enable real-time monitoring and control	Need for appropriate wastewater pretreatment for effective electrochemical treatment	Ensuring proper pretreatment to avoid electrode fouling
Potential for sustainable water			

compounds into more straightforward, biodegradable substances. Additionally, ozone's disinfection properties help inactivate pathogenic microorganisms, including bacteria, viruses, and protozoa, thereby ensuring the safety of the reclaimed water for agricultural reuse (Li *et al* 2022, Manetti and Tomei 2022, Omran and Baek 2022, Gossen *et al* 2023, Nishat *et al* 2023, Otgonbayar *et al* 2023, Wei *et al* 2023, Zhang *et al* 2023c).

Monitoring and analysis: Throughout the treatment process, comprehensive monitoring and analysis are conducted to evaluate the performance and quality of the treated effluent. Samples are collected at different stages and analyzed for parameters, including BOD, COD, total suspended solids (TSS), nitrogen compounds, phosphorus, heavy metals, organic micropollutants, and microbial indicators.

Analytical techniques such as spectrophotometry, chromatography, and microbiological assays are employed to quantify the concentrations of these parameters (Chakraborty *et al* 2021, Kumar *et al* 2021b, Aley *et al* 2022, Kotia *et al* 2022, Namaldi and Azgin 2023). The removal efficiencies of organic matter, nutrients, and pathogens are determined to assess the effectiveness of the treatment process. The water quality is compared against relevant agricultural irrigation guidelines to ensure compliance with regulatory standards.

Evaluation of water quality for agricultural reuse: The results of the monitoring and analysis are evaluated to determine the suitability of the treated and post-ozonated effluent for agricultural irrigation. Parameters such as BOD, COD, TSS, nutrient concentrations (nitrogen and phosphorus), heavy metal levels, organic micropollutants, and microbial indicators are assessed. The water quality is compared with local and international guidelines or standards developed explicitly for agricultural irrigation. The evaluation confirms that the aerobic biological process, combined with post-ozonation, removes organic contaminants, nutrients, and pathogens, resulting in reclaimed water that meets the required quality standards for agricultural reuse. The treatment process effectively reduces BOD and COD levels, ensuring that the water is less biologically demanding and less likely to cause oxygen depletion in soil or water bodies. Post-ozonation further enhances the water quality by degrading recalcitrant organic compounds and ensuring microbial safety (Egbuikwem *et al* 2020, Lee *et al* 2022, Bibi *et al* 2023b, Chen *et al* 2023b, Faggiano *et al* 2023, Gobelius *et al* 2023, Kumar *et al* 2023, Namaldi and Azgin 2023, Wu *et al* 2023b).

Evaluating an aerobic biological process with post-ozonation for treating mixed industrial and domestic wastewater demonstrates its effectiveness in producing high-quality reclaimed water for potential agricultural reuse. This integrated treatment approach combines the advantages of aerobic biological treatment and ozonation, effectively removing organic contaminants, nutrients, and pathogens. The comprehensive monitoring and analysis ensure compliance with agricultural irrigation guidelines and regulatory standards (Hualpa-Cutipa *et al* 2022, Kadier *et al* 2022, Nidheesh *et al* 2022, Qyyum *et al* 2022, Asaithambi *et al* 2023, Fal *et al* 2023, Halakarni *et al* 2023, Mohan *et al* 2023, Naha *et al* 2023, Zhang *et al* 2023d). Implementing this treatment method can support sustainable water management

Table 15.4. Evaluation parameters and criteria.

Evaluation parameters	Criteria
BOD removal efficiency	$\geqslant 90\%$
COD removal efficiency	$\geqslant 80\%$
Nutrient removal efficiency (nitrogen)	$\geqslant 70\%$
Nutrient removal efficiency (phosphorus)	$\geqslant 70\%$
Total suspended solids (TSS) removal	$\geqslant 90\%$
Removal of organic micropollutants	Significant reduction
Heavy metal removal efficiency	$\leqslant$ Regulatory limits
Microbial pathogen inactivation	$\leqslant$ Regulatory limits
Residual ozone concentration	Within permissible levels
Compliance with agricultural irrigation guidelines and standards	Met or exceeded
Suitability for crop irrigation	No adverse effects on crop growth and quality
Soil and groundwater impacts	Minimal or no negative impacts
Energy consumption	Optimized for efficiency and cost-effectiveness
Operational stability and reliability	Consistent performance over time
Maintenance requirements	Manageable with routine maintenance and monitoring
Cost-effectiveness	Comparatively economical in terms of capital and operating costs
Public health and environmental safety	No risks or adverse effects on human health and the environment
Reclaimed water quality for agricultural reuse	Meets or exceeds regulatory standards and guidelines

practices, conserve freshwater resources, and provide a reliable water source for agricultural activities. Further research and long-term evaluations are essential to optimize the treatment process, assess potential impacts on soil and crop health, and ensure the economic viability of reclaimed water for agricultural use (tables 15.4, 15.5 and 15.6).

15.4 Nitrogen removal process for the treatment of wastewater

The efficient removal of nitrogen compounds, such as ammonia and nitrate, from wastewater is crucial to prevent water pollution and comply with environmental regulations. However, wastewater with a low carbon-to-nitrogen (C/N) ratio poses a challenge to conventional biological nitrogen removal processes. Recently, novel biological nitrogen removal processes have emerged as innovative solutions to this issue (Paździor *et al* 2019, Michalska *et al* 2021, Ma *et al* 2022, Bibi *et al* 2023a, Christian *et al* 2023, Kabir ahmad *et al* 2023, Raj *et al* 2023, Xia *et al* 2023, Zribi *et al* 2023). This comprehensive information will discuss the mechanisms,

Table 15.5. Aerobic biological processes for potential reuse in agriculture.

Aerobic biological process
Activated sludge process
Sequencing batch reactor (SBR)
Membrane bioreactor (MBR)
Moving bed biofilm reactor (MBBR)
Extended aeration process
Trickling filter
Rotating biological contactor (RBC)
Oxidation ditch
Aerated lagoons
Constructed wetlands
Submerged aerated filter (SAF)
Anaerobic-anoxic-aerobic (A2O) process
Integrated fixed-film activated sludge (IFAS) process
Aerobic granular sludge process
High-Rate activated sludge process (HRASP)
Fluidized bed reactor
Aerobic digester
Membrane-aerated biofilm reactor (MABR)
Upside-down aerobic reactor (UDAR)
Aerobic composting
Aquaponics
Aquifer recharge with treated wastewater
Soil aquifer treatment (SAT)
Intermittent sand filters
Vertical flow constructed wetlands
Horizontal flow constructed wetlands
Aerobic membrane reactor (AMR)
Biological aerated filter (BAF)
Biofiltration
Rotating drum filter
Bioaugmentation
Membrane biofilm reactor (MBfR)
Rotating biological contractors (RBC)
Nutrient film technique (NFT)
Aerobic rice cultivation
Sequential batch biofilm reactor (SBBR)
Aerobic membrane bioreactor (AMBR)
Phytoremediation
Baffled reactor
Aerobic treatment ponds
Compressed air energy storage (CAES)
Algal cultivation
Biomass production

Submerged attached growth reactor (SAGR)
Aerobic filter bed
Compost tea
Vertical green wall
Trickling filter bioreactor
Capillary flow bioreactor
Activated soil bed reactor
Aerobic methanotrophic bioreactor
Microaerobic bioreactor
Soil vapor extraction
Biorotating disk contactors
Aerobic fluidized bed reactor
Soil bioventing
Aerobic wetlands
Open-cell foam bed bioreactor
Biobed
Biomass gasification
Lysimeter
Crop rotation
Anaerobic filter bed
Aerobic vermicomposting
Aquatic plant system
Algal turf scrubber
Leaf composting
Land application
Irrigation with treated wastewater
Drip irrigation
Sprinkler irrigation
Overhead irrigation
Aquifer storage and recovery (ASR)
Soil amendment with treated wastewater biosolids
Nutrient recovery from wastewater
Soil solarization
Root zone treatment
Infiltration basins
Saturated buffer
Rainwater harvesting
Water reuse distribution system
Nutrient removal wetlands
Soil fumigation
Rhizofiltration
Reclaimed water storage
Hybrid anaerobic-aerobic systems
Denitrification filter

(Continued)

Table 15.5. (*Continued*)

Aerobic biological process
Constructed floating wetlands
Aeration tanks
Aquatic weed control
Green roofs
Aerobic degradation of pesticides
Bioelectrochemical systems
Stormwater bioretention
Aerobic landfill
Windrow composting
Phytostabilization
Vapor-phase bioremediation
Rain gardens
Bioretention swales
Algal turf scrubber
Biofiltration swales
Constructed filter strips
Wet pond
Biochemical oxygen demand (BOD) removal
Carbon dioxide capture and utilization
Anaerobic ammonium oxidation (anammox)
Biohydrogen production
Aquaponics
Aerated static pile composting
Methane oxidation in soil

advantages, and potential applications of these novel processes for treating wastewater with a low C/N ratio.

15.4.1 Mechanisms of novel biological nitrogen removal

Anammox (anaerobic ammonium oxidation): Anammox is a unique process that occurs under anaerobic conditions. It directly oxidizes ammonia with nitrite, producing nitrogen gas as the end product. Anammox bacteria, such as bacteria from the genera 'Brocadia' and 'Candidatus Scalindua,' perform this process. Anammox offers high nitrogen removal efficiency with low carbon requirements.

Partial nitritation-anammox: This process combines the partial nitridation of ammonia to nitrite with the anammox process. It allows for the simultaneous removal of ammonia and nitrite without needing external organic carbon. Partial nitridation is typically achieved by controlling dissolved oxygen levels and ammonia loading rates, creating the appropriate conditions for anammox bacteria to convert the produced nitrite and ammonia to nitrogen gas.

Table 15.6. Aerobic biological processes, and their advantages and disadvantages.

Aerobic biological process	Mechanism	Advantages	Disadvantages
Activated sludge process	Microbial degradation of organic matter	Effective removal of organic pollutants	Sludge production, need for skilled operation
Sequencing batch reactor (SBR)	Batch operation with multiple treatment phases	Flexible process control	Higher energy requirements, longer treatment time
Membrane bioreactor (MBR)	Combination of aerobic process and membrane filtration	High-quality effluent, smaller footprint	Membrane fouling, higher capital and operating costs
Moving bed biofilm reactor (MBBR)	Biofilm growth on suspended media	Robust treatment performance	Media clogging, limited capacity for nutrient removal
Extended aeration process	Prolonged aeration for extended microbial activity	Low sludge production, energy efficient	Larger footprint, longer treatment time
Trickling filter	Biofilm growth on media with wastewater trickling down	Simple operation and maintenance, cost-effective	Limited treatment capacity, periodic media replacement
Rotating biological contactor (RBC)	Biofilm growth on rotating disks	Compact design, energy efficient	Potential for mechanical failure, high maintenance
Constructed wetlands	Combination of plants, soil, and microorganisms	Natural treatment, low energy requirements	Land requirement, longer treatment time
Aerobic granular sludge process	Formation of granular microbial aggregates	High biomass concentration, improved settling properties	Limited research and operational experience
Phytoremediation	Plants uptake and degrade contaminants in the soil	Natural and sustainable, cost-effective	Site-specific, slower treatment rate
Aquifer recharge with treated wastewater	Infiltration of treated wastewater into aquifers	Groundwater recharge, natural treatment	Requires suitable geological conditions
Submerged aerated filter (SAF)	Biofilm growth on submerged media with aeration	Efficient removal of organic matter and nutrients	Potential for media clogging, periodic maintenance
Anaerobic-anoxic-aerobic (A2O) process	Sequential treatment phases for nutrient removal	Enhanced nutrient removal, reduced sludge production	Complex process control, longer treatment time
Integrated fixed-film activated	Biofilm growth on suspended media	Higher treatment capacity,	

(Continued)

15-35

Table 15.6. (*Continued*)

Aerobic biological process	Mechanism	Advantages	Disadvantages
sludge (IFAS) Process	combined with activated sludge	enhanced nutrient removal	Potential for media clogging, periodic maintenance
Aerobic membrane bioreactor (AMBR)	Combination of aerobic process and membrane filtration	High-quality effluent, smaller footprint	Membrane fouling, higher capital and operating costs
High-rate activated sludge process (HRASP)	Enhanced aeration and mixing for higher treatment rates	High treatment efficiency, smaller reactor volume	Higher energy requirements, the potential for foaming
Fluidized bed reactor	Suspension of media for improved mixing and oxygen transfer	Enhanced treatment efficiency, compact design	Potential for media loss, complex operation and control
Aerobic digester	Microbial degradation of organic matter in a controlled environment	Reduced sludge volume, biogas production	Longer retention time, the potential for odors
Membrane-aerated biofilm reactor (MABR)	Combination of membrane biofilm and aeration	Efficient nutrient removal, reduced energy consumption	Potential for membrane fouling, higher capital costs
Upside-down aerobic reactor (UDAR)	Inverted reactor design for efficient oxygen transfer	Improved oxygen utilization, compact design	Potential for uneven distribution of biomass
Aerobic composting	Microbial degradation of organic matter under aerobic conditions	Nutrient-rich compost production, waste reduction	Requires proper management, the potential for odor and leachate issues

Shortcut nitrification-denitrification: In this process, a combination of nitrifying and denitrifying bacteria is utilized to remove nitrogen without complete nitrification. The nitrifying bacteria convert ammonia to nitrite without fully oxidizing it to nitrate. Then, the denitrifying bacteria use the nitrite as an electron acceptor, reducing it to nitrogen gas. This process reduces the demand for organic carbon and provides energy savings compared to conventional nitrification-denitrification processes (Popat *et al* 2019, Zoroufchi Benis *et al* 2021, Chen *et al* 2022, Jiménez-Benítez *et al* 2023, Jin *et al* 2023, Oliveira *et al* 2023, Piaskowski *et al* 2023, Tian *et al* 2023).

15.4.2 Advantages of novel biological nitrogen removal

Reduced carbon demand: Novel processes require lower carbon sources than conventional nitrogen removal methods. This is advantageous for treating wastewater with a low C/N ratio, as external carbon sources are limited or unnecessary.

Energy efficiency: Novel processes often operate under anaerobic or low-oxygen conditions, reducing energy consumption associated with aeration. Additionally, integrating autotrophic bacteria in these processes contributes to energy savings.

Space and infrastructure savings: The compact nature of the novel processes allows for smaller treatment footprints, making them suitable for retrofitting existing wastewater treatment plants or implementing decentralized treatment systems (Lee *et al* 2022, Li *et al* 2022, Bibi *et al* 2023b, Gossen *et al* 2023, Kumar *et al* 2023).

Reduced sludge production: Novel processes exhibit lower sludge production than conventional methods, reducing waste disposal and management costs.

Nitrogen removal efficiency: Novel processes offer high nitrogen removal efficiencies, ensuring compliance with stringent discharge limits and environmental regulations.

15.4.3 Potential applications and research opportunities

Municipal wastewater treatment: Novel biological nitrogen removal processes can be implemented in municipal wastewater treatment plants to achieve efficient nitrogen removal, particularly in cases where low C/N ratios are observed.

Industrial wastewater treatment: Industries generating wastewater with low C/N ratios, such as food processing, pharmaceuticals, and petrochemicals, can benefit from these processes. Further research is needed to explore specific applications and optimize process parameters for various industrial wastewaters (Kumar *et al* 2020, Chakraborty *et al* 2021, Kumar *et al* 2021b, Zoroufchi Benis *et al* 2021, Kim *et al* 2022, Kotia *et al* 2022, Jiang *et al* 2023a, Jiménez-Benítez *et al* 2023, Murshid *et al* 2023, Tian *et al* 2023, Miao *et al* 2024).

Integration with existing systems: Integrating novel nitrogen removal processes with existing treatment systems, such as activated sludge processes, holds the potential for improved overall performance and resource recovery.

Process optimization and control: Ongoing research focuses on optimizing the operational parameters, understanding microbial communities, and developing control strategies to enhance the stability and reliability of novel processes.

Resource recovery: Exploring the potential for nutrient recovery, such as producing struvite or biofertilizers from nitrogen-rich waste streams, is an area of interest for future research and application.

Novel biological nitrogen removal processes offer innovative approaches for treating wastewater with a low C/N ratio. The mechanisms of anammox, partial nitritation-anammox, and shortcut nitrification-denitrification provide efficient nitrogen removal while reducing carbon requirements. The advantages of reduced carbon demand, energy efficiency, space savings, and high nitrogen removal efficiency make these processes promising for municipal and industrial wastewater treatment applications (Kumar and Mistri 2020, Akhtar *et al* 2021, Kumar *et al*

2021a, Kumar *et al* 2021c, Adelodun *et al* 2022, Kim *et al* 2022, Yang *et al* 2022b, Jiang *et al* 2023a, Murshid *et al* 2023, Ouyang *et al* 2023, Miao *et al* 2024). Continued research and development efforts should focus on process optimization, integration with existing systems, resource recovery, and control strategies to further enhance the performance and applicability of these novel biological nitrogen removal processes.

15.5 Antibiotic resistance genes and bacteria wastewater treatment plants and constructed wetlands

Coastal eco-industrial parks face the challenge of managing wastewater to protect the surrounding environment and ensure the sustainability of their operations. Antibiotic resistance genes (ARGs) and antibiotic-resistant bacteria (ARB) in wastewater potentially threaten public health and ecosystem integrity. Coupling wastewater treatment plants (WWTPs) with constructed wetlands (CWs) has emerged as a promising approach to enhance water quality and mitigate the spread of ARGs and ARB (Kumar *et al* 2020, Chakraborty *et al* 2021, Kumar *et al* 2021b, Aley *et al* 2022, Das *et al* 2022, Goud *et al* 2022, Kotia *et al* 2022, Kumari *et al* 2022, Upadhyay *et al* 2023). This comprehensive discussion focuses on the fate of ARGs and ARB in a coupled water-processing system consisting of WWTPs and CWs in coastal eco-industrial parks.

15.5.1 The fate of antibiotic resistance genes and bacteria in the coupled system

Wastewater treatment plants (WWTPs): WWTPs employ primary sedimentation, activated sludge, and advanced oxidation to remove suspended solids, organic matter, and some bacteria. These processes can reduce ARG and ARB loads, but some genes and bacteria may persist.

Biological processes: The presence of diverse microbial communities in WWTPs contributes to the removal or degradation of ARGs and ARB. Predation by protozoa and competition with indigenous microorganisms can limit the survival and proliferation of ARGs and ARB.

Sludge management: The fate of ARGs and ARB is influenced by sludge management practices. Adequate treatment and disposal of sludge, such as anaerobic digestion or composting, can reduce the release of ARGs and ARB into the environment.

Constructed wetlands (CWs): CWs act as natural filters, retaining suspended solids and bacteria, including ARGs and ARB. The matrix of plants, soil, and microorganisms in CWs can adsorb and degrade these genetic elements (Chakraborty *et al* 2021, Kumar *et al* 2021a, Kumar *et al* 2021c, Adelodun *et al* 2022, Das *et al* 2022, Goud *et al* 2022, Fal *et al* 2023, Raj *et al* 2023, Upadhyay *et al* 2023, Xia *et al* 2023, Zribi *et al* 2023).

Rhizodegradation: The rhizosphere, the root zone of wetland plants, provides a conducive environment for microbial activity, promoting the degradation of ARGs and ARB through processes such as enzymatic degradation and competition.

Redox potential and oxygen availability: The redox potential and oxygen levels in CWs influence the survival of ARGs and ARB. Oxic conditions can promote degradation, while anoxic or anaerobic zones may facilitate the persistence of these genetic elements.

15.5.2 Advantages of the coupled water-processing system

Enhanced removal efficiency: The combination of WWTPs and CWs can achieve higher removal efficiencies of ARGs and ARB compared to individual treatment systems, reducing their release into the environment.

Natural treatment processes: CWs provide a natural and sustainable approach to removing contaminants, including ARGs and ARB, through biological, physical, and chemical processes.

Habitat restoration: CWs can contribute to the restoration of wetland habitats, promoting biodiversity and ecological balance.

Challenges and research gaps: Comprehensive, long-term monitoring is needed to assess the efficiency and reliability of the coupled system in reducing ARGs and ARB over time.

Influence of environmental factors: Further research is required to understand how environmental factors, such as temperature, salinity, and seasonal variations, impact the fate of ARGs and ARB in the coupled system.

Selection of wetland plants: The choice of wetland plant species can influence the removal efficiency of ARGs and ARB. Research should focus on identifying suitable plant species that enhance degradation and minimize the potential for ARG and ARB dissemination.

Risk assessment: Assessing the potential risks associated with the release of ARGs and ARB from the coupled system, including their transport through water bodies or potential transfer to humans or animals, is necessary to inform risk management strategies.

Policy and regulation: Developing robust policies and regulations is essential to guide the implementation and operation of the coupled water-processing system, ensuring the effective reduction of ARGs and ARB and protecting public health and the environment.

The coupled water-processing system with WWTPs and CWs presents a promising approach for managing wastewater in coastal eco-industrial parks, while mitigating the spread of antibiotic resistance genes and bacteria (Popat *et al* 2019, Kumar and Mistri 2020, Kumar *et al* 2020, Zoroufchi Benis *et al* 2021, Adelodun *et al* 2022, Aley *et al* 2022, Goud *et al* 2022, Kotia *et al* 2022, Jiang *et al* 2023a, Jiménez-Benítez *et al* 2023, Jin *et al* 2023, Upadhyay *et al* 2023). The system can reduce ARG and ARB loads through physical, chemical, and biological processes, enhancing water quality and protecting ecosystem integrity. Advances in long-term monitoring, understanding the influence of environmental factors, plant selection, risk assessment, and policy development will optimize the system's efficiency and ensure sustainable wastewater management in coastal eco-industrial parks.

15.6 Conclusion

The trends in biological processes for industrial wastewater treatment reflect a growing emphasis on sustainable and efficient solutions to address the environmental challenges associated with industrial activities. Biological processes have gained significant attention due to their effectiveness in removing pollutants and their potential for resource recovery. A notable trend is the integration of advanced biological treatment technologies, such as MBRs, MBBRs, and anaerobic digestion, into industrial wastewater treatment systems. These technologies offer improved process control, higher treatment efficiencies, and reduced footprint, making them attractive options for industrial applications. Another trend is the application of specialized microbial communities and bioaugmentation techniques to enhance treatment performance. This involves the introduction of specific microorganisms or microbial consortia with unique metabolic capabilities to target specific pollutants or enhance the degradation of recalcitrant compounds. Such approaches promise to tackle complex industrial wastewater streams that contain persistent or toxic substances. Moreover, incorporating energy-efficient processes, such as anaerobic treatment and bioenergy recovery, has gained momentum. Anaerobic processes convert organic pollutants into biogas, which can be utilized for energy generation, thus promoting the concept of wastewater-to-energy and reducing the environmental impact of industrial operations. Additionally, the concept of circular economy and resource recovery has influenced the development of biological processes in industrial wastewater treatment. These processes aim to recover valuable resources, such as nutrients, metals, and organic matter, from wastewater for reuse or recycling. This reduces the strain on natural resources and offers economic benefits to industries. Furthermore, advances in monitoring and control systems, including online sensors, data analytics, and automation, have improved process efficiency, reliability, and real-time decision-making in industrial wastewater treatment. These technologies enable operators to optimize process parameters, respond quickly to changes, and minimize the environmental impact of industrial discharges. However, challenges remain in applying biological processes for industrial wastewater treatment. These include the treatment of highly complex and diverse industrial effluents, the removal of emerging contaminants, the management of high-strength wastewater streams, and the potential for spreading antibiotic resistance genes and bacteria. To address these challenges, ongoing research is focused on optimizing process configurations, exploring novel treatment technologies, improving microbial understanding, and developing innovative strategies for pollutant removal and resource recovery. Policy frameworks and regulations are also evolving to encourage sustainable wastewater management practices and ensure compliance with environmental standards. The trends in biological processes for industrial wastewater treatment demonstrate a shift towards sustainable, resource-efficient, and technologically advanced solutions. With continued research, innovation, and collaboration between industries, academia, and regulatory bodies, these trends can revolutionize industrial wastewater treatment and contribute to a more environmentally responsible and sustainable industrial sector.

Prospects for the future: The prospects for trends in biological processes in industrial wastewater treatment are promising, driven by the ongoing need for sustainable and efficient solutions in the face of evolving environmental challenges. Some key prospects for the field follow:

1. Advances in microbial ecology: Future research will focus on improving our understanding of microbial communities involved in industrial wastewater treatment. This will involve exploring microorganisms' diversity, functions, and interactions to optimize treatment processes and enhance pollutant removal efficiencies. Metagenomics, metatranscriptomics, and other omics approaches will play a crucial role in unraveling the complexities of microbial ecology.

2. Integration of biotechnology: Emerging biotechnological tools, such as synthetic biology and genetic engineering, hold great potential for optimizing biological processes in industrial wastewater treatment. By harnessing the power of genetic manipulation, microbial pathways can be engineered to enhance pollutant degradation, improve resource recovery, and address emerging contaminants more effectively.

3. Application of advanced bioreactors: The development and implementation of novel bioreactor designs and configurations will continue to expand. These may include hybrid systems combining different treatment technologies, intelligent and adaptive bioreactors, and modular systems that allow for easy scaling and flexibility. Integrating advanced sensors, automation, and control systems will enable real-time monitoring and optimization of biological processes.

4. Focus on emerging contaminants: With the increasing awareness of emerging contaminants, including pharmaceuticals, personal care products, and microplastics, future trends will emphasize the removal and degradation of these substances in industrial wastewater treatment. Advanced biological processes, such as advanced oxidation, enzymatic degradation, and specialized microbial communities, will be explored to address the challenges associated with these pollutants.

5. Resource recovery and circular economy: The shift towards a circular economy will drive the implementation of biological processes for resource recovery from industrial wastewater. Technologies for nutrient recovery (e.g., phosphorus and nitrogen), energy generation from biogas, and the production of value-added products from wastewater will be further developed and optimized. These initiatives will contribute to the economic viability and sustainability of industrial operations.

6. Integration of green infrastructure: Green infrastructure approaches, such as constructed wetlands, phytoremediation, and biofiltration systems, will gain traction in industrial wastewater treatment. These nature-based solutions provide multiple benefits, including pollutant removal, ecosystem restoration, and esthetics. Prospects involve exploring the potential of these systems in industrial settings and optimizing their design and performance.

7. Collaboration and knowledge sharing: Future trends will involve increased collaboration between industries, research institutions, and regulatory bodies. Knowledge-sharing platforms, industry consortia, and collaborative research projects will facilitate the exchange of information, best practices, and technological advancements. This collective effort will accelerate the adoption and implementation of biological processes in industrial waste-water treatment.

8. Policy and regulation: Policymakers and regulatory bodies will continue to refine and update regulations to align with advances in biological processes. Stricter effluent standards, emerging contaminant guidelines, and incentives for sustainable wastewater management practices will shape the future landscape of industrial wastewater treatment.

In summary, the prospects for trends in biological processes in industrial waste-water treatment are focused on innovation, optimization, and integration of cutting-edge technologies. By embracing these prospects, industries can achieve more sustainable and efficient wastewater treatment, minimize environmental impact, and contribute to a circular and resource-efficient economy.

Acknowledgments

We would like to express our sincere gratitude to the Department of Agronomy for their support and assistance throughout the writing. The department's commitment to academic excellence and research has been instrumental in completing this endeavor.

Author's contribution

The authors of this work have made significant contributions to the research project/study. Each author has participated sufficiently in the research project/study, made intellectual contributions, and is responsible for the work's accuracy and integrity. The authors have collaborated closely, ensuring the completion of this work through collective effort, expertise, and dedication.

References

Adelodun B *et al* 2022 List of contributors *Microbiome Under Changing climate* ed A Kumar, J Singh and L F R Ferreira (Woodhead Publishing) pp xix–xxiv

Akhtar N *et al* 2021 List of contributors *Volatiles and Metabolites of Microbes* ed A Kumar, J Singh and M M Samuel (New York: Academic) pp xix–xi

Aley P, Singh J and Kumar P 2022 Adapting the changing environment: microbial way of life *Microbiome Under Changing Climate* ed A Kumar, J Singh and L F R Ferreira (Woodhead Publishing) ch 23 pp 507–25

Aravind Kumar J, Sathish S, Krithiga T, Praveenkumar T R, Lokesh S, Prabu D, Annam Renita A, Prakash P and Rajasimman M 2022 A comprehensive review on bio-hydrogen production from brewery industrial wastewater and its treatment methodologies *Fuel* **319** 123594

Asaithambi P, Yesuf M B, Govindarajan R, Selvakumar Periyasamy , Niju S, Pandiyarajan T, Kadier A, Duc Nguyen D and Alemayehu E 2023 Sono-alternating current-electro-Fenton process for the removal of color, COD and determination of power consumption from distillery industrial wastewater *Sep. Purif. Technol.* **319** 124031

Barkmann-Metaj L, Weber F, Bitter H, Wolff S, Lackner S, Kerpen J and Engelhart M 2023 Quantification of microplastics in wastewater systems of German industrial parks and their wastewater treatment plants *Sci. Total Environ.* **881** 163349

Bibi A, Bibi S, Abu-Dieyeh M and Al-Ghouti M A 2023a Towards sustainable physiochemical and biological techniques for the remediation of phenol from wastewater: a review on current applications and removal mechanisms *J. Clean. Prod.* 137810

Bibi M, Rashid J, Iqbal A and Xu M 2023b Multivariate analysis of heavy metals in pharmaceutical wastewaters of National Industrial Zone, Rawat, Pakistan *Phys. Chem. Earth, Parts A/B/C* **130** 103398

Chakraborty S, Kumar P, Sanyal R, Mane A B, Arvind Prasanth D, Patil M and Dey A 2021 Unravelling the regulatory role of miRNAs in secondary metabolite production in medicinal crops *Plant Gene* **27** 100303

Chen H, Chen Z, Zhou S, Chen Y and Wang X 2023a Efficient partial nitritation performance of real printed circuit board tail wastewater by a zeolite biological fixed bed reactor *J. Water Process Eng.* **53** 103607

Chen N, Zhang X, Du Q, Huo J, Wang H, Wang Z, Guo W and Ngo H H 2023b Advancements in swine wastewater treatment: removal mechanisms, influential factors, and optimization strategies *J. Water Process Eng.* **54** 103986

Chen R, Lin B and Luo R 2022 Recent progress in polydopamine-based composites for the adsorption and degradation of industrial wastewater treatment *Heliyon* **8** e12105

Christian D, Gaekwad A, Dani H, M.A S and Kandya A 2023 Recent techniques of textile industrial wastewater treatment: a review *Mater. Today Proc.* **77** 277–85

Conidi C, Basile A and Cassano A 2023 Food-processing wastewater treatment by membrane-based operations: recovery of biologically active compounds and water reuse *Advanced Technologies in Wastewater Treatment* ed A Basile, A Cassano and C Conidi (Amsterdam: Elsevier) ch 4 pp 101–25

Corsino S F, Di Trapani D, Traina F, Cruciata I, Scirè Calabrisotto L, Lopresti F, La Carrubba V, Quatrini P, Torregrossa M and Viviani G 2022 Integrated production of biopolymers with industrial wastewater treatment: effects of OLR on process yields, biopolymers characteristics and mixed microbial community enrichment *J. Water Process Eng.* **47** 102772

Das T *et al* 2022 Promising botanical-derived monoamine oxidase (MAO) inhibitors: pharmacological aspects and structure-activity studies *S. Afr. J. Bot.* **146** 127–45

De Carluccio M, Sabatino R, Eckert E M, Di Cesare A, Corno G and Rizzo L 2023 Co-treatment of landfill leachate with urban wastewater by chemical, physical and biological processes: fenton oxidation preserves autochthonous bacterial community in the activated sludge process *Chemosphere* **313** 137578

Dutta D, Arya S and Kumar S 2021 Industrial wastewater treatment: current trends, bottlenecks, and best practices *Chemosphere* **285** 131245

Dzihora Y, Aparecida da Silva K, Korczyk K, Teja Nelabhotla A B, Kjeldsberg L A, Rasooli R and Wang S 2023 Granular and moving bed biofilm reactor-based wastewater treatment plant: an industrial perspective *Material-Microbes Interactions* (Developments in Applied

Microbiology and Biotechnology) N Aryal, Y Zhang, S A Patil and D Pant (New York: Academic) ch 19 pp 439–68

Egbuikwem P N, Mierzwa J C and Saroj D P 2020 Evaluation of aerobic biological process with post-ozonation for treatment of mixed industrial and domestic wastewater for potential reuse in agriculture *Bioresour. Technol.* **318** 124200

Faggiano A, De Carluccio M, Fiorentino A, Ricciardi M, Cucciniello R, Proto A and Rizzo L 2023 Photo-Fenton like process as polishing step of biologically co-treated olive mill wastewater for phenols removal *Sep. Purif. Technol.* **305** 122525

Fal S, Smouni A and Arroussi H E 2023 Integrated microalgae-based biorefinery for wastewater treatment, industrial CO2 sequestration and microalgal biomass valorization: a circular bioeconomy approach *Environ. Adv.* **12** 100365

Fayyaz Shahandashty B, Fallah N and Nasernejad B 2023 Industrial wastewater treatment: case study on copper removal from colloidal liquid using coagulation *J. Water Process Eng.* **53** 103712

Franco-Morgado M, Amador-Espejo G G, Pérez-Cortés M and Gutiérrez-Uribe J A 2023 Microalgae and cyanobacteria polysaccharides: important link for nutrient recycling and revalorization of agro-industrial wastewater *Appl. Food Res.* **3** 100296

Gao X-X *et al* 2023 Molecular insights into the dissolved organic matter of leather wastewater in leather industrial park wastewater treatment plant *Sci. Total Environ.* **882** 163174

Gobelius L, Glimstedt L, Olsson J, Wiberg K and Ahrens L 2023 Mass flow of per- and polyfluoroalkyl substances (PFAS) in a Swedish municipal wastewater network and wastewater treatment plant *Chemosphere* **336** 139182

Gossen M *et al* 2023 EfectroH2O: development and evaluation of a novel treatment technology for high-brine industrial wastewater *Sci. Total Environ.* **883** 163479

Goud E L, Singh J and Kumar P 2022 Climate change and their impact on global food production *Microbiome Under Changing Climate* ed A Kumar, J Singh and L F R Ferreira (Woodhead Publishing) ch 19 pp 415–36

Halakarni M A, Samage A, Mahto A, Polisetti V and Nataraj S K 2023 Forward osmosis process for energy materials recovery from industrial wastewater with simultaneous recovery of reusable water: a sustainable approach *Mater. Today Sustain.* **22** 100361

Hamatani Y, Watari T, Hatamoto M, Yamaguchi T, Setiadi T and Konda T 2023 Greenhouse gas reduction of co-benefit-type wastewater treatment system for fish-processing industry: a real-scale case study in Indonesia *Water Sci. Eng.* **16** 271–9

Han M, Zhang C, Li F and Ho S-H 2022 Data-driven analysis on immobilized microalgae system: new upgrading trends for microalgal wastewater treatment *Sci. Total Environ.* **852** 158514

He Y, Li X, Li T, Srinivasakannan C, Li S, Yin S and Zhang L 2023 Research progress on removal methods of Cl- from industrial wastewater *J. Environ. Chem. Eng.* **11** 109163

Ho K C, Chan M K, Chen Y M and Subhramaniyun P 2023 Treatment of rubber industry wastewater review: recent advances and future prospects *J. Water Process Eng.* **52** 103559

Hu K *et al* 2023 Novel biological nitrogen removal process for the treatment of wastewater with low carbon to nitrogen ratio: a review *J. Water Process Eng.* **53** 103673

Hualpa-Cutipa E, Acosta R A S, Sangay-Tucto S, Beingolea X G M, Gutierrez G T and Zabarburú I N 2022 Recent trends for treatment of environmental contaminants in wastewater: an integrated valorization of industrial wastewater *Integrated Environmental Technologies for Wastewater Treatment and Sustainable Development* ed V Kumar and M Kumar (Amsterdam: Elsevier) ch 15 pp 337–68

Huang H, Ma H, Liu B, Yang S, Wei Q, Zhang Y and Lv W 2023a Bed filtration pressure drop prediction and accuracy evaluation using the Ergun equation with optimized dynamic parameters in industrial wastewater treatment *J. Water Process Eng.* **53** 103776

Huang L, Kong W, Song S, Quan X and Li Puma G 2023b Treatment of industrial etching terminal wastewater using ZnFe2O4/g-C3N4 heterojunctions photo-assisted cathodes in single-chamber microbial electrolysis cells *Appl. Catalysis* B **335** 122849

Irshad M A, Sattar S, Nawaz R, Al-Hussain S A, Rizwan M, Bukhari A, Waseem M, Irfan A, Inam A and Zaki M E A 2023 Enhancing chromium removal and recovery from industrial wastewater using sustainable and efficient nanomaterial: a review *Ecotoxicol. Environ. Saf.* **263** 115231

Jadhav A P *et al* 2022 Synchrotron x-ray assisted degradation of industrial wastewater by advanced oxidation process *Radiat. Phys. Chem.* **197** 110161

Jalali F, Zinatizadeh A A, Asadi A and Zinadini S 2023 A moving bed biofilm reactor coupled with an upgraded nanocomposite polyvinylidene fluoride membrane to treat an industrial estate wastewater *Chem. Eng. J.* **470** 144128

Jia Y, Shan C, Fu W, Wei S and Pan B 2023 Occurrences and fates of per- and polyfluoralkyl substances in textile dyeing wastewater along full-scale treatment processes *Water Res.* **242** 120289

Jiang D, Gao C, Liu L, Yu T, Li Y and Wang H 2023a Customized copper/cobalt-rich ferrite spinel-based construction ceramic membrane incorporating gold tailings for enhanced treatment of industrial oily emulsion wastewater *Sep. Purif. Technol.* **320** 124131

Jiang H, Chen H, Duan Z, Huang Z and Wei K 2023b Research progress and trends of biochar in the field of wastewater treatment by electrochemical advanced oxidation processes (EAOPs): a bibliometric analysis *J. Hazard. Mater. Adv.* **10** 100305

Jiménez-Benítez A, Ruiz-Martínez A, Robles Á, Serralta J, Ribes J, Rogalla F, Seco A and Ferrer J 2023 A semi-industrial AnMBR plant for urban wastewater treatment at ambient temperature: analysis of the filtration process, energy balance and quantification of GHG emissions *J. Environ. Chem. Eng.* **11** 109454

Jin L, Sun X, Ren H and Huang H 2023 Biological filtration for wastewater treatment in the 21st century: a data-driven analysis of hotspots, challenges and prospects *Sci. Total Environ.* **855** 158951

Kabir ahmad S F, Lee K T and Vadivelu V M 2023 Emerging trends of microalgae biogranulation research in wastewater treatment: a bibliometric analysis from 2011 to 2023 *Biocatal. Agric. Biotechnol.* **50** 102684

Kadier A *et al* 2022 A state-of-the-art review on electrocoagulation (EC): an efficient, emerging, and green technology for oil elimination from oil and gas industrial wastewater streams *Case Stud. Chem. Environ. Eng.* **6** 100274

Kim S Y, Park J W, Noh J H, Bae Y H and Maeng S K 2022 Potential organic matter management for industrial wastewater guidelines using advanced dissolved organic matter characterization tools *J. Water Process Eng.* **46** 102604

Kotia A, Rutu P, Singh V, Kumar A, Dhoke S, Kumar P and Singh D K 2022 Rheological analysis of rice husk-starch suspended in water for sustainable agriculture application *Mater. Today Proc.* **50** 1962–6

Kumar P, Devi P and Dey S R 2021a Fungal volatile compounds: a source of novel in plant protection agents *Volatiles and Metabolites of Microbes* ed A Kumar, J Singh and M M Samuel (New York: Academic) ch 6 pp 83–104

Kumar P, Kumar T, Singh S, Tuteja N, Prasad R and Singh J 2020 Potassium: a key modulator for cell homeostasis *J. Biotechnol.* **324** 198–210

Kumar P and Mistri T K 2020 Transcription factors in SOX family: potent regulators for cancer initiation and development in the human body *Semin. Cancer Biol.* **67** 105–13

Kumar P, Sharma K, Saini L and Dey S R 2021b Role and behavior of microbial volatile organic compounds in mitigating stress pp 83–104 ed A Kumar, J Singh and M M Samuel (Academic) ch 8 pp 143–61

Kumar R, Maurya A and Raj A 2023 Emerging technological solutions for the management of paper mill wastewater: treatment, nutrient recovery and fourth industrial revolution (IR 4.0) *J. Water Process Eng.* **53** 103715

Kumar V, Dwivedi P, Kumar P, Singh B N, Pandey D K, Kumar V and Bose B 2021c Mitigation of heat stress responses in crops using nitrate primed seeds *S. Afr. J. Bot.* **140** 25–36

Kumari P, Singh J and Kumar P 2022 Impact of bioenergy for the diminution of an ascending global variability and change in the climate *Microbiome Under Changing Climate* ed A Kumar, J Singh and L F R Ferreira (Woodhead Publishing) ch 21 pp 469–87

Lau P L and Trzcinski A P 2022 A review of modified and hybrid anaerobic baffled reactors for industrial wastewater treatment *Water Sci. Eng.* **15** 247–56

Lee Y-J, Lin B and Lei Z 2022 Nitrous oxide emission mitigation from biological wastewater treatment—a review *Bioresour. Technol.* **362** 127747

Li L, Liang T, Zhao M, Lv Y, Song Z, Sheng T and Ma F 2022 A review on mycelial pellets as biological carriers: wastewater treatment and recovery for resource and energy *Bioresour. Technol.* **355** 127200

Li R, Speed D, Siriwardena D, Fernando S, Thagard S M and Holsen T M 2021 Comparison of hydrogen peroxide-based advanced oxidation processes for the treatment of azole-containing industrial wastewater *Chem. Eng. J.* **425** 131785

Li W, Wei K, Yin X, Zhu H, Zhu Q, Zhang X, Liu S and Han W 2023 An extra-chelator-free fenton process assisted by electrocatalytic-induced *in situ* pollutant carboxylation for target refractory organic efficient treatment in chemical-industrial wastewater *Environ. Res.* **232** 116243

Liu F, Cheng J, Qian F, Zhang X and Zhang H 2023 Research on advanced treatment of phenolic chemical wastewater and carbon replacement by the multi-layer biological activated carbon filter *J. Water Process Eng.* **51** 103388

Ma C, Peng H, Chen H, Shang W, Zheng X, Yang M and Zhang Y 2022 Long-term trends of fluorotelomer alcohols in a wastewater treatment plant impacted by textile manufacturing industry *Chemosphere* **299** 134442

Ma W, Zhang X, Han H, Shi X, Kong Q, Yu T and Zhao F 2024 Overview of enhancing biological treatment of coal chemical wastewater: new strategies and future directions *J. Environ. Sci.* **135** 506–20

Manetti M and Tomei M C 2022 Extractive polymeric membrane bioreactors for industrial wastewater treatment: theory and practice *Process Saf. Environ. Protect.* **162** 169–86

Mao G, Han Y, Liu X, Crittenden J, Huang N and Ahmad U M 2022 Technology status and trends of industrial wastewater treatment: a patent analysis *Chemosphere* **288** 132483

Mao G, Hu H, Liu X, Crittenden J and Huang N 2021 A bibliometric analysis of industrial wastewater treatments from 1998 to 2019 *Environ. Pollut.* **275** 115785

Martín-González M A, Fernández-Rodríguez C, González-Díaz O M, Susial P and Doña-Rodríguez J M 2023 Open-cell ceramic foams covered with TiO_2 for the photocatalytic treatment of agro-industrial wastewaters containing imazalil at semi-pilot scale *J. Taiwan Inst. Chem. Eng.* **147** 104902

Miao S, Zhang Y, Men C, Mao Y and Zuo J 2024 A combined evaluation of the characteristics and antibiotic resistance induction potential of antibiotic wastewater during the treatment process *J. Environ. Sci.* **138** 626–36

Michalska K, Goszkiewicz A, Skalska K, Kołodziejczyk E, Markiewicz J, Majzer R and Siedlecki M 2021 Treatment of Industrial Wastewaters and Liquid Waste by Fungi *Encyclopedia of Mycology* vol 2 ed Ó Zaragoza and A Casadevall (Amsterdam: Elsevier) pp 662–682

Mohan K, Karthick Rajan D, Rajarajeswaran J, Divya D and Ramu Ganesan A 2023 Recent trends on chitosan based hybrid materials for wastewater treatment: a review *Curr. Opin. Environ. Sci. Health* **33** 100473

Mondal P, Nandan A, Ajithkumar S, Siddiqui N A, Raja S, Kola A K and Balakrishnan D 2023 Sustainable application of nanoparticles in wastewater treatment: fate, current trend and paradigm shift *Environ. Res.* **232** 116071

Murshid S, Antonysamy A, Dhakshinamoorthy G, Jayaseelan A and Pugazhendhi A 2023 A review on biofilm-based reactors for wastewater treatment: recent advancements in biofilm carriers, kinetics, reactors, economics, and future perspectives *Sci. Total Environ.* **892** 164796

Naha A, Antony S, Nath S, Sharma D, Mishra A, Biju D T, Madhavan A, Binod P, Varjani S and Sindhu R 2023 A hypothetical model of multi-layered cost-effective wastewater treatment plant integrating microbial fuel cell and nanofiltration technology: a comprehensive review on wastewater treatment and sustainable remediation *Environ. Pollut.* **323** 121274

Namaldi O and Azgin S T 2023 Evaluation of the treatment performance and reuse potential in agriculture of organized industrial zone (OIZ) wastewater through an innovative vermifiltration approach *J. Environ. Manage.* **327** 116865

Nguyen M L, Vo T-D-H, Dat N D, Nguyen V-T, Tran A T K, Nguyen P-T and Bui X-T 2023 Performance of low flux sponge membrane bioreactor treating industrial wastewater for reuse purposes *Bioresour. Technol. Rep.* **22** 101440

Nidheesh P V, Behera B, Babu D S, Scaria J and Kumar M S 2022 Mixed industrial wastewater treatment by the combination of heterogeneous electro-Fenton and electrocoagulation processes *Chemosphere* **290** 133348

Nishat A *et al* 2023 Wastewater treatment: a short assessment on available techniques *Alex. Eng. J.* **76** 505–16

Oliveira R, Silva R M, Castro A R, Rodrigues L R and Pereira M A 2023 Biological processes and the use of microorganisms in oily wastewater treatment *Advanced Technologies in Wastewater Treatment* ed A Basile, A Cassano, M R Rahimpour and M A Makarem (Amsterdam: Elsevier) ch 10 pp 257–288

Omran B A and Baek K-H 2022 Valorization of agro-industrial biowaste to green nanomaterials for wastewater treatment: approaching green chemistry and circular economy principles *J. Environ. Manage.* **311** 114806

Otgonbayar T, Pérez-Calvo J-F, Lucke M, Raiser T, Wehrli M and Mazzotti M 2023 Development and optimization of a novel industrial process solution for stripping of carbon dioxide and ammonia from bio-process wastewater *Chem. Eng. Res. Des.* **193** 810–25

Ouyang J, Miao Q, Wei D, Zhang X, Luo E, Zhao Z, Zhao Y, Li C and Wei L 2023 Biological treatment of cadmium ($Cd2+$)-containing wastewater with sulfate as the electron acceptor and its microbial community *Water Cycle* **4** 87–94

Ozyildiz G, Bodur M, Dilsizoglu-Akyol N, Kilicarpa A, Olmez-Hanci T, Cokgor E, Kilinc C, Okutan H C and Insel G 2023 Simulating the impact of ozonation on biodegradation characteristics of industrial wastewater concentrated with membrane filtration *J. Environ. Chem. Eng.* **11** 109286

Paździor K, Bilińska L and Ledakowicz S 2019 A review of the existing and emerging technologies in the combination of AOPs and biological processes in industrial textile wastewater treatment *Chem. Eng. J.* **376** 120597

Piaskowski K, Świderska-Dąbrowska R and Dąbrowski T 2023 Impact of cationic polyelectrolytes on activated sludge morphology and biological wastewater treatment in a Sequential Batch Reactor (SBR) *J. Water Process Eng.* **52** 103500

Popat A, Nidheesh P V, Anantha Singh T S and Suresh Kumar M 2019 Mixed industrial wastewater treatment by combined electrochemical advanced oxidation and biological processes *Chemosphere* **237** 124419

Qyyum M A, Ihsanullah I, Ahmad R, Ismail S, Khan A, Nizami A-S and Tawfik A 2022 Biohydrogen production from real industrial wastewater: potential bioreactors, challenges in commercialization and future directions *Int. J. Hydrogen Energy* **47** 37154–70

Raj S, Singh H and Bhattacharya J 2023 Treatment of textile industry wastewater based on coagulation-flocculation aided sedimentation followed by adsorption: process studies in an industrial ecology concept *Sci. Total Environ.* **857** 159464

Ribeiro J P and Nunes M I 2021 Recent trends and developments in Fenton processes for industrial wastewater treatment—a critical review *Environ. Res.* **197** 110957

Robles Á, Jiménez-Benítez A, Giménez J B, Durán F, Ribes J, Serralta J, Ferrer J, Rogalla F and Seco A 2022 A semi-industrial scale AnMBR for municipal wastewater treatment at ambient temperature: performance of the biological process *Water Res.* **215** 118249

Shah Maulin P 2020 *Microbial Bioremediation and Biodegradation* (Berlin: Springer)

Semaha P, Lei Z, Yuan T, Zhang Z and Shimizu K 2023 Transition of biological wastewater treatment from flocculent activated sludge to granular sludge systems towards circular economy *Bioresour. Technol. Rep.* **21** 101294

Singh N K, Yadav M, Singh V, Padhiyar H, Kumar V, Bhatia S K and Show P-L 2023 Artificial intelligence and machine learning-based monitoring and design of biological wastewater treatment systems *Bioresour. Technol.* **369** 128486

Shah M P 2021 *Removal of Refractory Pollutants from Wastewater Treatment Plants* (Boca Raton, FL: CRC Press)

Sun X, Jin L, Zhou F, Jin K, Wang L, Zhang X, Ren H and Huang H 2022 Patent analysis of chemical treatment technology for wastewater: status and future trends *Chemosphere* **307** 135802

Tian Y, Zhang H, Pan S, Yin Y, Jia Z and Zhou H 2023 Amine-functionalized magnetic microspheres from lignosulfonate for industrial wastewater purification *Int. J. Biol. Macromol.* **224** 133–42

Upadhyay S K, Devi P, Kumar V, Pathak H K, Kumar P, Rajput V D and Dwivedi P 2023 Efficient removal of total arsenic (As3+/5+) from contaminated water by novel strategies mediated iron and plant extract activated waste flowers of marigold *Chemosphere* **313** 137551

Wang G, Qiu G, Wei J, Guo Z, Wang W, Liu X and Song Y 2023a Activated carbon enhanced traditional activated sludge process for chemical explosion accident wastewater treatment *Environ. Res.* **225** 115595

Wang L, Xu Y, Qin T, Wu M, Chen Z, Zhang Y, Liu W and Xie X 2023b Global trends in the research and development of medical/pharmaceutical wastewater treatment over the half-century *Chemosphere* **331** 138775

Wei C *et al* 2023 Enrichment strategies of heavy metals in the O/H/O process composed of biological fluidized bed for wastewater treatment: a case study of Cu and Zn *J. Clean. Prod.* **411** 137334

Wu H, Li A, Zhang H, Gao S, Li S, Cai J, Yan R and Xing Z 2023a The potential and sustainable strategy for swine wastewater treatment: resource recovery *Chemosphere* **336** 139235

Wu Y, Gong Z, Wang S and Song L 2023b Occurrence and prevalence of antibiotic resistance genes and pathogens in an industrial park wastewater treatment plant *Sci. Total Environ.* **880** 163278

Xia P, Chen Z, Wang D, Niu X, Tang X, Ao L, He Q, Wang S and Ye Z 2023 Revealing the double-edged roles of chloride ions in Fered-Fenton treatment of industrial wastewater *Sep. Purif. Technol.* **319** 124035

Yakamercan E, Bhatt P, Aygun A, Adesope A W and Simsek H 2023 Comprehensive understanding of electrochemical treatment systems combined with biological processes for wastewater remediation *Environ. Pollut.* **330** 121680

Yang L, Xu X, Wang H, Yan J, Zhou X, Ren N, Lee D-J and Chen C 2022a Biological treatment of refractory pollutants in industrial wastewaters under aerobic or anaerobic condition: batch tests and associated microbial community analysis *Bioresour. Technol. Rep.* **17** 100927

Yang Q, Liu Y, Huang W, Liu Z, Guo R and Chen J 2022b Synchronous complete COD reduction for persistent chemical-industrial organic wastewater using the integrated treatment system *Chem. Eng. J.* **430** 133136

Yu H, Dou D, Zhao J, Pang B, Zhang L, Chi Z and Yu H 2021 The exploration of Ti/SnO$_2$-Sb anode/air diffusion cathode/UV dual photoelectric catalytic coupling system for the biological harmless treatment of real antibiotic industrial wastewater *Chem. Eng. J.* **412** 128581

Zahmatkesh S, Gholian-Jouybari F, Klemeš J J, Bokhari A and Hajiaghaei-Keshteli M 2023 Sustainable and optimized values for municipal wastewater: the removal of biological oxygen demand and chemical oxygen demand by various levels of geranular activated carbon- and genetic algorithm-based simulation *J. Clean. Prod.* **417** 137932

Zhang C, Quan B, Tang J, Cheng K, Tang Y, Shen W, Su P and Zhang C 2023a China's wastewater treatment: status quo and sustainability perspectives *J. Water Process Eng.* **53** 103708

Zhang C *et al* 2023b Critical analysis on the transformation and upgrading strategy of Chinese municipal wastewater treatment plants: towards sustainable water remediation and zero carbon emissions *Sci. Total Environ.* **896** 165201

Zhang M, Leung K-T, Lin H and Liao B 2021 Effects of solids retention time on the biological performance of a novel microalgal-bacterial membrane photobioreactor for industrial wastewater treatment *J. Environ. Chem. Eng.* **9** 105500

Zhang S, Jin Y, Chen W, Wang J, Wang Y and Ren H 2023c Artificial intelligence in wastewater treatment: a data-driven analysis of status and trends *Chemosphere* **336** 139163

Zhang X 2022 Selective separation membranes for fractionating organics and salts for industrial wastewater treatment: design strategies and process assessment *J. Membr. Sci.* **643** 120052

Zhang Y, Zhao Z, Xu H, Wang L, Liu R and Jia X 2023d Fate of antibiotic resistance genes and bacteria in a coupled water-processing system with wastewater treatment plants and constructed wetlands in coastal eco-industrial parks *Ecotoxicol. Environ. Saf.* **252** 114606

Zoroufchi Benis K, Behnami A, Aghayani E, Farabi S and Pourakbar M 2021 Water recovery and on-site reuse of laundry wastewater by a facile and cost-effective system: combined biological and advanced oxidation process *Sci. Total Environ.* **789** 148068

Zribi I, Zili F, Ben Ali R, Masmoudi M A, Karray F, Sayadi S, Ben Ouada H and Chamkha M 2023 Trends in microalgal-based systems as a promising concept for emerging contaminants and mineral salt recovery from municipal wastewater *Environ. Res.* **232** 116342